GEOMORPHOLOGY AND VOLCANOLOGY OF COSTA RICA

GEOMORPHOLOGY AND VOLCANOLOGY OF COSTA RICA

JEAN PIERRE BERGOEING
Retd. University of Costa Rica
Member of IPGH (Panamerican Institute of Geography and History)

ELSEVIER
elsevier.com

Elsevier
Radarweg 29, PO Box 211, 1000 AE Amsterdam, Netherlands
The Boulevard, Langford Lane, Kidlington, Oxford OX5 1GB, United Kingdom
50 Hampshire Street, 5th Floor, Cambridge, MA 02139, United States

Library of Congress Cataloging-in-Publication Data
A catalog record for this book is available from the Library of Congress

British Library Cataloguing-in-Publication Data
A catalogue record for this book is available from the British Library

ISBN: 978-0-12-812067-5

Publisher: Candice Janco
Acquisition Editor: Amy M. Shapiro
Editorial Project Manager: Tasha Frank
Production Project Manager: Anitha Sivaraj
Designer: Victoria Pearson Esser

Typeset by TNQ Books and Journals

Front cover image: Tarcoles River Mouth in the Pacific Ocean. Fluvial sediments in suspension that in contact with salt water particles spread and deposit on the bottom to form later step by step a new delta. Photo courtesy of CAVU.org.
Back cover image: Sunset in the Pacific Ocean near Dominical Beach taken by a Phantom Drone by Esteban Montealegre M.

Contents

9. Coastal Geomorphology of Costa Rica

10. Cocos Island Geomorphology

11. Costa Rica Quaternary Chronology Proposal

Biography of the Author

Jean Pierre Bergoeing, French geomorphologist and polyglot, studied at the Pontifical Catholic University of Chile. He continued his studies at the University of Aix–Marseille II, France, where he successively obtained a Master's degree in Physical Geography (1972), a Third Cycle Doctorate in Geomorphology (1975), and finally a State Doctorate in Letters and Human Sciences (1987). His career has developed on three continents, North and South America, Europe, and Africa. He has been a professor at the Pontifical Catholic University of Chile; Nantes University of France; Abdou Moumouni University of Niamey, Niger; and the University of Costa Rica. He has also worked for the French government as an international aid worker and later served a diplomatic career as a Scientific and Technical Cooperation Attaché. He is the author of several books, numerous publications in international journals, and geomorphic maps of Chile, Costa Rica, Central America, Africa, and Europe.

jegadana@gmail.com

506.86290712–506.22536318

Introduction

The first edition of *Geomorphology of Costa Rica* was published in 1998, in Spanish, by the Instituto Geografía Nacional of Costa Rica. In 2007 a second edition followed, a luxury and color print edition, in charge of the French Bookstore of Costa Rica. Both editions are now out of print, and thus a new book was essential. For this reason, the author wanted to re-form its context, making it more attractive and innovative in terms of content. This new English edition of *Geomorphology and Volcanology of Costa Rica* aims to renew the knowledge of that country accumulated in all these years, mainly of the ground. Geomorphology appears here as a science that allows the reader to understand the problems of a country with a complex relief and high altitudes that create stunning contrasts and which are equally the source of disasters of all kinds.

Costa Rica is a predominantly tropical country, only 10 degrees north latitude, with reliefs that reach 12,530 ft (Mount Chirripó) in a relatively small space, as the country covers only 19,691 square miles. We add that by its geological position, Costa Rica is subjected to the cortical thrusts of two tectonic plates, the Cocos and Caribbean plates, which border on Panama's Nazca plates. It is easy to understand the fragility of a country that borders an active abyssal pit, 9843 ft deep, in the Pacific, just 63 miles from its shores. The risks of tectonic and volcanic activity, mass landslides, and tsunamis are constant, and we must live with them. To these, we can add the consequences of climate change that will inexorably lead to a rise in the sea level over the next 100 years, and which may be 16–49 ft, according to the specialists' estimation, due to melting of the polar ice caps and Greenland.

Arenal Volcano *Jean Pierre Bergoeing photography (2010).*

This will directly affect Costa Rica's shores and the inhabitants who have built in the proximity of the coast. Also, it will affect agriculture, which will bring important changes in the crops, due to the decrease in rainfall, as well as urban areas, for which water is indispensable.

The text presented here, divided into 10 chapters, gives a regional vision of the country, with illustrations and original photographs that allow a better understanding of the ideas. The book is aimed at specialists, government and local officials, and students of the earth sciences, as well as the public generally eager to know the territory and where it is located or to have a deeper understanding of what is Costa Rica.

Finally, the author would like to express his thanks to colleagues and students who have worked in research and in the development of the illustrative color figures, as it is today in any publication.

Jean Pierre Bergoeing, Ph.D.

CHAPTER

1

Major Structural Units of Costa Rica

OUTLINE

Costa Rica, along with Nicaragua and Panama, forms the Central America Isthmus. They are mostly recent volcanic lands, created at the end of the Cretaceous, only about 70 million years ago. Although Costa Rica has a small area of 16,691 square miles, it has a wide variety of geomorphologic landscapes, products of its evolution throughout the Tertiary and Quaternary periods. The large structural units of Costa Rica are distributed from northwest to southeast. In the northern half, the units ranging from the Caribbean coast to the Pacific watershed of the country are as follows:

1. **Northern plains extension of the Nicaragua graben**. This corresponds to the vast plain basement of the Caribbean and passes to the

http://dx.doi.org/10.1016/B978-0-12-812067-5.00001-0

south and southwest, progressively shaping the Guanacaste and Central piedmont regions.

2. **The Plio-Quaternary volcanic ranges** (Guanacaste and Central ranges) in high tectonic position; this sector is the topographic and hydrographic axis of the country.
3. **The Tempisque tectonic depression**, traversed by Tempisque River, which corresponds to an inlet of the Gulf of Nicoya and constitutes the basal level of the Tempisque River. The volcanic Guanacaste Mountain Range brought piedmont deposits and above all deposits constituted of ignimbrites forming the great plateau of Liberia and serves as a transition from the previous tectonic unit.
4. **The Nicoya Peninsula**, which comprises mostly horst formations.
5. **The Tertiary mountain ranges**:
   a. **Tilaran volcanic mountain range** (Upper Miocene–Pliocene volcanism), which corresponds to Aguacate's geological formation)
   b. **Talamanca volcano–sedimentary mountain range**

Talamanca's mountain range occupies the south of Costa Rica, and the highest point of the country, Chirripó Mountain, is situated there at 12,530 ft. Only part of Talamanca is volcanic (Miocene volcanism). The base is formed by a Tertiary marine series, folded and faulted and pierced during the Miocene by granodiorite intrusions.

From the Caribbean coast to the Pacific we cross through the northern foothills and the axial zone of the Talamanca range, which is a mountainous region of difficult penetration covered by rainforest and dissymmetrical acute watersheds of the Pacific, contrasting with the slopes of the Caribbean side's broad development.

   c. **Littoral sedimentary mountains**. These are constituted by a partly folded monocline strip, limited in the Pacific by the tectonic Great Reverse Fault and which determines the coast path between Quepos and Palmar Sur.

6. **The tectonic depression of the General River-Coto Brus** is a faulted syncline, fed with piedmont deposits and Quaternary lacustrine deposits. These deposits are deformed by plate tectonics.
7. **Tectonic depression of Palmar Sur to the border with Panama**. This is a tectonic depressed area, situated west of the Great Reverse Fault and occupied by the basal levels of the Coronado and Coto Colorado rivers and the Golfo Dulce Bay.
8. **Osa Peninsula, a horst formation**.

1. NORTHERN PLAINS EXTENSION OF THE NICARAGUA GRABEN

Lake Nicaragua, or Lake Cocibolca, gradually formed during the Quaternary period. In effect the Tertiary-ending orogeny left exposed, folded (anticline) sediments of the Rivas formation (mainly sandstones and limestone) constituting an isthmus of some 13 to 19 miles wide, which definitely interrupts the contact of the southwestern region of Nicaragua with the Pacific Ocean. Inside, the tectonic depression of Nicaragua is occupied by a lake of marine origin whose waters became freshwater over the millennia thanks to the numerous tributaries that feed it. The lake's dimensions reduced considerably throughout the Quaternary period. Today the lake covers an area of 3,191,134 square miles and its major axis is 100 miles. Lake bathymetric studies indicate that the average depth is 43 ft, with a maximum between 19 and 230 ft, locate 8 miles to the southeast of Ometepe Island. The volume of the lake is estimated at 219,969,248.29909 gallons. At the beginning of the

Quaternary period the lake was salty, allowing a marine fauna, the *Carcharhinus leucas* shark, to slowly adapt to the conditions of lower levels of salt due to the surrounding river contributions from the Tertiary volcanic foothills (north–east) and Quaternary construction (west and south). During this period, communication with the Caribbean was broad. The intergraben sector, which extends to the south in Costa Rica, was fed with sediments of the Central and Guanacaste volcanic ranges, with contributions of lahars to the foot of the slopes and the construction of powerful alluvial fans that reached linear dimensions of 25 to 37 miles with a mainly northern orientation. The post-orogenic period corresponds to the Pliocene and Quaternary, characterized by an important development of tectonic fault system and big blocks that definitely make the major structural units. Particularly with consequences of tectonic activity in the sector of major shortcomings of the liminal, basins burier, as the depression of Nicaragua transformed into real grabben [sic]. These depressions were the centers of accumulation of the erosive elements of the mountain ranges which rose concomitantly with products of volcanic eruptions that occur at that time … (Butterlin, 1977).

A little developed volcanism (Tortuguero–Sierpe) also occurred, leaving small isolated cones in the middle of the flood plain under construction (Tournon, 1972). However, the intergraben sector also started to give signs of rising from the Middle Pleistocene. Flood rains that characterized the Upper Pleistocene period ended the construction of the powerful river fans, which took a multiconvex shape thanks to the consistent contributions of the volcanic mountain chains always in construction. The positive neotectonics was a constant from that moment. During the Upper Pleistocene, i.e., some 200,000 years ago, the southern part of Lake Nicaragua extended about 13 mi into the interior of Costa Rica in an area stretching from Los Chiles to Caño Negro, which was probably affected by the Eemian transgression. However, the communication with the Caribbean Sea was still partly open.

1.1 Alluvial Fans, Middle to Upper Pleistocene

Alluvial fans developed from the lower limit of lahar deposits produced by the volcanic mountain ranges covering the southern limit of the Nicaragua graben. They extend deeply northward, reaching linear distances of 25 to 37 miles. They are mainly made of decomposed ferruginous red clays. These alluvial fans contain upstream a matrix of some chaotic volcanic blocks (andesite, tuffs, and basalts) several meters in diameter, partially altered at the surface. This is the material that has better resisted the general breakdown of the lithic material swept away by river avalanches of this period alternating with periods of rexistasis and biostasis (Bergoeing, 1998). The alluvial fans are notched by paleochannels and modern channels giving a multiconvex modeling to this sector.

1.2 Fluvial–Lacustrine Terraces

From the Upper Pleistocene to the Holocene, the various rivers of Lake Nicaragua, including the Frío River, deposited sediments in the former paleolake, which contained the same clay genesis as the alluvial fans, because these eroded and the material was transported in suspension by the runoff. Between Los Chiles and the Nicaraguan border, an elongated NE–SW fluvial terrace rises about 32 ft above the general level of 131 ft above sea level. The wavy surface this terrace was shaped by the contributions of the Frío River, which currently flows 10.5 miles to the north, forming a lake delta a few miles from the drain of the San Juan River. The top of the terrace consists of decomposing clays, much leached and with whitish spots, whose thickness is about 32 ft, resting on a level of gray lake clay (base level) that does not exceed

FIGURE 1.1 Los Chiles fluvial–lacustrine terrace. Top: Fluvial deposits with small runoff courses in a red kaolin clay matrix. Bottom: Lake stratification. *Bergoeing (2006) photo.*

40 ft thickness and on a level of kaolin white lake clay (see Fig. 1.1).

The current paleolake level of Los Chiles is at 98 ft (the level of the surface of the lake, measured by the Instituto Nicaragüense de Estudios Territoriales of Nicaragua, was 101 ft above sea level). We can calculate that the sector covered by the Frío River, between Los Chiles and Lake Nicaragua, a distance of 10.5 miles, was settled only 17,000 years ago (if we calculate that deposited sediments in the middle lake gain approximately 0.62 mile of ground every 1000 years). It should be noted that the Flandrian transgression (−6500 years) probably interrupted the process of sedimentation; therefore the rate of sedimentation could be much faster and, in that case, it is fully Holocene.

The fluvial–lacustrine terrace of the Frío River is well exposed, on the left bank of the bridge of the highway between Los Chiles and Caño Negro, and is characterized by 33 ft of red leached clay deposit. It is a regional repository, which characterizes the low sector of Upala and Los Chiles. It corresponds to the upper reservoir of the lake terrace in Los Chiles. At the top of this deposit, the Frío River has built a composite fluvial terrace, by storied yellow lime–sandy sediments, which corresponds to the area of seasonal flooding caused by the river water. At the base, almost in contact with the current river level, red clays of the top level rest on a black clay stratum, about 0.6 to 1 ft thick, which can also be observed in the current margins of the Frío River in Los Chiles. Under this black layer appear clear lacustrine clay silt stratifications representing the base level.

1.3 Frío River Watershed

The Frío River originates, strictly speaking, from a number of tributaries originating in the Guanacaste volcanic range, between the Tenorio volcano slopes and Lake Cote. The basin has a high risk of erosion. Its mouth is formed by a delta only a few meters from the drain of the San Juan River. The amount of sediment that transports this water system into the upper part of the basin is very important.

It is inferred that during the Middle and Upper Pleistocene, particularly under a more contrasted climate, the basin underwent phases of rexistasis and biostasis, and the tectonic depression of Nicaragua was filled in quickly, reducing the lake's surface and conveying its coastline more northward.

2. GUANACASTE AND CENTRAL QUATERNARY VOLCANIC RANGES

2.1 Guanacaste's Volcano Range

The Guanacaste volcanic range is an eminently Quaternary continuation of a NW–SE fracture that continues into Nicaragua. It is characterized by acidic ignimbrite eruptions that created a plateau about 328–492 ft high, stretching west of the volcanic cones and going from the city of La Cruz northwest to Bagaces southeast.

2.1.1 *Orosí–Cacao Volcanic Complex*

Dominating the south of Lake Nicaragua, the Orosí–Cacao ensemble comprises the first volcanic cones that emerged, forming stratovolcanoes, during the Upper Pleistocene, and are composed of several cones and craters highlighted by the Orosí volcano (4593 ft above sea level), the Orosilito (3937 ft), El Pedregal (3609 ft), and Cacao (5443 ft). Seen from the north, the Orosí volcanic cone presents as a perfect, pointed cone coated on top with lush tropical vegetation, which demonstrates its relative youth, because tropical erosion has not fitted it with deep canyons. The craters of this set are closed mouth or open toward the southeast because of the effects of eruptions coupled with the northeastern trade winds that blow consistently. An explosion caldera structure separates the Orosi and the Orosilito craters.

Deep ravines covered by thick tropical vegetation, however, furrow the northern flanking cones. At the foot of the volcanic complex, an ignimbrite plateau extends north to Santa Cecilia village, where the natural modeling is flat-bottomed and formed by fluvial erosion into canyons and terraces modeled in the ignimbrites. From Santa Cecilia, the plateau is covered by lahars, creating a topographic plateau structure with multiconvex reliefs. Further to the south rises the Cacao volcanic cone, attached to the structure of the Orosi complex. This is characterized on the southwest flank by two very open and eroded craters facing the direction of the ignimbrite plateau. In the middle flank a third recent crater is imposed on its predecessors, where emerged an important lahar flow associated with pyroclastic rocks and large amount of pumice, forming a large cone of outfall fans ending in contact with the structural plateau. At the foot of the eastern flank of the Cacao cone, a vast circular depression that is without a doubt a collapsed caldera is partly fitted and drawn through the Pizote River, also known as the Niño River. A NE–SW tectonic fault borders this first volcanic complex and the Rincon de la Vieja volcano.

2.1.2 *Rincon de la Vieja Volcanic Complex*

The Rincon de la Vieja volcanic complex is seated between the Orosí–Cacao system and the Miravalles volcano. It is a Middle to Upper Pleistocene stratovolcano complex (Boudon et al., 1996 in Alvarado, 2000). Nine craters are aligned on the elongated top of this massif. The first is named Santa Maria (6286 ft above sea level), which houses a pluvial lake; next is the Rincon de la Vieja, the main active crater of this massif and whose south wall reaches 5906 ft above sea level. It presents great volcanic activity, the most recent of which dates to 1966–1970 and 1991–1992 to 2016, with intense fumarole activity. Northwest emerges the Von Seebach crater, which rises 2936 ft. This crater alignment is also characterized by the presence of a large explosion caldera at 5250 ft high, eroded and open toward the northeast, housing two craters. The edge is a caldera that has another much-eroded crater called Marmo, which reaches the altitude of 5381 ft. The western slope, though regular, presents in its sector average a great eroded circus shape, in which it is carved with deep river canyons. At the base emerge dacitic to rhyodacitic domes of the Lower Pleistocene (Bellon and Tournon, 1978), such as Fortuna (1572 ft), San Roque (1772 ft), Cañas Dulces (2149 ft), Góngora (2520 ft), and San Vicente (1969 ft) (Bergoeing, 2007). In addition, it is worth mentioning that at the Coyol site, bordered by the Colorado River and the Jaramillo Gorge, there are thermal springs and hot mud springs associated with the Rincon de la Vieja volcanism. The contact of the volcanic cone with the structural western plateau is the result of the spill of mainly successive ignimbrite flows and lahars during the Quaternary. The eastern slope, more regular, is covered by dense natural tropical vegetation and ends in a structural volcanic plateau that is inscribed in a former collapsed caldera, not recorded in the geological charts and unseen because of the permanent cloud cover of the sector, but which becomes evident when studying radar satellite imagery. The surface material of the past 300,000 years of this volcanic massif consists of cinder, lahars, and pyroclastic flow deposits of andesitic composition (Carr et al., 1986 in Alvarado, 2000; Chiesa et al., 1994 in Alvarado, 2000). The historical eruptive activity of this complex dates back to 1765 and has been characterized mainly by steam and ash columns. According to Boudon (1996) in Alvarado (2000), the Rincon de la Vieja can produce eruptions of lahars, spilling mainly toward the north; acid rain; volcanic avalanches; and ash fall.

2.1.3 *Miravalles Volcanic Complex*

The cusp of the Miravalles volcano reaches 7244 ft above sea level and the diameter of its crater is about 0.37 miles. It has six volcanic sources aligned NW–SE. This stratovolcano was rebuilt on several occasions during the Quaternary period (see Fig. 1.2). At its feet, in the northwest sector, stretches a vast topographic plateau, which corresponds to the bottom of the collapsed Guayabo caldera dating from the Lower Pleistocene (between 1.6 and 0.6 million years BP). At that time there were strong emissions of burning clouds (ignimbrites) and finally the subsidence or collapse of the Guayabo volcano (Gillot et al., 1994 in Alvarado, 2000). The surrounding La Montañosa hills represent the remaining rim. The collapsed caldera occupies an area of 77 square miles. Further, northeast, another collapsed paleocaldera, even greater, dominates the base of this colossus and separates it from

FIGURE 1.2 Miravalles volcano in the Guanacaste range, Costa Rica. *Photo courtesy of Gilbert Vargas (2008).*

the Rincon de la Vieja massif, and has been evidenced by satellite radar images. The current Miravalles cone has emerged on a vast collapsed caldera that is shared with the Tenorio volcano more to the south, whose evidence is shown once again by radar images. The Miravalles southwest flank is bordered by the Espiritu Santo and Gotas de Agua volcanic cones, which correspond to Lower to Middle Pleistocene volcanic activity. These volcanic hills are separated by a WSW–ENE tectonic fault that passes between the Tenorio and the Montezuma volcanic cones and disappears in the plain north of the Miravalles cone. Remains of less important calderas flank the northeast side of Miravalles. The Miravalles volcano is separated from the Tenorio–Montezuma volcanic complex by a SW–NE tectonic depression. The Miravalles volcano is currently active, showing fumaroles, solfataras, and thermal springs. Holocene andesitic lava flows cover the northwest flank, and the geothermal activity manifested has been used since 1946 by the Costa Rican Electricity Institute, still in operation today.

2.1.4 Tenorio–Montezuma Complex

Tenorio–Montezuma is the southernmost Guanacaste volcanic mountain range, if we take into account the gap between the Guanacaste range and the Arenal complex. It is composed of two Upper Pleistocene stratovolcanoes forming two main cones, the Tenorio (6286 ft above sea level) and its twin the Montezuma (5971 ft), separated by a SW–NE tectonic fault. Among them is equally drawn a small collapsed caldera of a third volcanic edifice, closed headed, whose remnants are a series of subequal-altitude structural plateaus. In the field, the hills present have a multiconvex morphology, being composed of pyroclastic material. Between the plateaus are hints of depressions that could be remains of small explosive craters. The Tenorio–Montezuma volcanic complex could be related to the modern building of the Miravalles cone because all are circumscribed in a peripheral caldera that could date back to the Middle Pleistocene and whose evidence can be seen in satellite radar images. Like Miravalles, the Montezuma volcano cone is separated, in the south, by a SW–NE tectonic fault, from another volcanic structure of the Upper Pleistocene, quite eroded, which, however, provides a glimpse of their original morphology. The Tenorio volcano dominates from the northwest the vast tectonic depression of Lake Arenal, which is a set of low mountain ranges of subequal altitudes, with volcanic dating to the Pliocene–Pleistocene and resting on a Tertiary sedimentary series (the Venado formation). In the northern sector of the Tenorio–Montezuma set a series of rivers and waterfalls associated with a hydrothermal volcanic activity gives a particular staining to the Celeste River due to the sulfur input from their waters.

2.2 Arenal–Chato–Los Perdidos

2.2.1 Modern Volcanic Complex

The Arenal, Chato, and Los Perdidos volcanoes (Upper Pleistocene–Holocene), of recent construction and nearby, south of the Guanacaste volcanic range, present as a new unit that serves as a transition with the Central Volcanic Range.

Arenal Volcano, 5436 ft above sea level, is a modern cone built on an old explosion caldera. It is estimated that Arenal has existed for only about 40,000 years and it would be the youngest volcano in Costa Rica. Historical eruptions of Arenal are unknown, but since 1937 it has shown violent activity. It is a stratovolcano composed of alternating ashes, slags, lapilli and lava blocks, and deposits of burning clouds. Volcanic products range from basalts to dacites (Borgia et al., 1988). Since 1968, eruptions have manifested it as a dangerous volcano emitting pyroclastic flows. Its activity did not cease until 2010, and this has attracted large numbers of tourists, mostly because of the spectacle that represents its unpredictable activity and its hot springs. For these reasons hotel facilities have been built at the foot of the volcano, in some cases just 1.8

to 3 miles away, as the crow flies, from the crater. It is the setting for an imminent disaster, legally authorized by the competent authorities, thus creating the conditions for a human catastrophe of major proportions. In the event of a catastrophic pyroclastic eruption, the red-hot avalanche would bury many hotels. The surrounding and nonradial road is completely inadequate for a quick evacuation.

2.2.2 *San Lorenzo Collapsed Caldera*

This description would be incomplete without mention of the San Lorenzo collapsed caldera. Indeed such a large caldera, perceived only through satellite images, probably fits in the Lower Pleistocene. This caldera marks the transition between the volcanic Central Range and the Guanacaste volcanic range.

2.3 Central Volcanic Mountain Range

The Central Volcanic Range is the result of the construction of five large volcanic edifices built gradually during the Quaternary period. They are the Platanar–Porvenir, Poas, Barva, Irazu, and Turrialba volcano complex. Each has its own structure and geomorphology, with twin cones, adventitious cones, "planezes," lava flows, lahars, and thermal and gaseous emissions. The current craters are aligned following a tectonic NW–SE fault direction.

The Central Valley is part of the Central Volcanic Range and is an extension of the same. It is a vast structural plateau in the western sector bounded by the Grande River and the Virilla River cuts. Ochomogo Pass separates the Central Valley into two sectors: the western area, where San José, Heredia, and Alajuela cities stand, and the eastern area, where Cartago and Turrialba cities are seated. The Central Plateau occupies a graben tectonic position, bordered by the horst of the Central Volcanic Range and the Talamanca Mountain Range, staggered by failures in this part. The Central Plateau is part of the basement of the Central volcanic system. Its first manifestations occurred during the Upper Pliocene and go back to the Lower Pleistocene. Its base consists of a powerful series of basaltic lava flows. The stratigraphy then evolves into a volcanic tuff series appearing clearly in the Virilla River Canyon, and finally, the tuffs are covered by important ignimbrite deposits, with an output 115–328 ft thick, divided into two units (Pérez, 2005). This bears witness to an increasingly important acidifying of the Central volcanic eruptive events. The Central Plateau is finally covered by numerous and important lahar deposits that descended from the southern flank of the Central Volcanic Range and are very visible, forming hills rows perpendicular to the Virilla River.

The eastern sector, where are seated Cartago and Turrialba cities, is constituted by smaller plateaus, but generally of the same nature as the western Central Plateau, such as the Paraiso and Juan Viñas plateaus. Between Turrialba and Lake Bonilla, the Reventazón River notch on its left bank contains marine sedimentary formations from the Eocene to the Miocene (Tournon and Alvarado, 1995), on which rest the oldest volcanic lava and lahars. The Central Volcanic Range stratovolcanoes, more modern chronologically than the Central Plateau, sit atop an understood margin between the Middle and the Upper Pleistocene. The current craters are mostly Holocene.

2.3.1 *Platanar and Porvenir Volcanic Complex*

Platanar Volcano is located northwest of the Central Volcanic Range. It is a composed stratovolcano, which could be tentatively dated from the Upper Pleistocene, consisting of the cone of the Platanar itself and, a little farther south, El Porvenir volcanic cone, separated from each other by 2 miles. The volcanic complex covers more than 44 square miles. Platanar Crater is located at 7162 ft above sea level, whereas Porvenir reaches 7437 ft.

On the northern flank of the Platanar, overlooking the Great Plains, is drawn the collapsed caldera of Palmera (Tournon, 1984), dating probably from the Lower Pleistocene, and covered in large part by lahars, arranged in large alluvial fans coming from the northern Platanar slopes. This filling can reasonably considered a constant stream, which covered the collapsed depression during all of the Quaternary. Farther east are nine small volcanic cones from the Upper Quaternary and Holocene, aligned north–south and known as the Aguas Zarcas volcanic region. They reach elevations no higher than 328–525 ft above sea level, but their cones are well preserved and some even show craters. They are the manifestation of the latest magma ascension of the sector. They consist of gabbros and dolerites, which correspond to Upper Pleistocene trace-alkaline volcanism. Some are Strombolian cones composed of lapilli, volcanic bombs, and alkali basalts (Tournon and Alvarado, 1995).

The Platanar and Porvenir cones are a modern construction (Upper Pleistocene) rebuilt over older structures where, notably, in the east appears the Chocosuela collapsed caldera (Alvarado and Carr, 1993 in Alvarado, 2000), which forms a canyon that can reach 1640–3280 ft deep and where the Aguas Zarcas River runs in the thalweg. If we consider this important, the structure rests on a bigger caldera that would go from the Toro River in the east to the San Lorenzo River in the west. Then the area takes on another dimension and the Platanar–Porvenir assembly fits within the framework of the largest calderas in the world, because the diameter of this structure would be about 22.3 miles, which would give a surface area of approximately 386 square miles.

2.3.2 *Poas Volcano*

Poas Volcano is a volcanic ensemble of the Central Mountain Range. It rises at 8884.5 ft above sea level, consisting of a main crater, more than 1.8 miles in diameter, containing an acid sulfuric lake with a pH of 1 and temperatures of 185°F. The lake also emits sulfurous gaseous emanations. A second volcanic crater is very close to the first one, but inactive, housing a pluvial lake, known as the Laguna de Botos. The whole is surrounded by dense mountainous tropical vegetation. Poas Volcano is also part of a tectonic system through which emerged, in a N–S line, several issues of magma. As well, a few miles north of Poas stands, at 6607 ft, the inactive Congo Volcano, whose crater is open to the northwest. Following the N–S tectonic alignment, farther to the north is the Hule collapsed caldera (Bergoeing and Brenes, 1977), a volcanic depression that houses three small lakes and in whose center stands a volcanic postcollapse cone. Finally, the complete tectonic–volcanic line ends north in the Laguna de Rio Cuarto gasmaar. Here is an explosive crater depression, covered with fluvial waters. This gasmaar is 1181 ft above sea level, measuring about 0,127414 square miles at the surface, and has a depth of 217 ft, and its walls are made of basaltic–andesitic lava (Alvarado, 2000).

Since 1828, Poas Volcano has had 39 eruptive episodes. On January 25, 1910, an eruption expelled a column of steam of 640,000 tons of ash. From 1952 to 1954, Poas Volcano experienced a new cinder cycle eruption with an emission of slag. In 1987 a phreatic explosion resulted in a modification of the lake level, and in 1989 the lake disappeared temporarily, exposing sulfuric "hot beds"; these were again submerged in 1990 when the lake returned to its "original" form. In 2006, Poas manifested again with phreatic explosions and fumaroles. Poas lavas are complex. The southern flank presents aphyric andesitic lava flows, a rare occurrence in calc-alkaline series, which can be acidic or basic (see Fig. 1.3).

Poas volcano lavas range from basalts to dacites. Congo volcanic cone lavas and the Hule collapsed caldera are mainly basaltic–andesitic (Tournon, 1983). Rivers running to the north are deeply encased in the volcanic slopes, including the Sarapiqui River, an important tributary of the San Juan River. The lava flows have allowed the creation of a choppy modeling, in which waterfalls and rapids are abundant. At the foot of the Poas cone powerful coalescing

FIGURE 1.3 Poas Volcano. Main crater hosting on the bottom a sulfuric acid lake. *Aerial photo Bergoeing (2012a).*

lahar fans of the Lower Pleistocene (Bergoeing, 2007) produce a transition with the northern flood plain. Finally, the Poas massif is border on the west by the Toro River, which forms part and shows the border of the collapsed caldera of the Platanar–Porvenir system.

2.3.3 *Barva Volcano*

Barva Volcano is separated from the Irazu–Turrialba volcanic complex by La Palma Pass and the Zurquí Hills volcanism relicts. Barva Volcano dominates the city of San José from its 9534 ft above sea level. It has about 12 eruptive outbreaks on its top and adventitious cones on its flanks. Three volcanic domes crown its summit, with a main crater occupied by the Barva crater lake at 8228 ft. To the north another crater is occupied by Lake Danta at similar altitude. Its historical eruptive activity is not known, although there are residual emissions of CO_2 (solfataras) and hot springs. Eruptive activity of Barva has been consigned at 6050 BC (Simkin and Siebert, 1994). Barva lavas are igneous andesites, basalts and some olivine basalts, and dacites (Alvarado, 2000). One of Barva Volcano's southern slope characteristics is the Los Angeles lava flow, a basalt–andesitic lava wash, contemporary with the Cervantes lava flow of the Irazu volcano, but its spill covers a much wider extension, reaching down to Barva City. This lava flow originated in the Cerro Redondo volcanic cone and, therefore, this is one of Barva Volcano's adventitious domes.

The western flank is dominated by the Guarari volcano structure (Protti, 1986), a Holocene explosive volcano that emitted andesite lava and pyroclastic flows, which came down to the village of Santo Domingo de El Roble. The crater is partly destroyed and open to the west. To the northeast Barva Volcano presents an explosion caldera structure, of large dimensions, but heavily affected by erosion and completely covered by tropical mountain vegetation.

Despite there being no historical records of Barva volcanic activity, this volcano is dangerous because there are traces of two Plinian eruptions and ignimbrite spills that show an explosive acid activity (Paniagua and Soto, 1988). The oldest known Holocene eruption dates back to 6050 BC ± 1000 years but the andesitic–dacitic tuffs exposed in the Tiribi River date back only 332,000 years and come from the Barva caldera (Venzke et al., 2008). Barva's northern slopes are also characterized by a series of calderas and domes and the Cacho Negro volcano. Despite being a Holocene cone it is very affected by erosion. From the bottom of this last crater is born the Puerto Viejo River.

The Cacho Negro stratovolcano rises to 7382 ft above sea level. It is located northeast of Barva Volcano. Its crater is fully northwest over mouth. On its flanks are observed recent radial lava flows (Holocene) and two adventitious cones (Alvarado, 1989). Lavas are predominantly andesitic and basaltic. It has no known historical activity. Arriving at the northern plain the volcanic limit is marked by the Santa Clara caldera, formed by two postcollapse cone inserts inside the caldera. The Barva Volcano northern slope can be divided into three sectors. The west sector corresponds to the cone of the Guarari volcano, the east to a series of volcanic domes that begin with the Zurquí Hills and culminate with the Cacho Negro volcanic cone, and at the center is a deep very regular slope that descends gradually and is interrupted by numerous rivers, which fit it vertically. This is due to the volcanic material, which is part of the northern slope, made from cinders interspersed with lava flows and tuffs.

It is perhaps in the northwestern part of Barva's north slope where you can better see the phenomenon of megalandslides, associated with the second and third landslide categories, products of a very thick altered mantle, formed by clay material coming from the decomposition of volcanic deposits. This material, deeply altered and overstretched by water under the tropical climatic conditions and favored by steep slopes, has allowed the mass landslides phenomenon and subsequent secondary landslides on the masses of material already removed.

2.3.4 *Zurquí Hills*

The Zurquí Hills are located between the Barva and Irazu stratovolcanoes. They are a volcanic group composed of the following cones: Chompipe (7411 ft), Turu (7018 ft), Caricias (6890 ft), Hondura (6716 ft), Tres Marias (5659 ft), and Mount Achiotillal (6174 ft).

They are in the most depressed part of the Central Volcanic Mountain Range and form the Desengaño Hill, easily penetrated by the humid flows from the Caribbean. The ensemble is much eroded despite its relative youth. An andesitic lava radiometric dating of these ancient andesitic volcanoes gave an age of 0.5 million years (Bellon and Tournon, 1978). The geological survey indicates that they were sources of emissions of basaltic lava flows, andesite, tuffs, breccia, ignimbrites, and pyroclastic materials that are exposed in the Braulio Carrillo highway and visible in the Zurquí tunnel. The hills are covered by significant thicknesses of altered ash.

2.3.5 *Irazu Volcano*

This stratovolcano has the highest active crater in Costa Rica at 11,260 ft above sea level. It is situated north of Cartago City and is composed of multiple Holocene craters. It has some of the oldest cones, aligned NW–SE, which end up in the destroyed crater of Las Nubes Volcano (Bergoeing, 1979). Irazu is the volcano with the best historical activity, because the colonial chronicles mention eruptions from the 18th century, with a cycle of 40–60 years, starting from 1723 and with some 23 eruptions since then. The 1963 eruption had as a complement a laharic flow, which buried the urban sector of Taras de Cartago, after heavy rains associated with the eruption of the volcano. San José and most of the Central Highlands of Costa Rica were then coated with ashes.

Tournon (1983) identifies basalt lava flows at the top as well as basic andesites in Cervantes lava flows and in Juan Viñas plateau. Cartago Plateau is composed of ignimbrites interspersed among the andesites. Irazu's southern slope is divided into two important morphological sectors:

The upper Irazu sector has an embossed wavy and multiconvex shape, born just from the crags of recent lava flows from the volcano, and upholds the southern flank to the hamlet of Cot. This sector has important thicknesses of much altered ashes and ancient andesitic–basaltic lava flows.

The lower Irazu sector goes from Cot to the volcanic plateau where Cartago sits on a series of four lahars coalescing into cones and occasional lava outcrops.

The northern sector of the city of Cartago was made particularly by lahars, highlighting two very specific generations, one ancient, at the bottom, with very altered ochre, made up of lapilli and pumice, and the other more recent, higher, with healthy rolled chaotic materials (Bergoeing, 2007).

The southern slope is also characterized by the Cervantes lava flow, which is an effusion of fluid lavas, basalt–andesitic, canoe type, which has been dated at 23,000 years BP by the method of $^{238}Th/^{232}Th-^{238}U/^{232}Th$ (Allegre and Condomine, 1976), but ^{14}C dating rejuvenated it to 20,000 years BP (Murata et al., 1966). The Strombolian eruption that caused the Cervantes lava flow (Tournon, 1984) came from the adventitious Pasqui Hill cone and resulted in a natural damming of the waters of the Reventazón River in today's Cachí Dam. This had to occur several times in the past and witnesses to this are the four lake terraces of the sector.

Irazu's northern slopes are covered with dense mountain primary tropical vegetation and this is why they are very different from the south side (see Fig. 1.4). "Lavakas," the Malagasy term designating megacircuses caused by erosion, characterize the upper part of the Irazu cone. In effect, the northern slope is modeled by scarps of 985–1640 ft of free fall, whose walls emit sulfuric gas because of their proximity to the volcanic chimney, a product of rainwater erosive wear. Lavakas are the first stage of a greater western phenomenon found on the Poas and Barva volcano slopes, which corresponds in reality to megaslides produced during the Middle Pleistocene, inside of which there are minor landslides, which are continuing today. It is partly the forest cover that has stopped the phenomenon, but it can be reactivated with the human colonization of the sector. Some major rivers, such as the

FIGURE 1.4 Irazu Volcano's northern slope, very eroded by the trade winds' orographic rainfalls that formed deep canyons in this stratovolcano. *Aerial oblique photo Bergoeing (2012a).*

Sucio River, have formed a river network that has undermined important river canyons with unevenness that can reach 1000 m. This is possible because the stratovolcanic material is very heterogeneous, comprising considerable thicknesses of cinder material, low welded tuffs, and lava flows that can fail where waters infiltrate.

2.3.6 *Turrialba Volcano*

Turrialba Volcano is the easternmost stratovolcano of the Central Volcanic Range of Costa Rica. It rises 10,958 ft above sea level and at its peak it presents a series of craters, of which the three main present activity from fumaroles and sulfur gas fumes. The volcano erupted several times in the 19th century (1853, 1855, 1864–1866) with pyroclastic and intense fumarole emissions. Its ashes arrived at Corinto Port in Nicaragua. An important volcanic activity started again in 2010 and continues today. The current top cone, like the others in the Central Volcanic Mountain Range, was built during the Holocene, presenting lavas ranging from basalts to dacites. It is estimated that in the past 3500 years this volcano has had at least five large eruptions with lava emission. In 2007, its activity increased dangerously, with sulfurous emissions affecting the local population (see Fig. 1.5).

Turrialba's top cone presents a series of planezes, which are structural plateaus formed of considerable thicknesses of lavas and that descend gently shaped skirts from the top volcano crater. They expand to the south, east, and north. To the east, the top of Dos Novillos Hill corresponds to a volcanic cone that is the easternmost of the Central Volcanic Mountain Range. The oldest Turrialba volcano lavas (basaltic–andesitic) go back 2.15 million years, placed at the end of the Pliocene. One of the most interesting washes of the Turrialba is the Peralta Lava Flow, on the southeastern slope, which draws a curved plateau at the end, reaching the altitude of 1969 ft. During the Pleistocene, it dammed the runway of the Reventazón River, leaving as witnesses the Bonilla Lakes. From that point and heading north, the lower part of the Turrialba cone is composed of powerful coalescent alluvial fans, consisting of important thicknesses of lahars associated with the volcanic activity of this massif. The Turrialba volcanic lavas are similar to those of the Irazu volcano, although basalts do not abound; they are mostly basic andesites (Tournon, 1983). Coalescing lahar cones descending from the Turrialba evolve gradually to alluvial fans and finally glacis that gradually merge with the great north plain (Nicaragua graben).

It is interesting also to note, in the Guapiles City sector, the Upper Pleistocene lava flows equally coming from Turrialba Volcano. In fact a few kilometers from this city stand three imposing andesitic lava flows, resting on old laharic fans that are part of the already mentioned coalescing volcanic fans and where surface underground water springs, which feed Guapiles City and would be

FIGURE 1.5 Turrialba Volcano, 2012 eruption. *Oblique aerial photo Bergoeing (2012a).*

able to provide water to all the Central Valley. In the southern crater, Turrialba City is seated on a major tectonic accident (NNW–SSE fault) covered with important lahar thicknesses. Turrialba Volcano and the nearby Irazu Volcano form the easternmost volcanic structures of the Central Volcanic Range. The southern volcanic slopes constitute the northern flank of the eastern Central Valley. In it sits an ancient explosion caldera structure whose remnants extend from Coliblanco to Juan Viñas.

3. TEMPISQUE RIVER TECTONIC DEPRESSION

The Tempisque River tectonic depression covers the northern Nicoya Peninsula. It is a tectonic subsidence zone pinched between the Guanacaste volcanic range and Nicoya Peninsula, both in horst position. The tectonic depression is oriented NW–SE and is traversed by the Tempisque River, which flows into it, leaving the ignimbrite plateau through a deep but short river canyon. The river with its sediments feeds the plain and this is how it has been partly sealing this depression. However, positive Quaternary tectonics affecting this sector of the country have raised regionally a vast sector, which was submerged in the sea during the late Miocene and formed a broad interior gulf; the pale remnant of it today is the Nicoya Gulf. Near the Tempisque River estuary rises a series of "mogots," or hills, which give testimony to the ancient coral reefs formed during the Miocene, and the rising plate tectonics are evident in the form of the isolated hills (Barra Honda hills). Inside, there are karstic caverns, the results of infiltration of the tropical rains into the last remnants of coral limestone. The Tempisque River depression of today is a vast plain, very fertile, exploited by agricultural activities. To the west of the Philadelphia sector is a series of isolated hills whose structures reveal a fluvial marine past formed by deposits of rolled stones and pebbles and are covered at the top by ignimbrites. These Pliocene–Quaternary acid deposits are witness to the maximum extent reached by the volcanic manifestations of the Guanacaste volcanic range.

4. TILARAN'S TERTIARY MOUNTAIN RANGE

The Tilaran Mountain Range, 31 miles long, is a mountainous unit with a much smaller area than the Talamanca or Central range. Its natural boundaries are Lake Arenal to the north; the San Lorenzo River to the northeast, which serves as a border with the Central Volcanic Mountain Range; and to the southeast the Aguacate Mountain Range, arriving in contact with the Talamanca Mountain Range. Tilaran extends south to the Tarcoles River, if we consider the Aguacate Mountains as forming an integral part of this unit. Indeed, by its geological history and the nature of the rocks and their ages, it is possible to include the mountain range that lies to the south of the Tarcoles River and part of Santiago de Puriscal. This part ends with the isolated Turrubares volcanic peak (5761 ft), a Cretaceous–Tertiary volcano overlooking the Pacific coast.

The Tilaran Mountain Range is the almost exclusive product of a Pliocene volcanic activity whose deposits have been regrouped under the name of the Aguacate group or formation (Madrigal, 1972), which includes a large number of effusive rocks. Erosion has eliminated most of the volcanic centers of this mountain range or made them unrecognizable. However, some domes, craters, necks, and caldera elements remain, such as the Palmares collapsed caldera or the volcanic unit Pelon–Mondongo–Tinajita (Bergoeing et al., 1978b), as well as ancient volcanic hot spots: Cerro Tilaran, Cerro Pelado–Delicias, Chopo (basaltic andesitic) (dated from 2.03 million years by the $^{40}Ar/^{39}Ar$ method by Alvarado and Gans, 2012), Cerro Perdidos, Poco Sol, Pelados–Herrera, Espiritu Santo (Mora, 1977), and the Cerro Macho

Chingo–Pelon (Paniagua and Tejera, 1980). Also in Santiago Alto Tajo, San Ramon, a basaltic lava dating gave a Pliocene age of 3.20 million years (K/Ar) (Bellon and Tournon 1978). Finally, the basalt of the quarry Tajo Alto, near Piedades de Santiago, gave an age of 4.04 million years ($^{40}Ar/^{39}Ar$) (data provided by MacMillan et al., 2004).

This volcanic range follows the general direction of the country, whose axis is NW–SE. It has two well-defined aspects. The northeast slope, or slope of San Carlos, pours its waters into the flood of the northern plain, where rivers, following a winding path, are tributaries of the San Juan River. The southwest side, or Pacific slope, is subdivided into three sectors:

1. the northern sector, between the Cañas and the Barranca rivers;
2. the southern sector, between the Barranca and the Tarcoles rivers;
3. the Turrubares volcanic hills sector.

In the western Central Valley, in the sector of Palmares, we find the most spectacular collapsed caldera of the sector. It is an ancient volcanic structure, from the Pliocene, which collapsed at the beginning of the Quaternary. Of this is the testimony of dikes on the rim. Post-collapse volcanism, however, continued giving rise to the volcanic cone known as Cerro Espiritu Santo (4439 ft). The history of the Palmares collapsed caldera extends over the Lower Pleistocene because it once housed a lake. In its waters were deposited volcanic ashes from the volcanic cones that are in the vicinity, in what would be the Central Volcanic Range (Espiritu Santo and Chayote volcanoes). The ashes were locked up in small spherical coffee-colored clayey matrices the size of marbles, by the effects of the rotating movements of the waters of the lake, and were deposited at the bottom. The lake, before being completely burst, emptied eastward, through the Grande River tectonic fault, which took advantage of the crushing of rocks to dig a deep canyon.

5. TALAMANCA MOUNTAIN RANGE

Gradually, Talamanca's mountain range has been hinting at its volcanic past with the intensifying research on its land. Thus, in its Caribbean watershed, Tournon (1980) for the first time described a series of explosion calderas along the Matama Row. The observations of satellite radar imagery have identified a series of volcanic structures. This did not escape the eye of Jean Tournon, PhD, from the University of Paris VI, Pierre-Marie Curie. This French geologist will be remembered for his Costa Rican geological map on a scale of 1:500,000, published in 1995 in Dieppe, France, in which the mentioned sector was punctuated by a series of basalt structures dated from the Pliocene, associated with intrusive formations mostly of diorite and quartz monzonite as well as granite and gabbro. The 2007 geological map by Percy Denyer and Guillermo Alvarado, at the scale of 1:400,000, published by the French Bookstore in San José, takes up part of the indicated information, which is of vast volcanic areas dating from the Miocene to the Late Pliocene that coexist with marine sedimentary formations composed by sandstone, shale, and conglomerates. These volcanic structures refer mostly to collapsed calderas of variable dimensions, ranging around 3.7 miles in diameter (see Fig. 1.6). More to the south, there are also old volcanic structures found in the Coastal Mountain Range as the Miocene–Pliocene alignment goes from the Mano de Tigre volcano through the craters of the Doboncragua to the China Kicha volcano. From the Paso Real sector to China Kicha, the Coastal Mountain Range on its border with the General River is characterized by a volcanic modeling in which are a series of structures very altered by erosion.

The geological sector known as the Paso Real Formation, composed of volcanic conglomerates, is actually a Pliocene volcanic complex (Bergoeing et al., 1978b) J.P. Bergoeing has already

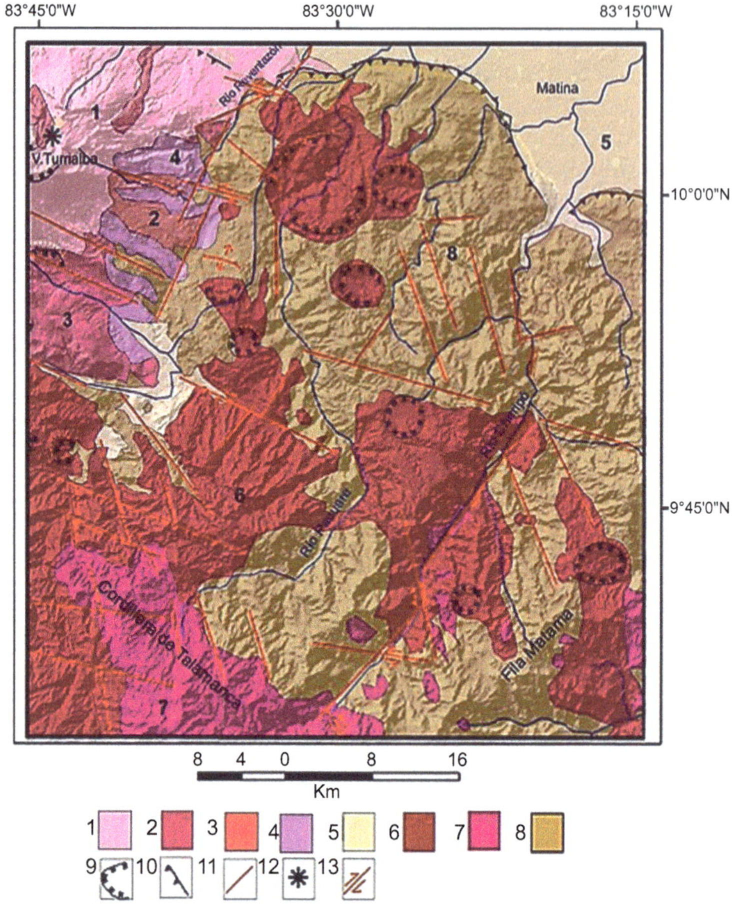

FIGURE 1.6 Neogene calc-alkaline magmatism and caldera structures on the Talamanca mountain range Caribbean slopes, which date back to the Pliocene and were formed in a marine environment before the orogeny. Modified by the author from Denyer's geological map (2007) and the satellite radar image (D: CR_RADAR\CR_HILLSHADES\). (1) Upper Pleistocene volcanism. (2) Middle Pleistocene volcanism. (3 and 4) Lower Pleistocene volcanism. (5) Quaternary, transitional coastal and continental sedimentation. (6) Tertiary volcanism of Talamanca (Miocene–Pliocene). (7) Granodiorite intrusive of Talamanca. (8) Tertiary sedimentary (Uscari and Senosri formations). (9) Volcanic calderas. (10) Reverse tectonic faults. (11) Undifferentiated faults. (12) Crater of the Turrialba. (13) Sliding faults.

recognized the structure of the Mano de Tigre volcano, which lies in El Brujo sector and has an age of 10 million years. From this position, there is a diversity of volcanic forms, remains of craters, calderas, and planezes associated with lahars in a much-altered state.

Further north is a particular volcanic complex that would correspond to the China Kicha volcanic cone, where it is possible to observe a crater in the sector of the town and two structures inside (visible only in satellite radar images).

Finally, the Talamanca Mountain Range, at its southern end, conceals a series of small volcanic cones that are nothing more than the extension of the volcanic system previously described between the Mano de Tigre and the China Kicha volcanoes. Talamanca's mountain range is characterized by two dissymmetrical watersheds. The first one falls to the Pacific, with strong but short and deep slopes, and the second one, with gentler slopes, goes to the Caribbean. It is an eminently volcanic area where stands the Cerro Fabrega caldera, open to the west, as well as Frantzius Hill (7001 ft), Cerro Pittier Hill (9331 ft), Cerro Gemelo (8865 ft) (volcanic cones), and other structures that are located more to the southwest.

These two recent volcanic cones could be associated with the postcollapse Fabrega caldera. The volcanism originated in the Upper Miocene and lasted until the end of the Pliocene and it is likely that it reached into the Lower Quaternary. The moment in which Talamanca's orogenesis began originated in the basaltic to andesitic rocks. The oldest rocks in the sector are sedimentary, formed of shale and conglomerates at the beginning of the Tertiary (Paleocene–Eocene). They are folded, forming synclines and anticlines, and in the area of contact are covered by volcanic deposits from the end of the Tertiary.

It is possible to access Irkibi Volcano through Alturas Farm. The volcanic cone is located to the east of the Frantzius volcano following an alignment of magma extrusion as already indicated. It is composed of several summits that leave one to assume many other craters where lava flowed: the Mount Chai cone of 6890 ft, with the Irkibi cone (7251 ft) (see Fig. 1.7), and further to the east the Cerro Bellavista cone (6719 ft), a volcanic package fitted only by the Cotilo River. At the foot of Mount Chai, lava flows abound, composed of dacites, according to an analysis by the Central American School of Geology. West, the Bellavista River separates the first described volcanic ensemble from a new volcanic system. It is the Cedro alignment, of 8995 ft, whose summit presents an old caldera. All these volcanic summits are limited to the south by the Coton River, and they are probably Pliocene. Among them are granodioritic rock deposits that would indicate a Pliocene volcanic extrusion coming from the granodioritic Cretaceous–Miocene Talamanca's batholith, and which would be extended to the northwest with the volcanic complex

FIGURE 1.7 Volcanic Irkibi Mount, 7218 ft high, covered by a dense tropical forest. *Photo Bergoeing (2011a).*

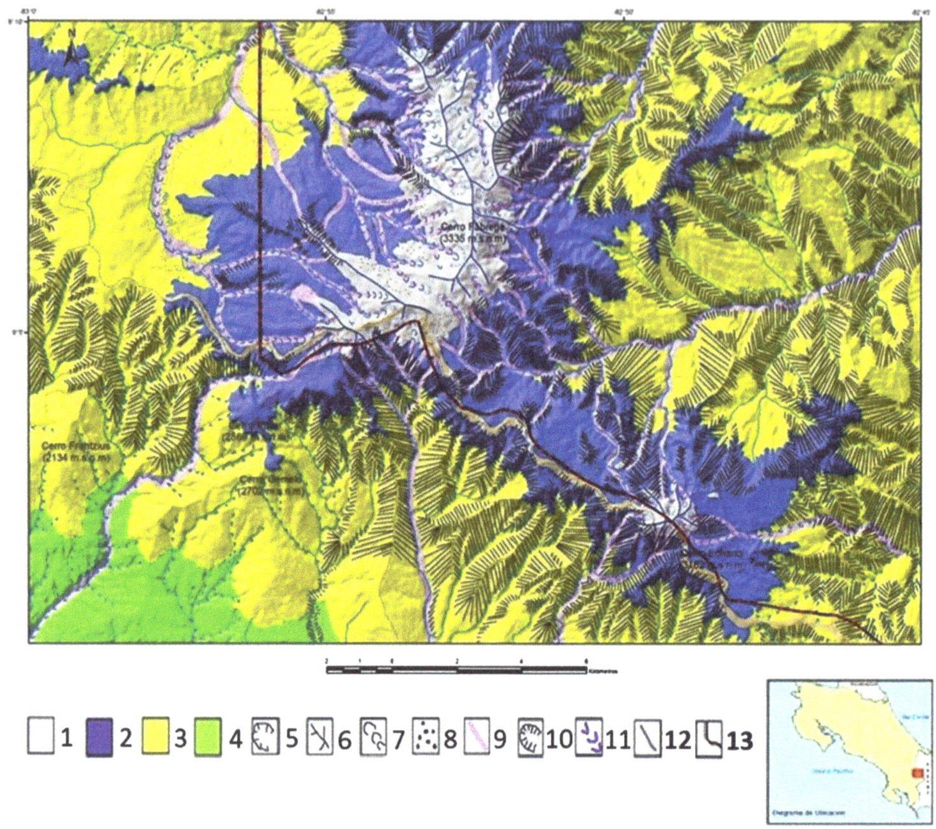

FIGURE 1.8 Maximum Würm–Wisconsinian ice age extension in Cerro Fabrega, Panama–Costa Rica border (Bergoeing, 2011a). (1) Paleo. The Riss and the Würm glacier parking. (2) Paleo. Snow parking during the Riss and the Würm. (3) Paleo. Páramo area. (4) Paleosavannas. (5) Paleoglacier circuses. (6) Edges of paleoglacier origin. (7) U-shaped paleovalleys. (8) Moraine deposits. (9) Paleoglacier valleys with moraines. (10) Slopes from erosion. (11) V-shaped valleys. (12) Divortium aquarium, or line of water partition. (13) International border Costa Rica and Panama.

Tigre–Doboncragua, hand dated to the Pliocene, of age 4–5 million years (K/Ar), by Kessel in 0 (in Alvarado, 2000).

In conclusion, we can say that Costa Rica is a country that is eminently volcanic and its structural construction is due to the clash of the Cocos and the Caribbean tectonic plates, which has resulted in the already described volcanism and orogeny that has raised the whole to a maximum altitude of 12,530 ft (Cerro Chirripó Mountain in Talamanca). This allowed the occurrence of the Riss and Würm ice ages during the upper Pleistocene (Bergoeing, 2011a) (see Figs. 1.8 and 1.9).

Indeed various studies in the field have allowed the determination of the presence of huge erratic blocks in the General valley that are the results of the Riss–Würm interglacial or glacial lakes, moraines, tarns, bolts, and U-shaped valleys, which are strong evidence of the glacial parking suffered by the high peaks in the upper Pleistocene when the orogeny exceeded altitudes higher than 9850 ft. The oldest volcanic and intrusive structures of the Talamanca

FIGURE 1.9 Tarns or paleoglacier lakes of Chirripó summit 12,500 ft above sea level surrounded by Würm–Wisconsinian deposits of moraines; Talamanca Mountain Range. *Photo Bergoeing (1982a,b).*

Mountain Range disappeared because of erosion, even through their relicts allow us to know its past, but Late Tertiary volcanic structures remain standing as a testimony to this igneous predominance in the country.

6. THE PACIFIC LITTORAL

The Pacific littoral of Costa Rica is composed of four peninsulas, Santa Elena, Nicoya, Quepos, and Osa, and Punta Burica. Here you will find the oldest volcanic outcrops of the country dating back to the Cretaceous. The Santa Elena Peninsula is a flake of the seabed made of green metamorphic rocks where the serpentine peridotite is predominant. The Nicoya Peninsula and Osa, as well as the promontory of Quepos, are the oldest remnants of a predominantly basaltic island volcanism at the end of the Cretaceous. At Montezuma Beach, at the tip of the Nicoya Peninsula, Bergoeing (2007) has dated, by ^{14}C in France, a lumachelle level that is located 7 m from sea level, by a strong positive neotectonism, and which marks the beginning of the Flandrian transgression (6620 $\pm$ 150 years BP).

The Central Pacific of Costa Rica is a both a tectonic and a climatic transition area. Indeed, south of the Tarcoles River the influence of the Cocos and Caribbean tectonic plates and the Panama subplate is evident. It is therefore an important seismic zone. The promontory of Quepos is the only notable accident that breaks the monotony of the Flandrian coastal cords of this sector of the Central Pacific. It is a sector where emerge, at the base of the coastal relief, the rocks of the end of the Cretaceous that are characterized by being basalt (pillow lavas) and pelagic limestone. However, here are large coastal plains where the Parrita and Naranjo rivers have built deltas covered by mangroves on both sides of this coastal promontory and there is no doubt they were built after the Flandrian transgression.

However, the country's largest mangrove is located at the mouth of the Grande de Terraba River, which despite its mouth, is a delta where the main channel has been running gradually northward because of the neotectonic process of the sector. The river mouth, covered by a vast mangrove swamp, was built up since the Flandrian regression period 6000 years ago.

The Osa Peninsula, less extended than the Nicoya, is mainly a plateau of sedimentary and volcanic rocks resting on Cretaceous basaltic basement. The Osa Peninsula is tilted toward the northwest as well as Punta Burica by the effects of the clash of the Nazca plate, which slides against the Cocos plate, and a result is that the greatest depths of the Golfo Dulce are in the northwest sector of the gulf and not in its mouth.

7. THE CARIBBEAN LITTORAL

The Costa Rican Caribbean coast, north of the Limon promontory and port, is characterized by a straight coastline that contrasts with the large bays, which were drawn to the south of this port by Flandrian coastal cords, aligned in parallel with the strong littoral drift. Inside is a series of navigable canals (Tortuguero canals) tipped by a dense tropical riparian vegetation that are silent witnesses to the recent emergence of this coast. In fact, we can say that the plain north of Limon, up to the level of 33 ft, was submerged by the Flandrian transgression until 6000 years ago and since then it has emerged, aided by positive neotectonics, as was been demonstrated by the Limon earthquake of 1991, which raised the field by 3–6 ft.

South of Limon, the coastline becomes narrower and forms an ancient reef fringe raised by the neotectonics giving origin to a coastal terrace bordered by a littoral cord to Cahuita Point, where is the only living reef of this sector of the coastal strip. From Cahuita the coral reef emerges from the coast and extends to Bocas del Toro, Panama.

Recent geomorphological studies of the northern sector of the Caribbean Costa Rican facade allow one to conclude that contact between Lake Nicaragua and the Caribbean Sea was definitely close during the Eemian. In fact, the Caño Negro sector constituted until the Holocene part of the Nicaragua paleolake, fed by constant contributions of volcanic material transported by the rivers from the Quaternary volcanic mountain ranges of Costa Rica. Currently the cones reach longitudinal extensions that vary from 18 to 35 miles and are completed by glacis in which run paleochannels containing remaining wetlands and wildlife (Bergoeing and Protti, 2007

FIGURE 1.10 Tortuguero channels. *Oblique aerial photography Bergoeing (2012a).*

in Bergoeing et al., 2007). The glacis reach 66 ft in altitude before contact, in some cases, with the margins of the San Juan River. During the Flandrian, the area was flooded to the Caribbean Sea, which occupied all areas below that altitude, taking into account the positive neotectonics, which corresponds to the Nicaragua graben.

The knowledge of the changes in the coastal relief of Costa Rica during the Holocene are interesting in particular not only to the country but to the international scientific community. During the Flandrian (7000 years BP) the sea penetrated deeply into the coastal low sectors of Costa Rica, making its printed mark in marine fossils (as lumachelles of Montezuma), in fossil deposits of sand (coastal cords) (see Fig. 1.10), and in lakes (today swamps) that are in process of drying. However, these testimonies are now in many cases at more than 33 ft of altitude and others are at 16 ft above sea level, which would be an anomaly, knowing that the Flandrian transgression did not surpass 6.5–13 ft, without the positive neotectonics very active in Costa Rica due to the collision of the Cocos and Caribbean tectonic plates. This knowledge would also allow the preparation of a paleotopographic map of the Costa Rican relief during that period and the interconnections that existed between Lake Nicaragua and a paleo-San Juan River much vaster than at present, with the consequent migration of lacustrine and marine faunas.

CHAPTER

2

Natural Shape Types in Costa Rica

OUTLINE

Costa Rica, a mountainous country, possesses two oceanic facades. Nestled in the humid tropical zone, it presents a diversity of shapes and models, which are the products of geological heritage (nature of rocks, volcanism, neotectonics, etc.) as well as the climatic oscillation games that occurred during the Quaternary (pluvial periods, or biostasis, and interpluvials, or rexistasis). The natural phenomena that have acted, and act still in our days, on this vigorous relief, inherited from the Plio-Quaternary orogeny, have built various models, which distribute at different altitudes. This is how large morphological climatic areas of Costa Rica distribute in altimetry order. To the north, the country is formed of large floodplains (ancient tectonic depressions of the Nicaragua graben and the Tempisque River plain, silted by Quaternary floods).

Geomorphology and Volcanology of Costa Rica
http://dx.doi.org/10.1016/B978-0-12-812067-5.00002-2

The Pliocene–Quaternary volcanic mountain ranges of Guanacaste and Tilaran separate the plains. These, as we will see, are differentiated among themselves by their different shapes.

The center of the country is represented by the Central Valley, a high basin bordered to the north by the Central Volcanic Range, in which certain volcanoes are in permanent activity (Poas, Irazu, and Turrialba) and by this one fact create and give great young volcanic forms.

To the south, the country has a greater variety of models, the Talamanca Mountain Range, including the imposing Chirripó Mountain, with an altitude of 12,530 ft. This impressive mountainous mass allows a complete distribution of forms at the different altitudes because of its elevation. Thus, we find shapes ranging from paleoglacial forms to modeled coastlines.

For this reason, we have elaborated seven fundamental morphologic types of models or shapes:

1. Paleoglacier and paleonival forms
2. Multifaceted shapes
3. Multiconvex modeling
4. Piedmont accumulation shapes
5. Large Quaternary volcano shapes
6. Floodplain shapes

1. PALEOGLACIER AND PALEONIVAL FORMS

Weyl (1955a) was the first to describe, in the Chirripó Mountain (12,530 ft) sector, paleoglacier forms of the Talamanca Mountain Range and speaks of "Pleistocene vestiges of Ice Age."

Indeed, in the Talamanca Mountain Range and approximately on Cerro de la Muerte Mountain, from 9186 ft of altitude, trees disappear progressively, giving way to a shrub vegetation and scrubland and then to a subalpine floor, leaving to appear a few extensive spaces of soil half covered by vegetation. Above, already on the top of the Chirripó, appears a wet alpine form, the páramo, progressing between 11,480 and 12,140 ft above sea level. This becomes a reality because average temperatures vary between 43 and 50°F. The summits are frequently subject to thick mist due to the clash of moist air masses coming from both oceans. On the Cerro de la Muerte, Chirripó, and Kamuk peaks located at more than 11,480 ft of altitude, frost is frequent at night. These peaks know equally sporadic precipitation of snow. At the Chirripó summit, in the Moraine Valley, Kohkemper (1968) in Weyl (1980) measured a temperature of 17.6°F (see Fig. 2.1).

Geomorphological implications of ancient colder periods are visible only in the peaks of the Talamanca Mountain Range. In other sectors, and particularly in the volcanic cone of the Irazu (11,260 ft), or the Turrialba (10,955 ft), there is no evidence, because of volcanic eruptions (Irazu's most recent eruption was in 1963).

1.1 Chirripó's Paleoglacier Shapes

Mainly on the Chirripó and Kamuk summits, it is possible to see small glacier circuses, well developed, whose bottoms are at 11,811 ft altitude. A wider ones is located upstream of the Talari River and extends over an area covering 0.6 square mile. The circus walls have a slope ranging from 500 to 650 ft. At the feet of the circuses, small valleys of concave shape have "locks" and "navels," which are in the sector reaching up to touch small circular lakes or tarns. The concave valleys, or U valleys, develop for a distance of only 0.6 to 1.2 miles. The circuses are surrounded by sharp and well-preserved edges, modeled in a rocky material, eminently granodiorite in the area of Chirripó Mountain, and a material of volcanic origin with respect to Kamuk. Some U valleys are suspended. The Lake Valley of Cerro Chirripó Grande is furnished with small lakes of glacial erosion, or some are located just behind small underdeveloped terminal moraines (Moraines Valley, Uran mountainous row). The strong diurnal temperature fluctuations create effects of cryoclastism

FIGURE 2.1 Talamanca Mountain Range. Tarns, or paleoglacier lakes, appear in old glacier circuses. Observe the outcrops of granodiorite rocks shaped by the erosion of the glacier and the moraines that border the extent of the lake. *Photo Bergoeing (1982a,b).*

(rocks broken by ice) translated by small accumulations of angular debris (debris cones) mostly located at the foot of the rocky walls.

> Everything tends to prove that glaciers organisms of little expansion, but very active, modeled the massif of Mount Chirripo. Without a doubt it was glaciers, quite thick and relatively quick, susceptible to develop deep cirques, U well marked valleys and navels of undermining, at an altitude between 3 800 and 3 500 meters if the merger was probably very fast, the snowy power must have been exceptionally important. ***Beaudet et al. (1982a).***

On the tops of both Chirripó and Kamuk, the forms are very well preserved. Between the two summits, the continental water divide retains equally fresh forms of snow parking, particularly on Durika's mountaintop (10,761 ft) and Utyum Mountain (9990 ft). These same forms are obvious on the Cerro de la Muerte (11,473 ft). The glacial forms of the high peaks of Talamanca are very fresh, as the described material remains intact despite the aggressiveness of the humid climate of the current altitude. This allows us to deduce that the ice age that affected Chirripó was recent. On the other hand, C. Schubert (1984) studied the paleoglacial forms of the Dominican Republic and Venezuela, put them in relationship, and attributed them to an Upper Pleistocene age (Merida glaciation).

The high peak modeling of Talamanca can be divided into two sectors according to altitude:

- Paleoglacier forms sector: Located between 9850 and 12,470 ft and in which traces of one or two recent glaciations are evident (Chirripó, Kamuk)
- Paleonival forms sector: Located between 8530 and 9850 ft, where the forms linked to old snow deposits become evident (Cerro de la Muerte, Cuerici, Durika, Utyum, etc.)

1.2 Cerro de la Muerte Paleonival Forms

The Cerro de la Muerte sector presents small accumulations of snow ("niches"), probably seasonal, favoring cryoclastism at their edges, and the melting waters created downstream deposits of freezing debris surrounded by a clay—fine matrix. This notched the volcanic rocks, altering their shape, modeling them by snow accumulation in small depressions that have a flat bottom and are quite scattered. On these rocks a humus soil was subsequently developed. In some other sectors a shrublike vegetation followed.

In Costa Rica, the highest mountains, over 9200 ft in altitude, therefore met with specific glacial and snow conditions, which have today disappeared. Traces of the cryoclastism, however, are rare, which suggests that it was a rather extremely cold snow paleoclimate.

2. MULTIFACETED SHAPES

This modeling exists where there are important high massifs, subjected to a deep ferruginous alteration, and therefore at the feet of Talamanca's glacial and periglacial modeling. Especially where the volcanic massifs are sufficiently old, a profound change was able to develop (as, for example, in the Tilaran Mountain Range). This type of modeling occurred under conditions of dense forest cover, and steep slopes, which fits perfectly with the high mountains of Costa Rica. The rocks that constitute the nature of this type of modeling are of various origins, and rocky outcrops without alteration are rare. Rocks of plutonic origin predominate in summit sectors, whereas downstream are above all volcanic sedimentary rocks. Costa Rica is therefore a region dominated by schist and volcanic–plutonic rocks. Here the runoff of water plays a big role. These rocks are changed into clays, but as muscovite is very rare in clay, the iron oxides do not abound (in Costa Rica the formation of ferruginous shields has not been observed). This allows, therefore, these clays to be very sensitive to the phenomenon of solifluction.

This phenomenon is widespread in the mountainous areas of the country, but starts and becomes gradually evident in the lower floors at 9186 ft altitude. Multifaceted shapes cover much of the mountainous sectors, places where the rivers deeply fit into rock, digging powerful thalwegs and modeling in facets of triangular-type sheds. The multifaceted topography is therefore ordered by the sizing of the rivers, the steep slopes, and the abundant rains on both sides (Caribbean and Pacific) that reach more than 2500 mm per year. These steep slopes are responsible for cleaning the rocks (altered or healthy), because of a very active orogeny coverage, which continues today, and very active neotectonics. They play fundamental roles in the preparation of the material. Multifaceted shapes exist generally between 9420 and 9186 ft of altitude (without taking into account the large Quaternary volcanic cones), gradually leaving the place to other forms of modeling, in particular those produced by solifluction, with decreasing altitude. It is also true that the slopes have important values: 30, 40, and sometimes up to 45 degrees! Therefore, the multifaceted modeling can occur in two major sectors.

2.1 Forest Sector

The forest sector is situated between 6560 and 9186 ft above sea level (Fig. 2.3). It is a sector populated by large endemic trees and where on-site slopes are altered, leaving features with forms of triangular facets. Rivers at their headwaters start the work of the incision of deep thalwegs. However, thick vegetation covers the slopes of these rivers, which allows the constitution of an altered coverage, little affected by erosion, except when suffering major anthropogenic deforestation (Fig. 2.2).

2.2 Solifluction Sector

The solifluction sector is situated mainly between 3280 and 6560 ft above sea level, but is also present outside these altitudinal levels. Although the slopes are still sometimes acute, the forest has a tendency to disappear because of uncontrolled deforestation, on one hand, and the devastating effects of some exceptional floods during the rainy season. Here the phenomenon

FIGURE 2.2 Chirripó Summit, Talamanca Mountain Range. Multifaceted shape under moor (páramo) vegetation. *Photo courtesy of Karol Fallas (2010).*

FIGURE 2.3 Altitudinal forest sector, Kamuk Mountain in the Talamanca Mountain Range. *Photography courtesy of Guillermo Artavia (2010).*

FIGURE 2.4 Coastal mountain range. Landslides due to creeping and solifluction create a transition of multifaceted to multiconvex shapes. *Photo Bergoeing (2009c).*

of solifluction appears clearly in net-carried red clay soil. The top sectors are affected by laminar solifluction, which leaves many times an outside sound rock, but retains the multifaceted shape. Downstream, a mass solifluction appears, with big erosion shapes, sometimes on the order of more than 30 ft. Solifluted material covers the marshy bottoms of these basins (see Fig. 2.4).

3. MULTICONVEX MODELING

From 4265 to 3280 ft altitude, and downstream, the mountainous Costa Rican sector is transformed into a hill landscape of rounded shapes. Rocky outcrops become scarce, except in the thalwegs, where the weathered layer disappears completely, from the effects of erosion, exposing the unaltered rock. These rocks are predominantly volcanic (andesite) or sedimentary (flysch).

An ombrophilous forest usually covers this modeling and weathered layers can be several meters thick (see Fig. 2.5). Red clays, which constitute the outer part of the hills, are often of montmorillonitic origin and hydrate easily and rapidly in the humid tropical climate. In these hills, however, one feels the presence of andesitic blocks more or less used, which testifies that the weathered layer is not very thick; however, it is the result of an in situ evolution. On the steeper slopes, these clays are the result of material transported from the top (torrential lava, solifluction). Indeed, this modeling seems to be mostly where a cyclic evolution has been able to develop (multiple levels at altitudes, often difficult to recognize). It is a privileged, human-intervention sector that leads to areas of pasture where there are still some patches of original forest, on both banks of the thalwegs, or on the tops of the hills.

Here the water flows only during the rainy season, with the exception of the rivers. These multiconvex forms are very evident in Costa Rica, particularly in the hamlet of Pavones, on the path to Siquirres (East Central Valley). The residual volcanic relief (Aguacate Formation, Pliocene) leaves, little by little, a wavy, planted landscape of pejibayes (*Gulielma utilis* Oerst), where pace Brahman-type cattle, and for which

FIGURE 2.5 Sarapiquí sector multiconvex shape (Bergoeing, 2011a).

such lands were deforested. Other examples are possible to see on the hills, where ripples the Siquirres–Limon route. On the Pacific slope, the examples are not scarce, in the Esparza piedmont, following the route to Puntarenas, as well as in the Parrita piedmont following the Parrita–Quepos route, and in the Coastal Mountain Range road following the San Isidro–Dominical route. Finally, this shape is also present in the east piedmont of the Guanacaste Volcanic Range, where clay deposits of the slopes reach considerable thickness (Colon River sector).

4. PIEDMONT ACCUMULATION SHAPES

The landscapes of low altitude in Costa Rica, out of earlier forms of multiconvex shapes, characterize the construction of large alluvial fans, glacis, and river levels.

4.1 Alluvial Fans

There are large cones that form the base of the Talamanca Mountain Range (especially in the Pacific watershed and particularly in the General Valley) (see Fig. 2.6), as well as appearing in the Quaternary volcanic mountain ranges. There are powerful coalescing alluvial fans, with sometimes acute slopes (25–30 degrees), products of rains and river deposits during the Quaternary. Clay is the matrix of the lithic material, in blocks from 1 to several feet in size, but well worn. Lahar layers (chaotic masses of ashes, clay, blocks of lava, and mostly pumice) coat some alluvial fans, next to recent volcanic emission centers, such as the Irazu and Platanar volcanoes.

The most representative modeling of this type that can be observed is in the General Valley (southwestern sector of Talamanca), the big northeast piedmont of the Central Volcanic Range, the Tilaran and Guanacaste mountain ranges, the southwest sector of the Tilaran and Aguacate mountains, and finally the Parrita–Quepos area of contact with the littoral plain.

4.2 Accumulation of Glacis

The large alluvial fans built by debris from the Quaternary volcanic ranges come into contact with the northern plain, imperceptibly becoming forms of glacis accumulation from the moment in which the descent slows (Fig. 2.7). The material that covers the surface of the glacis is generally thick upstream and is made up of boulders that

FIGURE 2.6 Quaternary mega-alluvial fans or dejection cones of the General Valley. *Photo Bergoeing (2008).*

FIGURE 2.7 Piedmont filling of alluvial fans becoming glacis in the northern region, Sarapiquí. *Photo Bergoeing (2011a).*

rest on a sand and silt matrix. Downstream, the material becomes increasingly fine until the boulders disappear completely. In these glacis, torrential-regime rivers, as in the case of the Toro Amarillo River, which plunges by the northern slopes of Irazu, frequently form vagrant riverbeds during the rainy season. The fluvial network forms braided channels with the creation of river barriers. Some are destroyed and rebuilt according to the vagaries of the river, modeling them. It is only once the general level of the Great Plain (328 ft above sea level) is reached that the rivers lose all their competition and consequently describe numerous and large meanders.

4.3 Fluvial Terraces

The third type of modeling of the sector is therefore characterized by numerous river levels, which we find everywhere on the riverbanks. Generally there are two or three levels, many times encased, consisting of well-worn pebbles with a degree of alteration in the surface, directly related to the antiquity of the level. These floors are described in detail in the chapter devoted to the great morphological units; however, we can say that in Costa Rica, there are three fundamental river levels that are very well represented on the river banks of the

FIGURE 2.8 Orosí Valley showing two cyclic fluvial terrace levels produced by the melting of the ice cap of Talamanca 12,000 years ago and where Orosí Town stands. *Photo Bergoeing (2006).*

Reventazón, Tarcoles-Virilla, Chirripó del Pacífico, Barranca, and others (see Fig. 2.8).

The Pacific piedmont of the Tilaran Mountain Range shows equally varied deposits, which characterize alluvial fans that overlap one another, laterally passing to fluvial terrace levels. As an example, we will cite the case of the Ciruelas River.

The alluvial fans are assembled in the fluvial terraces, downstream, where they dilute into the coastal plain, which, incidentally, is very recent (Flandrian). The fluvial terraces present three very different levels, which are visible on the banks of the Guacimal, Ciruelas, and Aranjuez rivers. Let us then see the description of the terrace levels of the Ciruelas River, on the road to the Port of Chomes.

T1 level: It is 49–65 ft from the T2 level, formed by boulders very well worn and with important volume, 0.0032 to 0.016 ft in diameter, with some blocks of rolled stone. The rolled material has little surface alteration. The largest riverbed is covered by very well worn boulders forming braided islets.

T2 level: This is a high well-developed terrace, consisting of rolled pebbles of 0.065 to 0.10 ft in diameter, little changed on the surface. Boulders rest on an array of red clays. The material presents a significant wear and is well tamped.

T3 level: This is the top level attached to the alluvial fan of outfall fans. At 65 ft above the T2 level, this level consists of a heterometric material. The lithic material is very mixed and rests on a dark red clay matrix.

The Tilaran coalescent alluvial fans, where the river level appears, come to die in the littoral, becoming in turn a plain of much flattened alluvial fans passing to glacis. Finally, the youngest alluvial fans give rise to currently deltaic forms covered by mangroves (*Rhizophora mangle*). Only major accidents break the monotony of the inclined plane, formed by the alluvial fans and glacis out of the incision of rivers and fluvial deposits of volcanic or sedimentary origin from isolated hills, which have not been completely submerged by Quaternary alluvial spills.

These torrential washes, as well as the river levels, are directly related to the climatic crisis that occurred during the Quaternary period, with repetitive characteristics, and they eroded the mountain, taking a considerable amount of material, because sometimes spills covered a surface of approximately 39 square miles, between the extreme limits of the Lagarto River west and the Barranca River east. On the other hand, the average distance between the peaks of Tilaran, which stand at an average of 3300 ft altitude, and the mangrove coast is only 14 miles. The easy erosion of the slopes due to floods can explain this watershed, in part, in addition to torrential rains, taking into account also the igneous material already predisposed to profound alterations under the tropical wet climate.

5. LARGE QUATERNARY VOLCANO SHAPES

Modeling of Quaternary volcanic mountain ranges is different from one mountain range to another, and it depends on the volume of the volcanic sets, as on the value of the slopes and their position with regard to the northeast trade winds, as well as the value of the human presence, which plays an important role in the degradation of slopes.

5.1 The Central Volcanic Range

This is higher than the Guanacaste Volcanic Range (Irazu's summit reaches 11,329 ft against the 6654 ft of the Miravalles summit). Therefore, in the volcanic peaks of Irazu and Turrialba, we find the moor level. The mountain range, which is directed toward the Caribbean, has two very asymmetric slopes, from both the topographic and the shape modeling points of view.

5.1.1 The Northern Slope

The northern slope is covered by dense mountain tropical forest, with no real dry season. Deep ravines (a more than 984-ft difference between the base of the volcanic lava flows and the thalwegs) erode it. At the summit, between the craters of Turrialba and Irazu, the northern side shows very active erosion, which gives rise to lavakas. By contrast, the northern Poas and Barva volcano slopes, with larger masses, found the material needed to build a large piedmont, highly developed, where the phenomena of canyon formation and mass solifluction are checked on a daily basis. Volcanic rocks are present with an advanced surface alteration.

The Platanar—Porvenir north piedmont has the same characteristics as the previous ones, with deep notches of fluvial origin, material that eminently failed because of strong volcanic activity (Chocosuela's caldera is inserted into a series of concentric calderas indicating the repeated collapse of this volcanic complex). It is on the northern slope of the Central Volcanic Range, where are the main rivers of the northern plain, and goes directly to the San Juan River or the Caribbean Sea. They are responsible for the large piedmont accumulations (coalescing alluvial fans and glacis) that run from Siquirres southeast to Ciudad Quesada northwest.

5.1.2 The Southern Slope

Protected from the northeast trade winds, the southern slope presents as a "drier" slope, due to a dry season of 4 months. Because of this, humans have exploited it since the beginning of their occupation. Only the volcanic peaks, dominated by the altitude forest or the "páramo," or moor, are not exploited, because they are under "Natural Parks" regime protection.

From 4600 ft altitude, these slopes present strong human activity that translates primarily as grazing areas in the form of "bocage" (closed green fields), as well as open fields dominated by European-type crops (potatoes, onions, artichokes, ornamentals, and irrigated crops). From a physical point of view, these slopes are under a thick cover of volcanic ashes, due to the northeast exposure to winds. These ashes have evolved into brown highly fertile soils, giving rolling sheds toward 6560 ft altitude, carved by the main rivers that delve deeply into their runways, sometimes making canyons more than 300 ft deep. Downstream, the piedmont is much more rugged because of the presence of fractured volcanic terminal lava flows and the presence of an important tectonic fault line escarpment, much eroded (Poas and Ciruelas rivers sector). This escarpment is the result of a tectonic fault located between Naranjo and Heredia, which in some quarters is fitted by the rivers that descend from the volcanic piedmont, forming a hydrographic subparallel network. At the current level of the Central Valley, discharges produce accumulations forming important and predominantly lahar cones.

Today, these cones present a revival of the erosive phenomenon.

In short, the southern slope of the Central Volcanic Mountain Range is constituted mainly of lahars, and is covered by thick layers of volcanic ash. This material often hides important old andesitic lava flows. The most recent were found only approximately by the crater where they are exposed forming cornices.

5.2 Guanacaste Volcanic Range

Like the previous one, this mountain range has a northeast slope directly exposed to the humid trade winds from the Caribbean and which, therefore, is very different from the southwest side. In addition, the volcanic range presents as a series of cones or isolated volcanic ensembles, which are disposed in a rosary form, separated by low-lying hills.

5.2.1 The Northeast Watershed

The northeast watershed is constituted by three different sectors:

- **The sector of volcanic peaks** is fitted by deep streams of a torrential system and covered in part by a dense forest of altitude, of difficult access.
- **A clay plateau** formed by a thick material (cast mud), lahars, and solifluction material from the higher sector is next. It is a very compact brown-clay color material with little pebbles, chaotic, with large elements (diameter greater than a meter).
- **An area of contact with the floodplain** follows. The northern contact is frank, because the chaotic mass of the plateau passes to the Paleocene–Eocene sedimentary hills. To the south, contact is made by means of alluvial fans, sometimes major, such as those located between the Tenorio and the Arenal volcanoes. The rivers that descend from these summits have high capacity because of the acute slopes (between 27 and 32 degrees); however, reaching the floodplain, the rivers depose the entire thick load, carrying it to the foot of the slopes.

5.2.2 Southwest Watershed

The Pacific Guanacaste Volcanic Range slope enjoys a prolonged dry season (5–6 months) affecting all the Guanacaste region. To this is added the effects of foehn winds and the northeast trade winds, when they exceed the hills and peaks of the Guanacaste Volcanic Range. The volcanic peaks downstream are transformed in piedmont clay, which quickly disappears, coated by an important ignimbrite deposit (more than 300 ft). This southwest side can be divided into several sectors:

- **Open craters area to the west**: They are covered by their original important vegetation and have not been exploited by humans, except in the case of the Rincon de la Vieja volcano; because of the recent volcanic eruptions there the vegetation is in a phase of reconstitution. Hillside, dense vegetation disappears to leave room for altitudinal grazing areas.
- **Altitudinal grazing area**: This is an old dense Mesophyle jungle area, with soils predominantly covered by lava flow (andesite) outcrops between this sector and the plateau, where it disappears under the most recent ignimbrite flows. This sector is also the headquarters of the first lahars and an altered ash soil landslide area (for example, the lahar flows of Cacao volcano).
- **First ignimbrite spills in contact with the plateau** were scattered during the Pliocene and Low Quaternary, and downstream they continue to constitute a vast topographic plateau.
- **The ignimbrites plateau**: This is of considerable dimensions, as it covers an area 62 miles in length by 13 miles in width. It is located at the foot of the volcanic mountain

FIGURE 2.9 Guanacaste ignimbrite plateau with xerophyte vegetation. In the foreground is the collapsed Guayabo caldera. *Photo courtesy of Gilbert Vargas Ulate (2009).*

range on the Pacific slope (see Fig. 2.9). Its Great Plain, lake levels, and fault escarpments characterize it. There are various sectors at the heart of the plateau:

- **a high external escarpment** (because of the active neotectonics in this region), in contact with the Tempisque River plain; contact corresponds to a tectonic fault line escarpment;
- **a more depressed area, the center of the plateau**, with the presence of lacustrine terraces (estates: Ciruelas and Pelon de la Bajura);
- **an area of small ignimbrite cones and erosion terraces** fitted in the same ignimbrite, in the rivers area in contact with the plateau.

Similarly, the ignimbrite plateau possesses various sectors that we will develop in the following chapters. They are:

- **Bagaces to Liberia sector**: A former paleolacustrine dam area
- **Santa Rosa sector**: The highest point of the plateau, which has a former Tempisque River lake area
- **La Cruz sector**: Characterized by its staggered fluvial erosion terraces and its northward draining

6. FLOODPLAIN SHAPES

In Costa Rica, well-developed floodplains are especially in the Pacific watershed, without omitting of course the Great North Plain. In the Pacific area, we can distinguish four major sectors subject to weather conditions that confer their origin.

- The Great North Plain
- The Tempisque River tectonic basin
- The littoral plain of Parrita–Dominical
- The littoral plain of Diquís–Coto Colorado.

The morphological characteristics of floodplains in Costa Rica due to the climatic differences of each sector are associated with the type of rocks that sometimes emerge on the same plain, forming small isolated hills. In effect, the Great North Plain and the Diquís–Coto Colorado sector present a greater precipitation (more than 4000 mm per year), and the dry season is very short (a month), compared to the rest of the country. The Parrita–Dominical sector ranks as a transition area, with annual rainfall between 2000 and 3000 mm, and the dry period is longer (2 to 3 months). This sector is also a biological sempervirens species and deciduous transition zone. Finally the Tempisque River plain is the most relevant area, with a prolonged dry season (5 to 6 months) and weak annual rainfall (1600 mm).

6.1 The Great Northern Floodplain

It is a vast expanse of lowlands, which originates at the foot of the volcanic ranges of Guanacaste, Tilaran, and Central, after a progressive contact with large Quaternary piedmont alluvial fans. It extends, to the north, to the edges of Lake Nicaragua and the right bank of the Río San Juan, and into the interior of Nicaragua. To the east it is bordered by the Caribbean Sea. To the south it narrows considerably, after the Limon promontory, because of the presence of important side spurs of the Talamanca Mountain Range. The Great North Plain covers an area of 4633 square miles, which represents 25% of the national territory.

The plain has several names according to the sector. In the far north the country adopts the name of Guatusos Plain (the native name for the place). It is a depressive and flooded area, characterized by Lake Caño Negro, a vast marsh of shallow waters growing disproportionately with the episodic floods of the rivers and leaving some sectors isolated (see Figs. 2.10 and 2.11). Therefore, it is an area of difficult access, with roads unusable during the rainy season; however, it is also an area of great commercial value to the country, as it produces the red bean (*Phaseolus vulgaris*), one of the main elements of the daily diet of Costa Ricans.

In the center, the plain is known as San Carlos Plain. It is a relatively narrow sector, because between the contacts with the Central Volcanic Mountain Range and the San Juan River, there is a distance of only 10 miles in a straight line. The rivers that bathe the San Carlos Plain are the largest rivers (San Carlos, Sarapiquí rivers). In the margins, despite the absence of observable river levels, the beds of the rivers have rocks of reduced dimensions (0.016 to 0.065 ft diameter), which are part of a compact sandy–clayey material. These rivers are navigable, in their lower courses, by small boats, allowing them to reach the San Juan River. Similarly, San Carlos Plain

FIGURE 2.10 Great northern floodplain in Caño Negro sector. *Aerial photo, courtesy of Francisco Solano Mata (2010).*

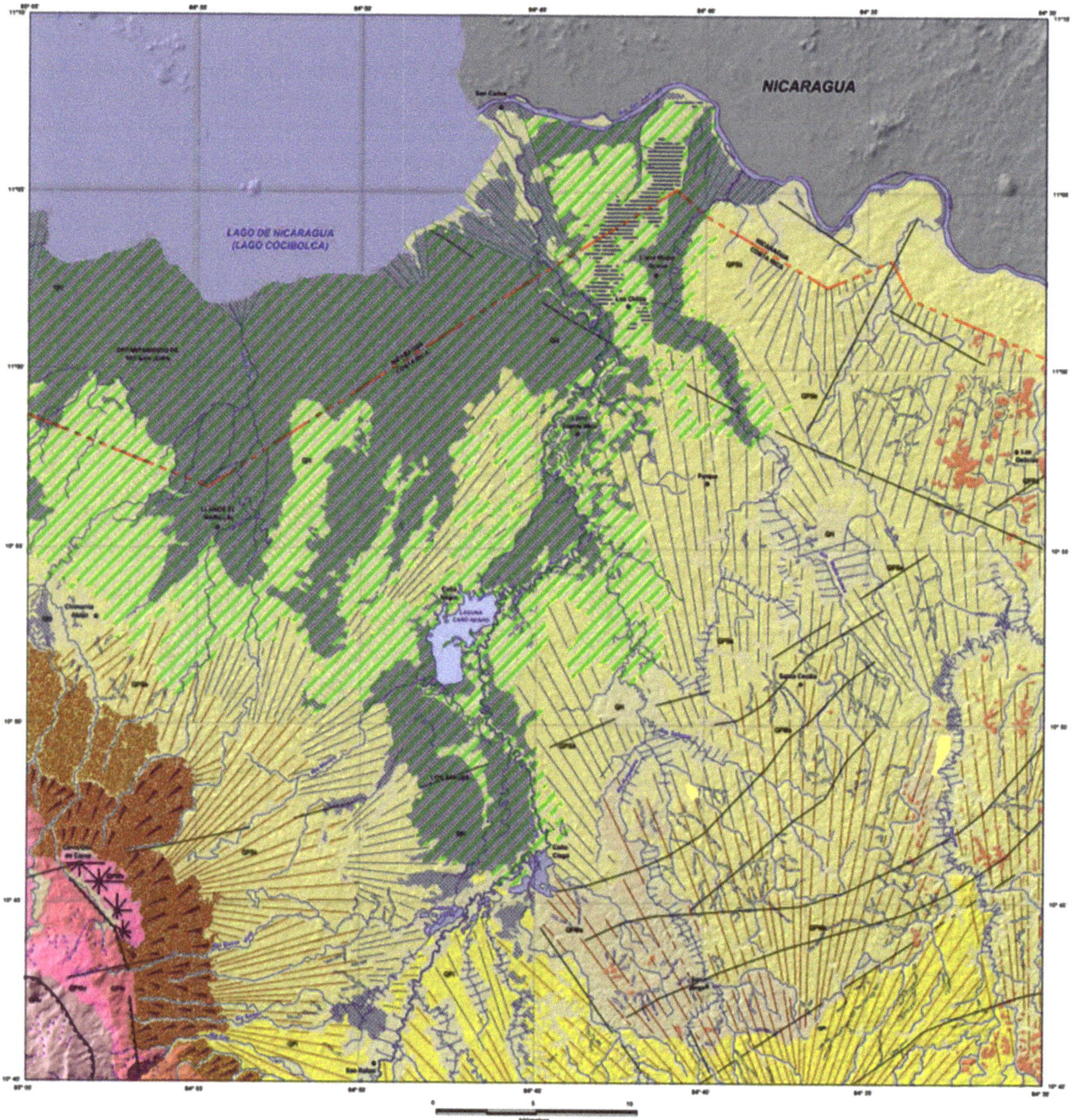

FIGURE 2.11 Northern floodplain in Caño Negro bordering Lake Nicaragua (Bergoeing, 2016). Geomorphology Atlas of Costa Rica–UNA.

is characterized by some isolated, residual hills of volcanic origin, specifically, Tortuguero calc-alkaline volcanism (Tournon, 1972). Some tops may reach an altitude of 660 ft, which break the flat monotony of the sector (see Fig. 2.12).

The lower course of the San Juan River until its delta sector marks the northern limit of the third sector known as the Tortuguero Plain, which runs south to the Parismina River. It is a riparian, very moist, and frequently flooded

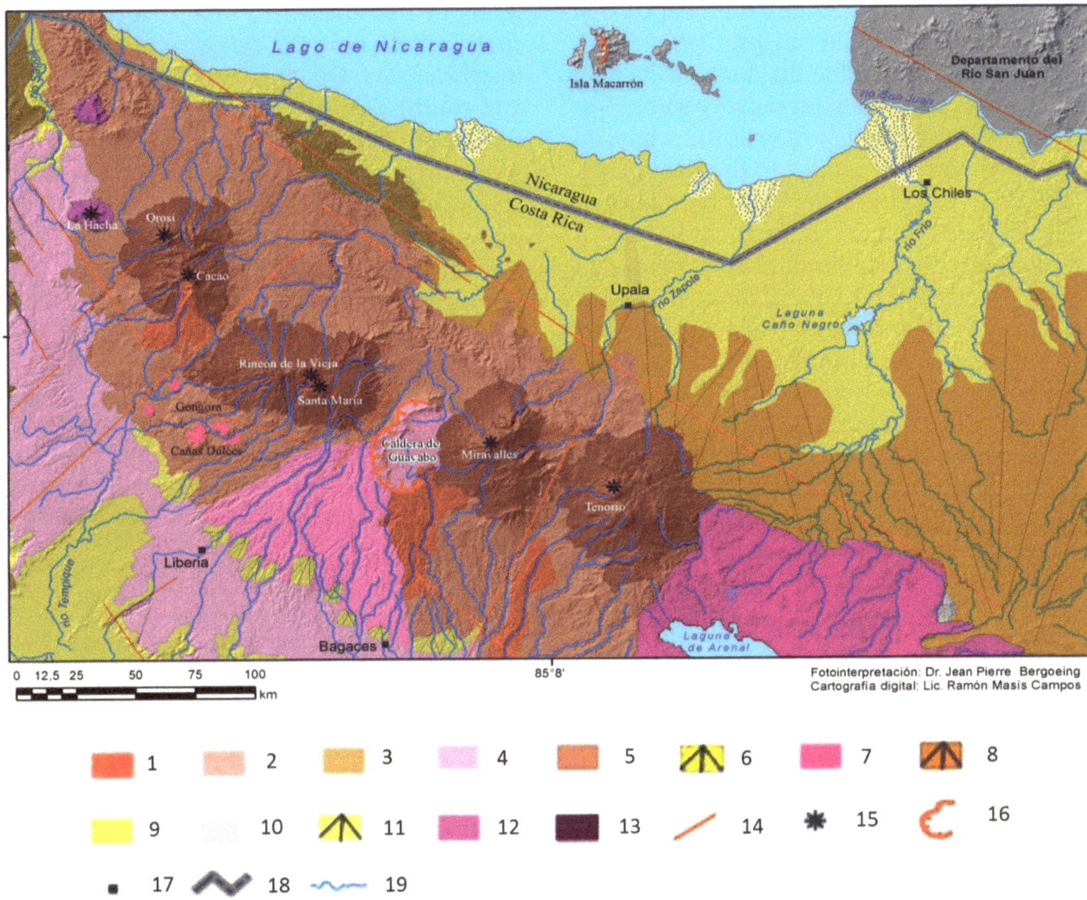

FIGURE 2.12 Geomorphologic map of Costa Rica–Nicaragua intergraben sector (Bergoeing, 2012b). (1) Lahar fields. (2) Quaternary altered lava flows. (3) Paleocene–Eocene sedimentary hills. (4) Guayabo's caldera ignimbrite cone. (5) Middle Pleistocene volcanic cones and craters. (6) Lacustrine deltas. (7) Pliocene Cañas Dulces dacitic domes. (8) Pleistocene big alluvial fans. (9) Alluvial plain. (10) Liberia ignimbrite plateau. (11) Little alluvial fans. (12) Pliocene volcanic remains. (13) Pliocene volcanism. (14) Tectonic faults. (15) Volcanic craters and cones. (16) Volcanic calderas. (17) Urban centers. (18) International border. (19) Fluvial network.

area, with wetlands and coastal lagoons. However, two sectors can distinguish this frequently flooded green set: the sector of the Lomas Azules or Lomas de Sierpe hills (1293 ft) and Lomas de Colorado and the small peak of the Tortuguero volcano cone (Tournon, oral communication). This old volcanic sector consists of basanite lavas whose age is Pliocene as they oscillate between 3.6 and 3.07 million years (Saginor et al., 2011 in Alvarado and Gans, 2012).

In this last Tortuguero sector, the rivers have lost all their force. Therefore, they describe a number of highly sinuous meanders. Finally, the coastline is characterized by a series of Flandrian coastal cords (Battistini and Bergoeing, 1983).

The fundamental morphological features of this Northern Great Plain are the following:

- **Subsident area**: This area corresponds to the extension of the Nicaragua graben, filled in part during the Quaternary with deposits of the volcanic mountain ranges.

- **Great Plain**: Located to the north of the plain, the Great Plain is altered only by some residual isolated reliefs of volcanic origin; it is oriented NW–SE parallel to the Central Volcanic Mountain Range (northern border end of the Nicaragua graben).
- **Frequently flooded land**: The overall level is situated at 328 ft because of frequent rainfalls brought by the northeast trade winds and by a weak dry season (unfolded: October and January).
- **Rivers with important flows**: Forming meanders in the lower courses, these rivers give rise to coastal lagoons due to the presence of Flandrian cords. They are often communication waterways.

6.2 Paleolacustrine Geomorphology: Southern Sector of Lake Nicaragua

Lake Nicaragua, or Cocibolca, gradually formed during the Quaternary period. In effect the Ends–Tertiary orogeny left exposed folded (anticline) sediments of the Rivas Formation (mainly sandstone and limestone) constituting an isthmus some 65–100 ft wide, which definitely interrupts the contact of the southwestern region of Nicaragua with the Pacific Ocean. Inside, the tectonic depression of Nicaragua was occupied by a lake of marine origin whose salty waters would become fresh over the millennia thanks to the numerous tributaries that feed it. The dimensions of the lake reduced considerably throughout the Quaternary period. Today the lake covers an area of 3191 square miles and its major axis is 100 miles. Bathymetric lake studies indicate that the average depth is 42 ft, with maximum depths between 197 and 230 ft located 0.8 mile southeast of Ometepe Island.

At the beginning of the Quaternary period there was a salty lake where lived marine fauna, such as the shark *Carcharhinus leucas*, which would slowly adapt to the conditions of lower levels of salt created by the surrounding river contributions from the Tertiary volcanic foothills (northeast) and Quaternary in construction (west and south). During this period, communication with the Caribbean was broad. The intergraben sector that extended to the south, in Costa Rica, would be fed with sediments from the Guanacaste and Central volcanic ranges, and also with contributions of lahars to the foot of the slopes, and the construction of powerful alluvial fans would reach linear dimensions of 25 to 37 miles going to the north.

> The post-orogenic period corresponds to the Pliocene and Quaternary, characterized by an important development of tectonic fault system and big blocks that definitely make the major structural units. Particularly with consequences of tectonic activity in the sector of major shortcomings of the liminal, basins burier, as the depression of Nicaragua transformed into real grabben [sic]. These depressions were the centers of accumulation of the erosive elements of the mountain ranges which rose concomitantly with products of volcanic eruptions that occur at that time (...). ***Butterlin (1977).***

Also present was an underdeveloped volcanism (Tortuguero–Sierpe), leaving small isolated volcanic mountain ranges in the middle of the floodplain under construction (Tournon, 1972). However, the intergraben sector also started to give signs of rising from the Middle Pleistocene. Flood rains that characterized the Upper Pleistocene period would end the construction of the powerful alluvial fans, which would take a multiconvex shape thanks to the consistent contributions of the volcanic mountain chains always in construction. The positive neotectonics were a constant that must be considered from that moment. During the Upper Pleistocene, i.e., some 200,000 years ago, south of Lake Nicaragua the extension went about 13 miles into the interior of Costa Rica in an area between Los Chiles, Caño Negro, and Upala, which was surely affected by the Eemian transgression, again giving communication with the Caribbean Sea (see Fig. 2.13).

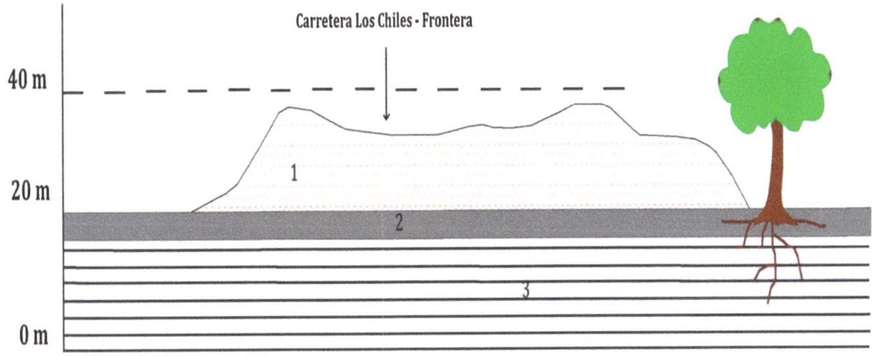

FIGURE 2.13 Fluvial–lacustrine terrace of Los Chiles (Bergoeing, 2006). (1) Lixiviated laterite–kaolin clays. (2) Lacustrine gray stratification melted with red clays of the upper level. (3) White lacustrine stratification probably of diatoms.

6.3 Middle to Upper Pleistocene Alluvial Fans

These alluvial fans start from the lower limit of the volcanic lahars coming from the volcanic mountain ranges. Those deposits cover the southern limit of the Nicaragua graben, which extends deeply northward, reaching linear distances of 30 to 40 miles. They are mainly made of decomposed ferruginous red clays. These alluvial fans contain upstream some chaotic volcanic blocks (andesite, tuffs, and basalts) several feet in diameter within this matrix, which are partially altered on the surface. This is the material that has better resisted the general breakdown of the lithic material swept away by river spills from this period, succeeded by alternate periods of rexistasis and biostasis (Bergoeing, 1998). Paleochannels and modern channels in that sector of multiconvex modeling carve the alluvial fans.

6.4 The Fluvial–Lacustrine Terraces

From the Upper Pleistocene to the Holocene, the various rivers of Lake Nicaragua, including the Frío River, deposited sediments in the former paleolake, which were of the same genesis clay as the alluvial fans, because these eroded materials are transported in suspension by runoff. The Los Chiles lacustrine terrace cut allows us to see its genesis. Between Los Chiles and the border with Nicaragua arises a fluvial terrace of elongated NE–SW orientation. It rises about 33 ft over the elevation of 131 ft above the sea surface. The wavy surface of this terrace was shaped by the contributions of the Frío River, which flows currently 11 miles to the north, ending in a delta in Lake Nicaragua only a few miles away from the drain mouth of the San Juan River. The top of the terrace consists of decomposition clays, very leached with whitish spots, whose thickness is about 33 ft, resting on a gray clay level (base level) that does not exceed 3 ft thickness, resting on a lake level of white kaolin clay.

The current paleolake level of Los Chiles (see Fig. 2.14) is therefore situated at 100 ft (the mirror, or lake surface, measured by the Instituto Nicaragüense de Estudios Territoriales of Nicaragua, giving a value of 103.31 ft above sea level). We can calculate that between Los Chiles and the lake sector, the Frío River distance is 11 miles, being constructed for only about 17,000 years (if we calculate that deposited sediments in the middle lake gain approximately 0.62 mile of ground every 1000 years). This indicates that the Flandrian transgression (−7000 years) probably interrupted the sedimentation process; therefore the rate of sedimentation could be much faster and in that case be fully Holocene.

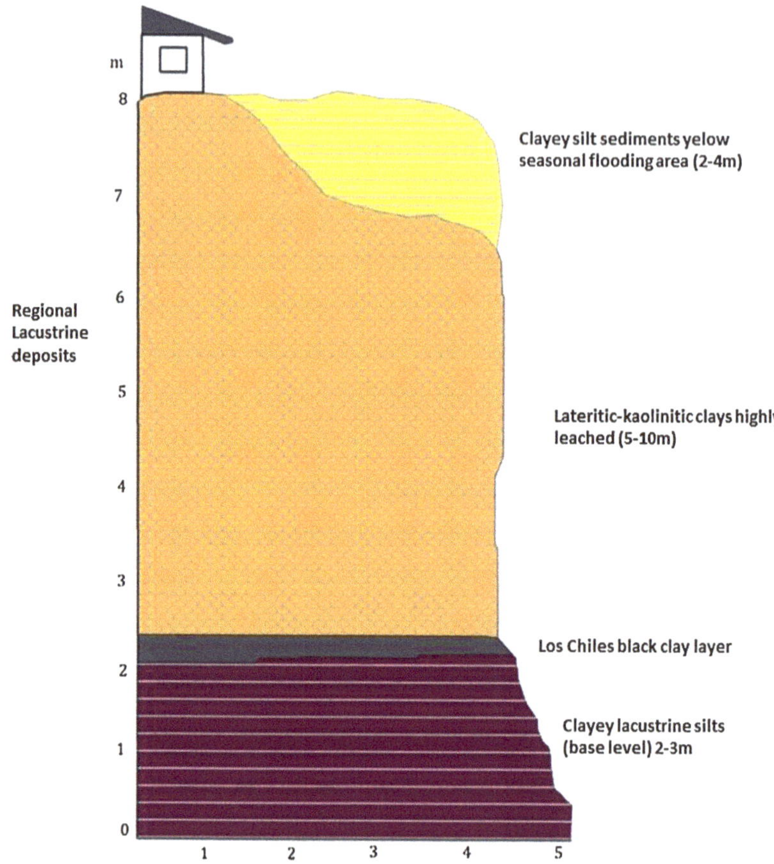

FIGURE 2.14 Fluvial terrace in the Frío River bridge on the Los Chiles–Caño Negro Road (Bergoeing, 2006).

6.5 The Fluvial–Lacustrine Terrace of the Frío River

The Frío River fluvial–lacustrine terrace is well exposed on the left bank of the bridge of the highway between Los Chiles and Caño Negro that follows it (Fig. 2.14). The Frío River fluvial–lacustrine terrace, characterized by 33 ft of leached red clay deposits, is a regional repository, which characterizes the Upala low sector and Los Chiles. It corresponds to the lake terrace upper reservoir in Los Chiles. At the top of this reservoir, the Frío River has built a fluvial terrace composed of stratified yellow slime–sandy sediments, which correspond to the seasonal flooding of the area by water carried by the river. At the base, almost in contact with the current river level, it is possible to observe red clay resting on top of black clay, a stratum about 1 ft thick, which we can also observe in the current margins of the Frío River in Los Chiles. Under this black layer appears a lacustrine clayey clear silt stratification, which represents the base level.

6.6 Frío River Basin

The Frío River originates, strictly speaking, from a number of tributaries, which originate in the Guanacaste Volcanic Range, between the slopes of Tenorio Volcano and Lake Cote. The basin has a high risk of erosion. Its mouth is formed by a delta less than a mile from the drain of the San Juan River. The amount of sediment transported by this water system in the upper part of the basin was demonstrated by the following study:

> The specific sediment yield calculated in **Guatuso station** was of 298 tons/km^2/year, and in the **Venado station** 181 tons/km^2/year of lesser magnitude than the contributions of the basin of the San Carlos basin river, but with the same implications on turbidity. *UNEP-OAS (1997)*.

This study not only demonstrated the amount of current sediment transported in suspension and dilution by the Frío River, but also the immediate consequences in the high speed of the silting up of the wetlands in the areas flooded, particularly during the summer period. It is inferred that during the Middle and Upper Pleistocene, particularly under a climate in which it underwent phases of rexistasis and biostasis, the tectonic depression of Nicaragua was filled quickly, decreasing the lake surface and transporting the lake coastline north (see Fig. 2.15).

6.7 Drilling of Deep Wells

The Costa Rican Oil Refinery Company (RECOPE) carried out a drilling of 23 deep wells from 1922 to 1988, associated with international consultants such as Pemex, Sinclair, Union Oil, Gulf Oil, and Elf. Two of these perforations are a matter for the study area: piercings practiced in Pataste and Tonjibe (basin of San Carlos). The lithological column results reveal that the observed rocks belong to the geological Venado Formation and Aguacate Group, where it follows that the intergraben sector of the studied area was first occupied by marine sediments and then by volcanic deposits associated with the orogeny in this sector of Central America. Here are the details of the drills:

> **Pataste Well No. 1**: The sequence began with 984 ft of pyroclastic rocks and volcanic material correlated with the Cote and Buena Vista Formations. The 984- to 3373-ft range is composed of shale, sandstone tuffs, limestone, and coal horizons corresponding to the Venado Formation, and finally a section of 3373–6200 ft is equipped with rocks correlated with the Aguacate Group. Ages, assigned by Bujak Davies, are Quaternary, for the rocks found up to 984 ft, based on the discovery of the palynomorphs *Bombacacidites baculatus* and *Pediastrum simplex* alga, and Middle Miocene, Late Miocene, and Pliocene for the Venado

FIGURE 2.15 Geological profile of the Liberia plateau through the Miravalles Volcano to Lake Nicaragua by Bergoeing (2006). (1) Quaternary stratovolcano volcanism. (2) Pliocene–Quaternary Liberia ignimbrite. (3) Pliocene–Pleistocene Guayabo ignimbrite. (4) Lower Pleistocene, volcanic flows. (5) Miocene, Venado sedimentary formation. (6) Paleocene, Rivas sedimentary formation. (7) Miocene–Pliocene comagmatic Aguacate Formation. (8) Tectonic faults. (9) Graben border.

Formation sequence, given the occurrence of the planktonic foraminifera *Globoquadrina baroenoenensis* and *Globorotalia (Jenkinsella) mayeri*; the dinoflagellates *Diphyes latiusculum*, *Lejeunecysta* cf *Hyalina*, and *Selenopemphix nephroide*; and the palynomorphs *Bombacacidites* cf *ciroleonsis*, *Echitricolporites spinosus*, and *Polypodiaceoisporites potoniei* (Pizarro, 1993). **Tonjibe Well No. 1**: This goes through a sequence of 3822 ft of Venado Formation rocks, under volcanic materials corresponding to the Aguacate Group. The age of the first according to Bujak Davies is Late Miocene–Late Pliocene based on the discovery of the palynomorphs *B. baculatus*, *E. spinosus*, *Hemiteli* sp., *Mauritia* sp., and *Pachydermites diederixi*. The Aguacate Group is regarded as Lower Miocene. The column concludes with 889 ft of shale, sandstone, and tuffs correlated with the Loma Chumico Formation, whose age according to Bujak Davies is Late Albinianus, and finally 633 ft of serpentinite correlated with the Santa Elena ultrabasic complex (Astorga, 1992).

6.8 Lake Caño Negro

It is a vast freshwater lake depression covering some 3 square miles, which is situated about 13 miles in a straight line south of Lake Nicaragua. The Monico and Frío Rivers feed it. Both rivers originate in the eastern foothills of the Quaternary Tenorio–Montezuma volcanic complex. The shallow lake is also the natural habitat of many species, both fowl and aquatic, and received National Wildlife Refuge recognition with an area of 38 square miles. We can say that about 100,000 years ago, i.e., in the Upper Pleistocene, the region of Laguna de Caño Negro was part of the southern coast of the former paleo-Lake Nicaragua.

The lacustrine depression of Caño Negro (see Fig. 2.16) is located in the intergraben zone of Lake Nicaragua and is silting up owing to river sediments from the water supply, which was formed in the Middle Pleistocene, starting from the volcanic ends of the Guanacaste Volcanic Range. The construction of stratovolcanoes, already mentioned, associated with a humid tropical climate and a continuous neotectonic rising, allowed the construction of vast alluvial fans

FIGURE 2.16 Caño Negro wetland and remnant lakes. *Aerial photo courtesy of Francisco Solano M.*

passing surreptitiously to glacis before being confused with the floodplain and the water supply. The Caño Negro wetland has a great capacity to accommodate both migratory and local fauna and therefore is listed as a RAMSAR.2 site; nonetheless, it is threatened by the sedimentation of the Frío River, the invasion of exotic species, and the surrounding agricultural pollution, which leads to a cross-border impact. The importance of the conservation of the wetland lies in the protection of habitat for migratory species. Today its waters are the natural habitat of a living fossil, the *Atractosteus tropicus*. This fish, known better by its common name "Gaspar," cohabits with other fish such as the guapote, *Cichlasoma dovii*, and reptiles such as caiman and crocodile, *Caiman crocodilus* and *Crocodylus acutus*.

The study area lies within the tectonic pit or graben of Nicaragua. Deep deposits show volcanic activity (Early Miocene—Aguacate Group) followed by a period of calm with sedimentary deposits in the middle sea (Venado Formation). In the Pliocene was a volcanic wake that would fill the trench gradually followed by the sediments of volcanic origin of the Guanacaste Volcanic Range, found in its evolutionary phase during the Quaternary. The area was occupied by the sea to at least the Lower Pleistocene, and later by the shallow waters of the paleo-Lake Nicaragua. It was during the Middle and Upper Pleistocene that the tectonic pit was filled, giving the current landscape of wetlands and the lakeshore, thanks to the contributions of the basin.

CHAPTER

3

Regional Study of Large Morphological Units and the Coast

OUTLINE

1. VOLCANIC MORPHOLOGY

1.1 Talamanca Mountain Range

The Talamanca volcano-plutonic mountain range (Butterlin, 1977) is the backbone of the country. From this axis was established the isthmus bridge between southern and northern Central America (Dengo, 1962a). The watershed line (divortium aquarium) is therefore the southeast route, which corresponds to the general direction of the country, and whose position also serves as the layout for large tectonic accidents, in particular the Reverse Tectonic Fault of Costa Rica (Pacific coast) (see Fig. 3.1).

The Talamanca Mountain Range was formed by two very well defined watersheds. The Caribbean slope is well developed and corresponds to

Geomorphology and Volcanology of Costa Rica
http://dx.doi.org/10.1016/B978-0-12-812067-5.00003-4

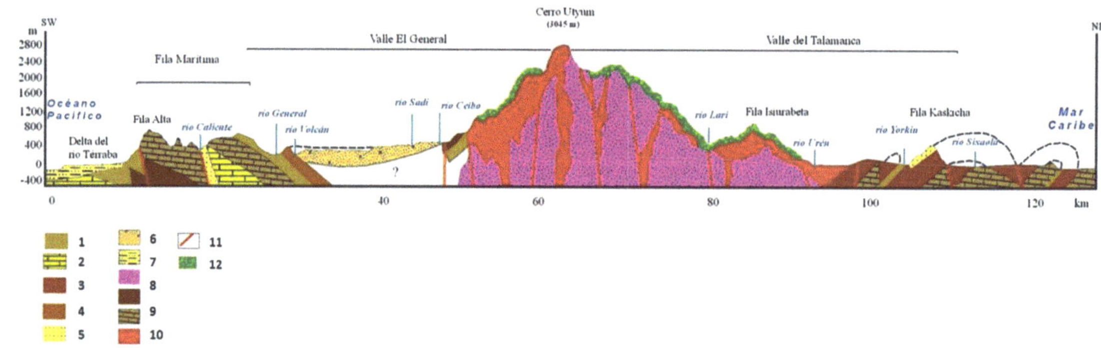

FIGURE 3.1 Schematic of a geological section of the Talamanca Mountain Range from the Pacific Ocean to the Caribbean Sea (Bergoeing, 2012b). (1) Sandstone and lutites from the Gatun Formation, Upper Miocene. (2) Shales, Brito Formation, Upper Eocene. (3) Shale and sandstone, Río Banano Formation, Upper Miocene. (4) Continental conglomerates, Paso Real Formation, Pliocene. (5) Conglomerate, Suretka Formation, Pleistocene. (6) Coalescent alluvial fans and piedmont deposits, Quaternary. (7) Delta deposits. (8) Talamanca's intrusive Tertiary batholiths, stocks, and dikes. (9) Lutites and sandstone, Terraba Formation, Oligocene–Miocene. (10) Lutites, Uscari Formation, Middle Miocene. (11) Tertiary volcanism of Talamanca, lava flows, aglomerates, and tuffs. (12) Tectonic faults.

a raised block (the Pacific slope corresponds to the movement sector of tectonic depressed blocks), however, a volcano–sedimentary series at the foot of the slope is placed, in general, in a position of folds created by lateral compression of the massif.

This set was lifted up during the Pliocene. On this side of the Caribbean, rivers have much farther to travel compared with those of the Pacific slope; however, they have paths ordered by tectonics, because they exploit the large alignment faults, as in the patent cases of the Atlantic Chirripó and Telire rivers. In the case of the Telire River, the fault alignment that allowed it to notch its thalweg continues more northward, cutting the Atlantic Chirripó River, and joins a second E–W fault system known as La Candelaria Fault (to the south of the western Central Valley). By this fact the Telire–Candelaria Fault becomes one of the biggest tectonic accidents in the country, because it divides Costa Rica into two sectors: one northern and another southern (geological map of Talamanca, de Escalante, 1968a; geologic map of Costa Rica, Denyer and Kusmaull, 2000).

Talamanca's volcanism is closely related to the genesis of the country, because it was from the volcanic archipelago (internal arc) that this mountain range began to develop, finally building the aforementioned isthmus bridge. The first products of the volcanic eruptions of Talamanca were mainly basalt, andesite, olivine, and tuffs, slightly more acidic, associated with pyroclastic material, which indicates phases of more intense volcanism (Boudon, 2003a,b). The volcanic buildup started in a marine environment, which is evident because such volcanic deposits alternate with marine sedimentary deposits of radiolarians, foraminifera, and mollusks, which give the series a thickness of more than 6600 ft (Dengo et al., 1970).

It was during the Late Miocene, and especially during the Pliocene, that the Talamanca Mountain Range was lifted up. This orogeny continues today. The volcanism of Talamanca, in that period, was closely related to an intrusive activity that translates today for multiple outliers, ranging from gabbro to granodiorite. There are some radiometric data: Siberia, Talamanca, leucogranite granophyre = 10.15 MY; División,

Talamanca, quartz diorite = 8.5 MY; División, Talamanca, andesite with hypersthene = 9 MY (Tournon in Saenz, 1981). This shows that the Talamanca batholiths ended their construction during the Miocene. Talamanca Pliocene volcanic activity in the Panamanian sector is represented by the Chiriquí volcano isolated cone and the volcanic remains of Las Perlas Archipelago. In the Costa Rican sector persist sketches of some old craters. However, there are volcanic domes that have better resisted weather erosion, and these are visible in Kamuk Mountain (11,693 ft), Cerro de la Muerte, and the sector of Santa Maria de Dota.

The Pacific watershed, in contrast to the Caribbean slopes, has stronger slope values (up to 45 degrees), because of a general tilting due to regional tectonics. The principal tectonic axis of Costa Rica is NW–SE (Great Reverse Fault of Costa Rica). In parallel to this great tectonic accident, other tectonic faults drawn on this slope form a staggered series (Mora, 1979). The erosion during the Quaternary, aided by the tectonic repaving of plutonic rocks, allowed cleaning of much of this material, depositing it at the foot of the slopes (large deposits of the General Valley), thus allowing the outcrop of the granodioritic bedrock on some high peaks, the sector of Santa María de Dota and Pittier and Echandi Peaks.

Talamanca's eruptive material from the Middle Pliocene covers a vast range of magmatic types, from basalt with olivine to the riodacites. The lavas are therefore in general basic, whereas the pyroclastic material has a net tendency to acidity (Weyl, 1956).

1.1.1 Talamanca's Caribbean Watershed

Talamanca's Caribbean watershed is much more important than its Pacific watershed, its surface being almost three times that of the Pacific. This feature can form two groups from the Cuerici Summit:

Between Cuerici Hill and La Carpintera Row, Talamanca is the southern part of the Central Eastern Valley basin.

Between Cuerici Hill and Pittier Peak, representing the Caribbean slope, the great rivers arrive directly at the Caribbean Sea. The large hydrographic systems of this slope, outside the Reventazón River, are the Pacuare, Atlantic Chirripó, and Estrella rivers and the large hydrographic complex of the Telire–Sixaola rivers, where two areas arise: the High Talamanca and the Low Talamanca.

The summits, as well as intermediate sections of the slopes of the Caribbean, are largely composed of igneous rocks, which have allowed the creation of important plateau sectors, particularly downstream of the Kamuk. According to Von Ballman (1976) in Alvarado (2000), and cited by Tournon (1972): In the axis of the Talamanca (Lari River) exist extrusive dacitic domes spanning the granodiorite batholith presenting, by their morphology, a Plio–Quaternary age. The rivers that fit the middle part of the slope created deep thalwegs, aided by the presence of a little cohesive igneous material and preconditioned by plate tectonics producing faults. However, the tropical vegetation is present everywhere in exuberance, because it is one of the areas that have been preserved from colonization by the white man. The lower slopes of the Talamanca Valley are outlier sectors of much-folded sedimentary rocks, which were formed in layers by lateral compression, as well as faulty folds and the presence of structures in horst or graben positions.

1.1.1.1 TELIRE RIVER BASIN

The Telire River is the main collector of several tributaries such as the Coen, Lari, Yorkin, and Uren rivers. All of them are born in the Chirripó or the Kamuk summit (see photo of Telire, Yorkin, and Coen river basin confluence, Fig. 4.2). They form upstream a large hydrographic circus where the waters are products of the abundant rains of Talamanca (High Talamanca) brought by the northeast trade winds. These waters run in deep ravines, transformed into canyons to reach the course average. Downstream, all these rivers meet in the Low

Talamanca Valley. These rivers built in this place, during the Quaternary period, creating powerful alluvial fans, sometimes coalescent, right at the mouths of the canyons (see Fig. 3.2).

The Telire River is the main collector from the middle course, rolling in a WNW–ESE tectonic fault alignment (Telire Fault) that extends westward (Candelaria Fault), as well as to the east, because the tectonic accident gradually disappears under the alluvium of the coastal plain. It is from the hamlet of Bratsi that the Telire River changes its name to the Sixaola River. Both the Low Talamanca valley and the river lock of Bratsi–Bribri, where the Telire runs through a narrow gorge, are of tectonic origin. The Low Talamanca valley is a tectonic depressed area, and at the same time, the sector breaks down the slopes of rivers that descend the steep slopes of the Talamanca Range, which explains the constitution of the powerful coalescent alluvial fans.

In the Low Talamanca valley, the subsiding area allows the Telire River to create its course through sedimentary rocks, taking advantage of two tectonic fault alignments; the first runs WNW–ESE and is nothing more than the extension of the Telire Fault, where suddenly the river changes direction, forming an elbow, following the second tectonic fault aligned ENE–WSW. It is in this sector where it joins the Yorkin River.

River levels recognized in this narrow corridor are fluvial terraces consisting largely of puddingstone, very compacted, whose red clay matrix cemented volcanic and plutonic elements of the high mountain (fractured andesite, breccia, sandstone, and shale), whose diameters range between 0.001 and 0.8 ft; there is also the presence of some blocks of 3.2 ft diameter (see Fig. 3.3). It is almost a certainty that this heterogeneous material forming the puddingstone must be contemporary with the large alluvial fans of the General Valley, dating from the Villafranchian (Beaudet et al., 1982a).

Sergio Mora, a geologist, in his technical study on the construction of the Telire River hydroelectric dam, takes the deposit of puddingstone as part of the Suretka Formation and therefore that it goes back to the Upper Pliocene–Quaternary (Taylor, 1975). This material is tilting toward the northeast from the effects of neotectonics.

FIGURE 3.2 Telire, Yorkin, and Coen rivers in the Low Talamanca confluence sector. *Photo courtesy of Roberto Castillo Vásquez (2008).*

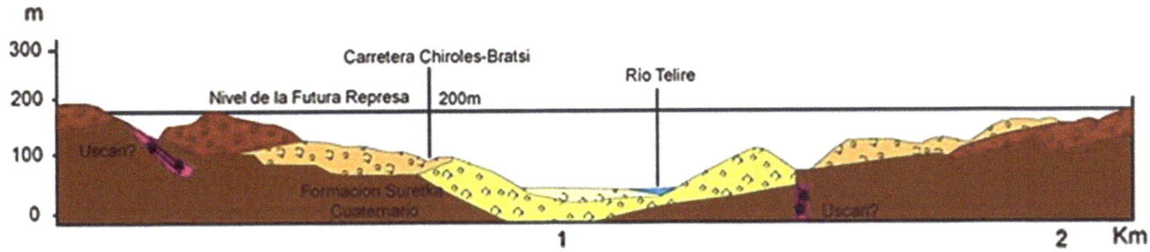

FIGURE 3.3 Four fluvial terrace levels of the Telire River (Bergoeing, 2007).

Indeed, the Telire River, before reaching the coastal alluvial plain, describes a bayonet course, down the gorge already described, where the river previously eroded its thalweg, exploiting the crushed areas. It is at that place where the dam project of the Costa Rican Institute of Electricity (ICE) should have been completed. It is important to note that the Telire River has an average expenditure of 52,854 gal/s together with its tributaries. However, in May 1973 and April 1975 minimum runs of 161,938 and 135,256 gal/s were measured. On the other hand, the exceptional flooding of November 1973 and December 1974 allowed the river to reach records of 486,076 and 560,045 gal/s, respectively, not to mention the spate of 1970, which saw the water level rise up to 23 ft, flooding the hamlet of Bribri, located at the mouth of the Telire River Gorge (Mora, 1978b). It is by watching the current river expenditure that it is possible to understand the importance of gross contributions, so far transported in the lower half, as well as deposits of the river, modeled from the Villafranchian levels contrasted most likely with one fluvial capacity increase, due to changing climatic periods during the Quaternary. In the field, it is possible to observe the following Telire River terrace levels:

Level T1: This is a low terrace of 16–33 ft relative height. It is constituted by healthy boulders of intermediate size (0.065 to 0.13 ft diameter) resting on an ochre sandy clay matrix. The bed, wholesale or flood, wide and flat, is located at 164 ft altitude. During exceptional flooding, as we have already been able to see, the Telire overflows the T1 level. During the period of decline, the largest bed presents itself as a sector of great sandy flats with deposits of well-worn, healthy boulders of large size (0.16 to 0.20 ft diameter).

Level T2: This is situated at 328 ft altitude and exposed on the route going from Bribri to Shiroles. It is not present everywhere.

Level T3: This is located between 430 and 500 ft altitude. T3, as well as T2, is present in levels swiveled by the effects of neotectonics. The material consists of puddingstone, very altered on the surface and wrapped in a bright red clay matrix.

Level T4: This level is located between 690 and 722 ft above sea level. This layer consists of very compacted puddingstone. The rolled material is much altered on the surface.

Sergio Mora studied the narrow part of the Telire River, where the dam should be, in short, which shows four well-defined river levels. Are these levels tectonic or climatic? It is evident that neotectonics has played, during the Quaternary period, an important role in Costa Rica and particularly in this aspect of the Talamanca area. The concrete proof is the swiveled upper levels of the fluvial terraces, which rise to the east. On the other hand, between Bribri and Home Creek, rivers show fossil conglomerates (Tertiary) that are in staggered forms. Such staggered series are little eroded by the rivers, which, however, carry an

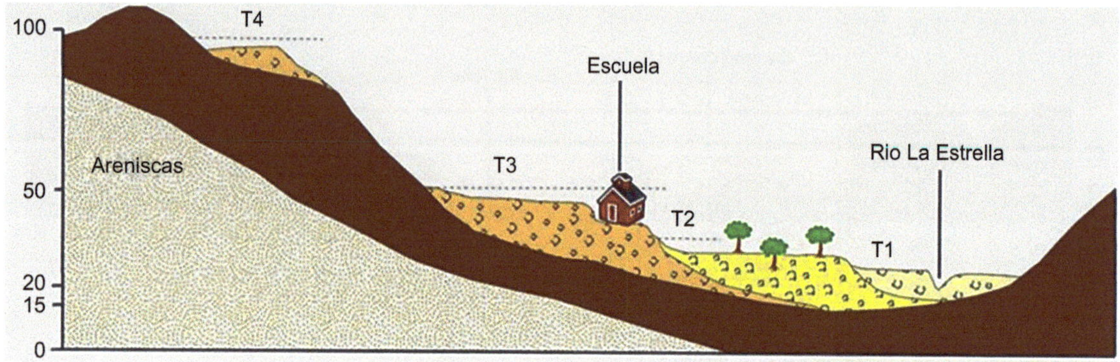

FIGURE 3.4 Estrella River fluvial terraces in Pandora (Bergoeing, 2007).

important well-rolled material and drain in river jumps and waterfalls. This suggests that these tectonic forms are relatively young and that neotectonics manifests itself steadily, regularly rejuvenating the reliefs (see Fig. 3.5).

In regard to the climate origin of the terraces, there are no doubts that the T1 level may be associated with the last high-mountain cold event [there are boulders that have stretch marks and some Costa Rican geologists (Madrigal) attribute them to the Würm–Wisconsinian glaciation of Talamanca]. Concerning the higher levels, it is evident that more dry periods repeated, and must have occurred during the Middle Quaternary and higher, allowing the recoil of the ombrophilous jungle coverage, which allowed the flood rains of short duration to intensely erode the sheds. The material of the T3 and T4 levels, a red clay matrix strongly cemented in segregation levels, is very similar to the material of the old alluvial fans of the General Valley (T4 level), which were dated as Low Quaternary (Beaudet et al., 1982b). We think therefore that the T4 level of the Telire River can be associated with the T4 level of the General River and therefore belongs to the Low Quaternary.

Finally, the geological profile allows us to say the following:

- **Telire River (T4) upper terraces** are contemporaries of the alluvial fans of the General Valley and therefore are from the Low Quaternary.
- **The T2 and T3 levels** are in a range that covers the Middle to the Upper Quaternary. The T3 level, by its consistency and its tectonic deformations, must be quite old (perhaps contemporary with the Pacific Chirripó River T3 level), whereas T2 may be a contemporary of the Eemian period.
- **The T1 river level**, by the freshness of the material, is contemporary with the thaw of Talamanca (terminal Würm–Wisconsinian).

At the mouth of the Telire Canyon, strongly influenced by plate tectonics, the river changes its name and is called the Sixaola, which flooded a vast coastal plain describing numerous meanders. The lower course also serves as the international border with Panama, which poses an international problem with the riverbed, which wanders frequently in the plain. This sector of Talamanca is also home to the last indigenous groups of Costa Rica, living in complete freedom. They are the Bribri and Cabecar tribes of the lower and upper Talamanca Range. It is also in this area where Costa Rica, with the help of Mexico, started oil exploration, which so far has been unsuccessful. However, the volcano-sedimentary intermediate mountain land has saved important copper resources.

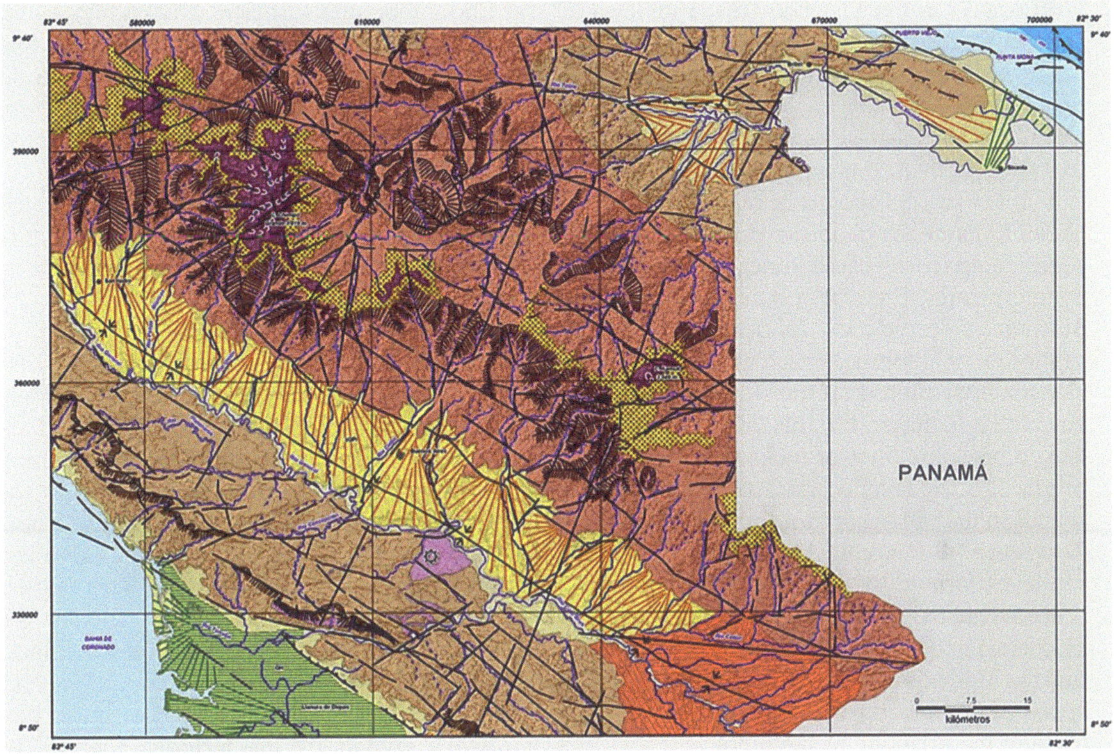

FIGURE 3.5 Talamanca Mountain Range showing the asymmetry between the Caribbean and the Pacific watersheds, the latter characterized by powerful accumulated alluvial fans over the Quaternary. The peaks higher than 9850 ft are the last witnesses to two ice age episodes (Riss–Illinoian and Würm–Wisconsinian). *Geomorphology Atlas of Costa Rica Talamanca, Sheet 1: 350,000, Bergoeing et al. Costa Rican Institute of Electricity, ICE (2010).*

1.1.2 Estrella Valley

The Estrella Valley, located in the north of Low Talamanca, is a small tectonic depression (see Fig. 3.4). The valley relies on the southeast Matama row flank, Muchilla Peak (2989 ft) and Mount Dichibeta (5003 ft). The basin consists of sedimentary slopes, folded, faulty, and deeply altered. The Estrell a River is the main collector of this depression, almost closed. As in the case of the Telire River, the Estrell a River flows into the Caribbean Sea, after crossing an epigenic narrow gorge. On the banks of the lower course of the Estrella River, close to the hamlet of Pandora, we also find four river levels:

Level T1: This corresponds to the bottom of the valley terrace and is made of pebbles of medium size (0.065 ft diameter), well blunted, whose lithology is similar to that of the other levels. However, the boulders are healthy. They rest in a red clay matrix, similar to the T2 and T3 levels. The Estrella River runs in a bed, undermined by antecedence through the hills of the coast, finally reaching the narrow coastal plain before flowing into the sea.
Level T2: This level is the product of ancient axial fills from the bottom of the valley (river beaches crisscross beds) and is between 33 and 66 ft from the T3 level and raised 16 ft

with respect to the T1 level. This large level constitutes the bottom of the valley and is partly covered by impenetrable bamboo vegetation.

Level T3: This is approximately 164 ft lower than the T4 level. It is an important level of fluvial terrace, composed of medium-sized (0.065 ft diameter) pebbles formed by sedimentary rocks (sandstone, limestone, conglomerates, breccia) and also plutonic–volcanic rocks (andesite, granodiorite, gabbro), which are upstream. The material altered on the surface is easily cut with a hammer. On this level, there is at least a predominance of rocks of igneous origin. It is the level where stands the public school in the village of Pandora (see Fig. 3.4).

Top level T4: This coincides with the general hill level, formed by thin Neogene sandstone and covered by boulders of small size (0.030 ft diameter) in a terrace grounded on a sandy matrix. This higher level is testimony to the speed of uplift of the sector and therefore the force of the regional neotectonics.

The Estrella River basin, less developed than the Telire River basin, is, however, important as we glimpse the same problems already exposed in reference to the river's levels and its origin. The Estrella River's lower levels (T1 and T2) are of interest in its development because of the freshness of the pebbles that testify to two recent climatic crises that allowed its placement. As for the Telire River basin, we easily associate level T1 with the terminal Würm–Wisconsinian of Talamanca. For T2, it is possible to associate this level with a likely phase of Riss–Illinois, which no doubt existed in Talamanca, but for which we do not have more than assumptions. In regard to the upper levels, T3 and T4, allow us to go back to the Quaternary. T3 is a much altered level and therefore older than the lower levels. Throughout Costa Rica it is present in the high-level rivers (El General Valley, Central Valley). Altered elements have not yet undergone one sufficient breakdown, as for which they are not recognizable. The clayey matrix is usually red ochre. By analogy with the T3 levels in the Central Valley and its elevated position, we associate them with the Middle Quaternary. The T4 level, often consisting of puddingstone, including a much-consolidated clayey matrix, is the most altered, and therefore we associate it with the Lower Quaternary.

1.1.3 Atlantic Chirripó River Basin

The Atlantic Chirripó River begins at the summit of Chirripó Mountain (12,530 ft) and flows in a nearly straight path (tectonically charted by tectonic fault alignment), following the crushing of a large accident in the NE–SW area. In the middle of its course, the river is cut by a new set of southeastern tectonic faults that give it a zigzag path (detachment faults). This makes the northern block, downriver, slip toward the northwest, whereas the southern block, upstream, slips to the southeast.

The Atlantic Chirripó River, whose linear plot is always guided by the tectonic, changes its course, meeting the northern flood plain. Having lost most of its water force, it forms meanders. The tributaries of the Chirripó River follow a path that instructed by plate tectonics, which gives them a reticulated general network. The right bank slope of the Chirripó River is influenced by the Matama row, an important mountain spur of Talamanca, which moves toward the Caribbean and whose average height is 7218 ft. This row is directed toward the northeast (see Fig. 3.6).

On its summit line, Talamanca harbors numerous witnesses of the Pliocene volcanic activity (collapsed calderas or explosion calderas, craters, dikes, semidestroyed, partly by the same Pliocene volcanic activity and the Quaternary erosion) (Tournon, 1972; Escalante, 1968a). This region is a "hydropower reserve" for a future dam.

Given the mountain slopes and the mountain rainforest covering the slopes of the Atlantic

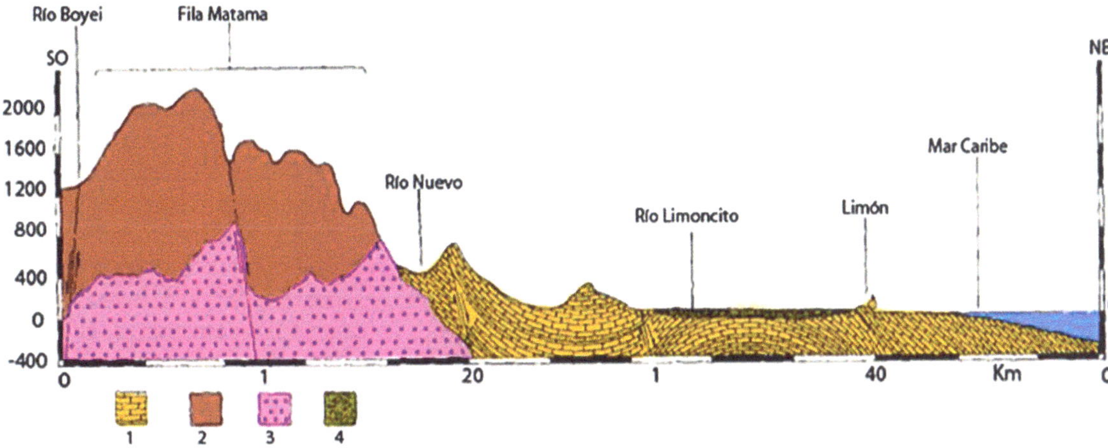

FIGURE 3.6 Geological profile from Matama row to the Caribbean Sea (Bergoeing, 2007). (1) Miocene, Río Banano Formation, sandstones, shale, conglomerates, and clays. (2) Tertiary Talamanca volcanism. (3) Talamanca's intrusive comagmatic group. (4) Quaternary piedmont deposits.

Chirripó this is of very difficult access, especially Matama row, being the major east–west spur of the Talamanca Mountain Range. Matama row is a Tertiary (Miocene) volcanic row, with peaks where some authors describe volcanic forms such as calderas and craters (Escalante, 1968a). The intrusive comagmatic group of Talamanca emerges on the west bank of the Nuevo River. From that point, there is the Tertiary sedimentary series of the Río Banano Formation creating a relief inversion, because the folded series in contact with the plain is in syncline position, whereas the folded series of the coastal plain arises in anticline position. The Limoncito River covers the plain with its modern fluvial sediments where there are boulders of both sedimentary and volcanic type whose origin lies in the Matama row. The only important accident of the coast is Puerto Limon Promontory, being an old coral reef dating from the Miocene.

Contact of the Talamanca Caribbean slope with the coastline is made through a series of spurs or hills whose average altitude does not exceed 656 ft. These sedimentary hills are arranged parallel to the coastline. The same are also witness to stages of rexistasis. Indeed, their slopes have been excavated, resulting in small V-shaped valleys of torrential origin, amid a necessarily drier climate than the current one, being a moister phase, creating a series of concave valleys due to colluvial accumulation coupled with a low capacity of the current rivers.

This is impressive in the Home Creek sector, on the slopes right off the road from Puerto Viejo to Bribri. In this sector, we have examples of detachment scars and solifluction packages at the base, which give the region a general modeling that shows the coastal hills to have evolved from a multifaceted type to a multiconvex type.

The role of neotectonics is not negligible. In this same sector, between Home Creek and Bribri, the rivers form small stepped waterfalls with 16–33 ft of free fall. Here sedimentary rocks (sandstones, travertine, shale) cut by a series of parallel faults give rise to the landscape in a staggered fashion. On the coast, between Punta Mona and Punta Cahuita, develops a beautiful coastal terrace. Punta Cahuita is witness to an old tombolo, or coastal sand spit, which today rises 10 or 13 ft above the high tide and is colonized by a tropical vegetation. The old coastal line formed during the Flandrian transgression is in the contact area of the low hills (Fila Carbon) and the current coastal terrace. This is at least what can be concluded from the observations made by Battistini and Bergoeing (1980) of the coastal sector of Cahuita—Puerto Viejo. In this sector, the road that goes along the coast between Puerto Limon and Puerto Viejo cuts the contact area between the Fila Carbon rank and the littoral terrace. This contact is not free, because a series of colluvial or talus cones fossilizes the slope; however, it becomes very visible in the flat areas where flourish cocoa plantations.

1.2 Talamanca Mountain Range Volcanic Calderas

Satellite radar imagery observations have identified a series of volcanic structures. This did not escape the eye of Jean Tournon of the University of Paris VI, Pierre-Marie Curie. This French geologist recorded in his geological map at 1:500,000 scale, published in 1995 in Dieppe, France, sectors that are punctuated by a series of basalt structures dating from the Pliocene, associated with intrusive structures formed mostly by diorite and ledge monzonite, as well as granite and gabbro. The geological map of Percy Denyer and Guillermo Alvarado, at 1: 400,000 scale, published in 2007 by the San José French Bookshop, takes up part of the indicated information. It is a vast volcanic area, dating from the Miocene to Late Pliocene, coexisting with marine sedimentary formations composed of sandstone, shale, and conglomerates and turbidites dating from the Paleocene to the Eocene known as the Tuis, Uscari, Senosri, and Suretka Formations, to mention the most important.

The aforementioned volcanic structures refer mostly to collapsed calderas of variable dimensions, which are around 4 miles in diameter. We have been able to determine this in the field and, thanks to the vision given by satellite imagery radar, in a region covered by thick jungle, collated with the aforementioned geological maps. Here, we would be in the presence of an oceanic volcanism, probably of isolated volcanic islands that developed throughout the Tertiary, but which finds its apotheosis at the end of the Pliocene, when the Talamanca's orogeny starts. In that moment a series of volcanic structures ceased their activity by the movement of the tectonic plates, exhausting the magmatic sources and along the way the magma migrating toward the northeast, giving rise to the construction of the Central Volcanic Mountain Range as a vast basaltic plateau. Tournon concludes that the presence of recent alkaline lavas is relatively common in the regions of the Caribbean from Honduras to Costa Rica. Alkaline basalts are relatively important in Costa Rica, and in the region of Siquirres, the lava would be associated with intrusions of teschenite of the Upper Tertiary from where it is concluded to have a common origin with the basalts.

An article by E. Gazel completes the picture, saying that the "Sarapiqui Arc (22,2M.y. to 11,4M.y.) represents a segment of the magmatic arc from the Miocene of Costa Rica, whose magma associations ranging from basalts to rhyolites." Basalts have a chemical composition that demonstrates a primary character; the fusion mechanism is interpreted as a decompression of the mantle in response to an extensive tectonics, which subsequently presented a contribution of the fluids of the subducted plate. "Within this unit are two partial melting trends: the first evidence of low degrees of partial melting and fluids of the subducted plate limited contribution and the second presents a greater degree of partial melting, in response to an increase in the contribution of the fluids of the subducted plate" (Gazel et al., 2005 in Alvarado, 2000).

In terms of lithological appearance and geologic history, a sample of rock taken from the Quebrada Siuta River (Moravia collapsed caldera) was classified by the petrology laboratory at the Central American School of Geology as a dacite with hornblende (transition to acid andesite) with the following textural composition: hypocrystalline glomeroporphyritic serial with phenocrystals of plagioclase, hornblende, augite, hypersthene, and olivine altered within a matrix of intersertal texture. The section contains a few holocrystalline inclusions with a maximum diameter of 3 mm, composed of plagioclase, augite, and magnetite; it is probably a cumulate from the magmatic chamber.

This last aspect indicates that the sample is therefore a material of volcanic origin in which exists contamination of the basic magma. This marks a transition between a marine volcanism and other fringes of volcanic islands off the coasts, because of the degree of variation in the chemical composition of the magma rising and cooling, allowing further development, and a variety of crystal content has formed. The dacites are Tertiary or post-Tertiary effusive volcanic rocks, passing laterally by a decrease of the quartz to andesite or by an increase of the quartz to liparite. In any case their volcanic origin is proven, and also their relative age. The presence of quartz in any case corresponds to a magmatic differentiation mechanism, linked to a greater rise of magma from the original sources, so it is easier to relate the training to the process of orogeny.

2. BARBILLA VOLCANIC SECTOR

Located on the Caribbean slopes of Costa Rica, the volcanic sector of Talamanca is mentioned for the first time by Tournon and Alvarado (1995) as a sector of the Pliocene basaltic volcanism. In fact, Tournon represents in his geological map at 1:500,000 scale (1995) a series of basaltic outcrops ranging from Matina sector up to the foot of the Turrialba volcano in contact with the Reventazón River. The geological map of Costa Rica by Percy Denyer (Denyer and Kusmaull, 2000) takes the information from Tournon and adds an important element, which is the reverse faults in contact between the foothills and the alluvial plain. Field investigations begun in 2007 by Jean Pierre Bergoeing et al (2007) in this sector made it clear that the Barbilla sector is eminently volcanic, and a collapsed caldera of about 4 miles in diameter occupies a vast depression surrounded by a basaltic plateau. The volcanic material rests 1312 ft above the sedimentary Uscari and Senosri Miocene Formations and in a series of silted lake depressions corresponding to explosion craters probably concomitant with the collapse of the Barbilla caldera (see Figs. 3.7 and 3.8).

In addition, it was seen that the upper terraces of the Pacuare River (T3 level) were about 1312 ft in altitude and that their current position is probably due to a regional positive neotectonics.

In Costa Rica the three river levels are generally well represented and correspond to climate levels associated in part with the last two ice ages, whose glacier shapes are witnesses that have remained in the high peaks of Talamanca exceeding 9850 ft altitude: Chirripó, Kamuk, and Cerro de la Muerte (Bergoeing, 2007).

The road works made by the ICE and property owners have provided recent observations in the field. These works have unraveled the abundant tropical vegetation, exposing a lot of basalt rocks that carpet the floor, cleaned with tractors, as well as the volcanic material from the rim, ranging from basalt to dacite, with a strong degree of alteration that in some sectors is from hydrothermal alteration. Chemical analysis of the igneous material collected was performed in 2007 (Bergoeing et al., 2007). At the caldera's bottom, there are occasional fossil sedimentary rocks of Uscari and Senosri Formations (Miocene) that would show, on one hand, that the Pliocene–Quaternary basaltic volcanism was an eruptive phase that upholstered the underlying sedimentary formations, becoming discordant between the Miocene and the Pliocene–Quaternary and completed by an explosion crater. On the other hand, they would show that the sector affected by the positive neotectonics was in an intense erosive phase that

FIGURE 3.7 Silted lake of volcanic origin near the Barbilla collapsed caldera. *Photo Bergoeing (2006).*

cleared the upper volcanic deposits on the caldera floor, leaving exposed the underlying sedimentary layers.

Within the crater, we saw for the first time a basaltic postcollapse cone that rises 984 ft from the bottom of the caldera, with important slopes that so far have been supported by the abundant vegetation covering it, but when it disappears this will produce generalized solifluction, immediately beginning an erosive work in clay soils with thicknesses ranging from 3 to 66 ft or more (see Fig. 3.9). The cone is sectioned by an east–west fault that appears clearly on the radar satellite image. The study of satellite images not only allows corroboration of the observations and studies in the field but also determines that it is a vast volcanic area, very dissected, that goes from the left bank of the Chirripó River to the banks of the Reventazón River, where a series of structures sit inside of this sector. It in addition is a volcanic area associated with basal volcanism in the Central Volcanic Mountain Range. Volcanic ashes found in the Reventazón and Pacuare rivers and attributed to the Turrialba volcano in reality would be remnants of the Pliocene–Quaternary volcanic activity of the exposed sector (see Fig. 3.10).

3. MORAVIA DEL CHIRRIPÓ CALDERA

This caldera is located between coordinates 9°45′ and 9°53′ north latitude and 83°25′ and 83°30′ west longitude. It is a vast circular depression about 4 miles in diameter that draws a crescent depression around Quetzal Hill, 4872 ft above sea level (Fig. 3.11). Quetzal Hill is a volcanic postcollapse cone. To the southeast stretches, a vast volcanic plateau, with an average altitude of 4900 ft, very eroded at its edges and particularly in the southeast edge by a faulting line scarp of 2300 ft, which corresponds to the great SW–NE fault of the Chirripó River, which runs linearly in the direction of the Caribbean. The plateau is probably the remnant of the ancient volcanic edifice that was established where today is the collapsed depression of Moravia. To the north, the volcanic depression of Moravia is drained by the Moravia River, which is directed toward the north by a canyon going to Lake Ayil, and then continues its course through homonymous falls to finally surrender as a tributary of the Chirripó River. This entire sector is strongly faulted and it highlights the parallel orientation of NW–SE faults among

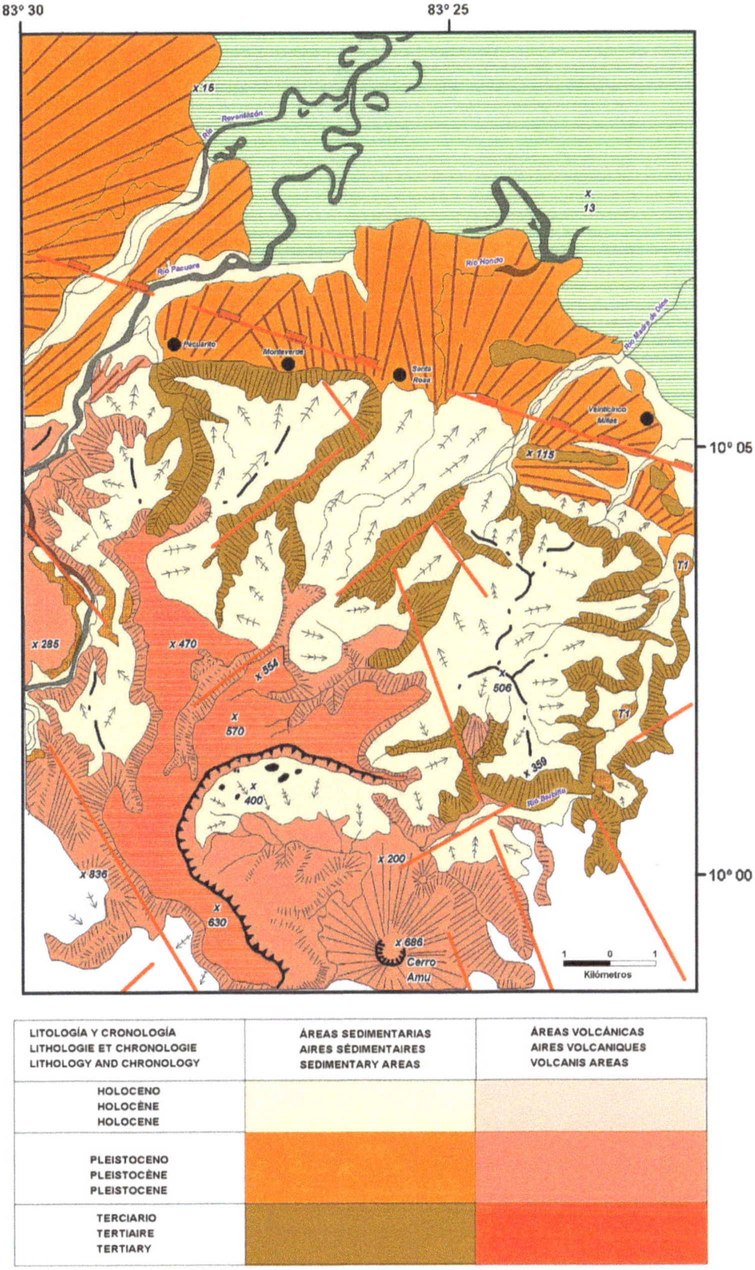

FIGURE 3.8 The Pliocene Barbilla collapsed caldera developed under a sea environment when the Talamanca Mountain Range orogeny began (Bergoeing, 2006).

FIGURE 3.9 Barbilla caldera, basaltic postcollapse cone. *Photo JP Bergoeing (2006).*

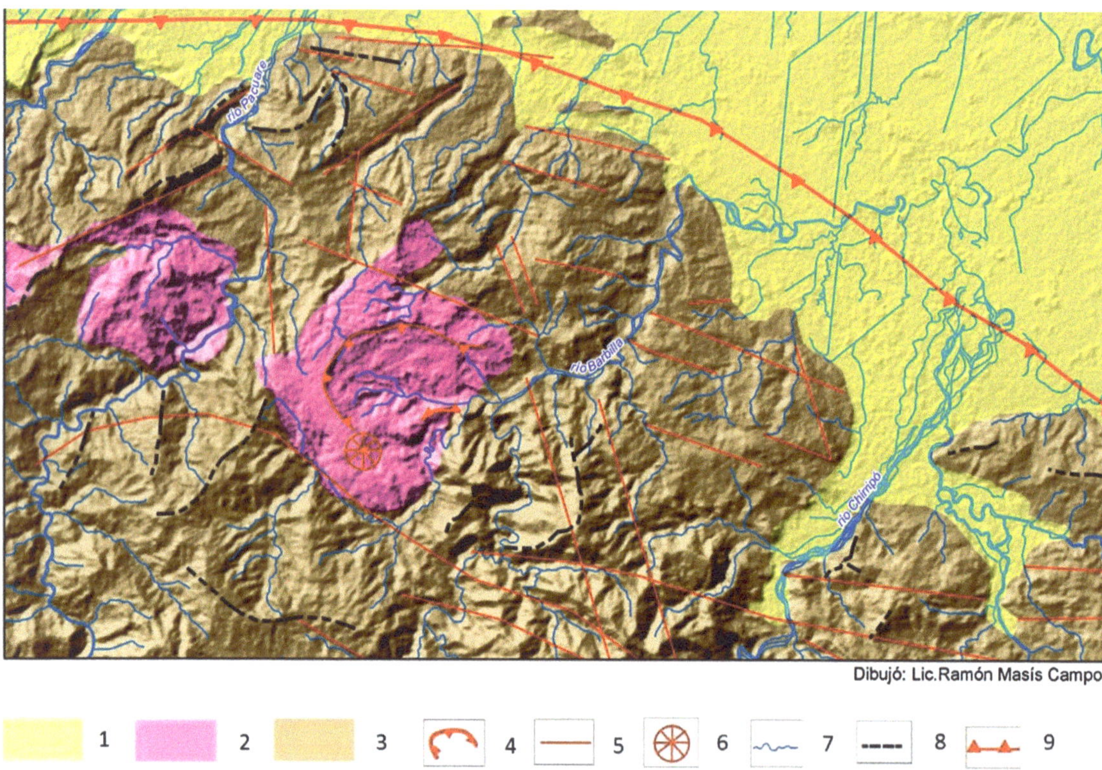

FIGURE 3.10 Barbilla's geomorphology sector through a 2000 satellite radar image, in which it is possible to see Barbilla's caldera and the Quaternary Caribbean North Depression (Bergoeing, 2007). (1) Holocene fillings. (2) Pliocene volcanism. (3) Miocene–Pliocene sedimentary outcrops. (4) Collapsed caldera. (5) Volcanic cone. (6) Tectonic faults. (7) Fluvial network. (8) Water dividing line. (9) Tectonic inversed fault.

FIGURE 3.11 Rim and bottom of the Moravia collapsed caldera. *Photo taken from the south by Bergoeing (2009c).*

which we can see the Shara River, which cuts the Moravia caldera at its southern limit.

The caldera rim is composed of Pliocene basalt (Tournon). The surface of the rim is altered, with thicknesses going from 7 to 40 ft of decomposition red clays. From the center and south of the caldera is situated a postcollapse cone, therefore younger, of semicircular shape, whose center is found in the main crater and which is probably from the Lower Pleistocene. The summit is dominated by Mount Quetzal (4872 ft), covered by a dense tropical forest that hides one of the secondary craters of this cone. A multiconvex shape dominates the lower part of the cone, where tropical forest has disappeared because of anthropogenic action into grasslands. Numerous tributary rivers of the Moravia River rush toward the depression to the north (See Fig. 3.11 and 3.12).

Small alluvial fans of one linear mile built by these rivers may date to the Middle to Upper Pleistocene. Some alluvial fans have traces of river shape passing to terrace fans, like that of the Moravia Church. The caldera depression shows a perfectly flat bottom through which the Moravia River runs SW–NE and disappears abruptly in a river canyon near the northeast rim of the caldera. The sector is also modeled by landslides, which have been partly saved by reforestation, because forestry dominates this sector owing to the strong influence of the Cabecar natives.

Around the caldera structure and especially in the south, a vast basaltic plateau developed with substantial thicknesses of clay alteration (see Fig. 3.12). It is there where is born the Moravia River, through its tributaries the Shara River and Quebrada El Seis River, who has trained at least two levels of fluvial terraces between the villages of El Seis and Grano de Oro. More southeast the Shara River is deeply encased in an altered terrain of basalt origin. The Shara structural plateau has a rectangular appearance and is abruptly bordered by the Chirripó River fault line escarpment, 2300 ft high, between the plateau surface and the Chirripó River thalweg. The Chirripó River fault line escarpment is one of the largest in Costa Rica and developed on both sides of the river. In this sector the fault strike–slip covers about 13 miles from southwest to northeast of the Chirripó River, adopting a bayonet tectonic path. The Shara structural plateau is probably a volcanic core of the preexisting building of the collapsed caldera of Moravia. This area is crossed by a NW–SE fault orientation, perpendicular to the direction of the Great Fault of the Chirripó River, and the Shara River is encased by one of them until the hamlet of Grano de Oro.

Quebrada Suita River volcanic samples collected from the postcollapse caldera cone were analyzed by the Central American School of Geology laboratories at the University of Costa Rica and were revealed to be dacite with hornblende (transition to acid andesite).

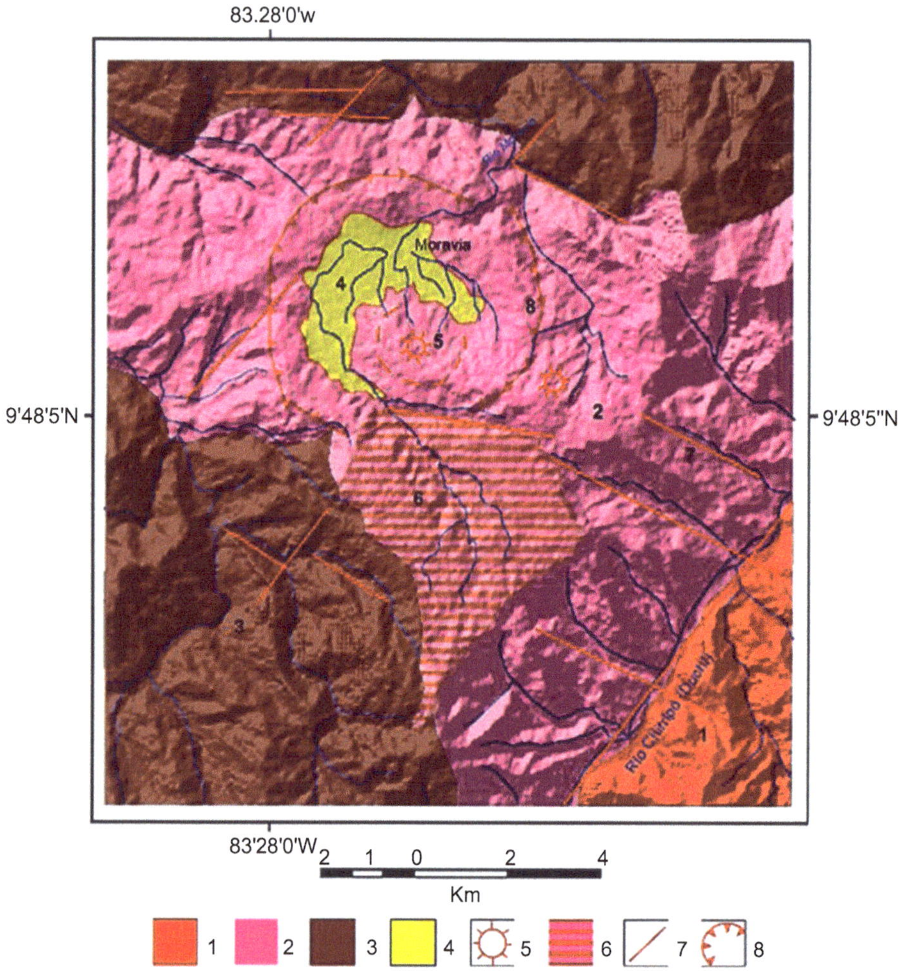

FIGURE 3.12 Radar satellite image of the Moravia caldera and postcollapse volcanic cone. A southern structural volcanic plateau is cut by the faulting line and scarp of the Chirripó River (Bergoeing, 2009c). (1) Talamanca's Miocene volcanism. (2) Talamanca's Pliocene volcanism. (3) Talamanca's Tertiary sedimentary series. (4) Moravia's caldera Upper Pleistocene–Holocene sediments. (5) Craters and cones. (6) Structural volcanic plateau. (7) Faults. (8) Volcanic calderas.

According to Dr. Siegfried Kusmaull, "volcanic hornblende rocks are not very common in Costa Rica and so I know only hornblende rocks from the Pliocene" (oral communication). This confirms that we are in the presence of a volcanic postcollapse caldera cone probably from the end of the Pliocene.

The geomorphology landscape as well as the lithology of the studied places shows that we have volcanic collapsed structures dated to the end of the Pliocene by Jean Tournon. Such structures are predominantly basalt with alternating andesitic and dacitic material, which shows us that there was a transition of the ocean environment to the continental environment, leaving as testimony the emerging volcanic cones inside of postcollapse calderas (see Fig. 3.13). Volcanic depressions cover the mountainous sector between

FIGURE 3.13 Postcollapse volcanic cone of Moravia's caldera. A multiconvex shape dominates the slopes, and alluvial fans going to the bottom of the caldera are dissected by rivers showing basalt rocks on their thalwegs. *Photo Bergoeing (2007).*

the Reventazón and the Chirripó rivers and, extending even beyond Matama row, those forms extruded through marine sedimentary formations dating from the Paleocene to the Miocene. The sector is also crossed by a dense network of faults and guidelines with predominant NW–SE orientation, which highlights the Great Fault of the Chirripó River.

The Quaternary orogeny raised these formations to their current positions where they are subject today to intense weathering effects of the tropical mountain climate, taking in some cases profound alterations in situ and widespread solifluction sheds due to the disappearance of the protective forest, often by anthropic effects.

3.1 The Pacific Watershed of Talamanca

The Talamanca Mountain Range has been arbitrarily limited to the NORTH by the Río Grande de Candelaria, which separates the Talamanca Range from the small volcano-sedimentary mountain range that serves as a southern boundary of the western Central Valley. The great west accentuates this border to the east Candelaria tectonic fault, whose river more or less follows the path. In the crushing area, it is easy to see how exposed is the Terraba Formation flysch series of the Oligocene–Miocene, partly covered at the top with Pliocene volcanic material.

Fig. 3.14 shows various geological formations from the summits of Talamanca to Quepos Port Promontory, which rises to 330 ft above sea level and rests on the Nicoya Complex Formation (Cretaceous), formed by basaltic rocks (pillow lavas) and sedimentary deposits of limestone and sandstone of the Rivas and Sabana Grande Formations interbedded with shale of the Terraba Formation (Eocene–Miocene) (see Fig. 3.14).

The important NW–SE Costa Rica Inverse Fault, a product of the clash of the Cocos and Caribbean tectonic plates, passes near the coast (see Fig. 3.15). The coastal plain in this sector is little developed, limited to the construction of deltaic river cones created by the sedimentary input of the Naranjo River, before flowing into the sea describing numerous meanders that sometimes flood the coastal plain out of its channel. The mountainous area corresponds to the

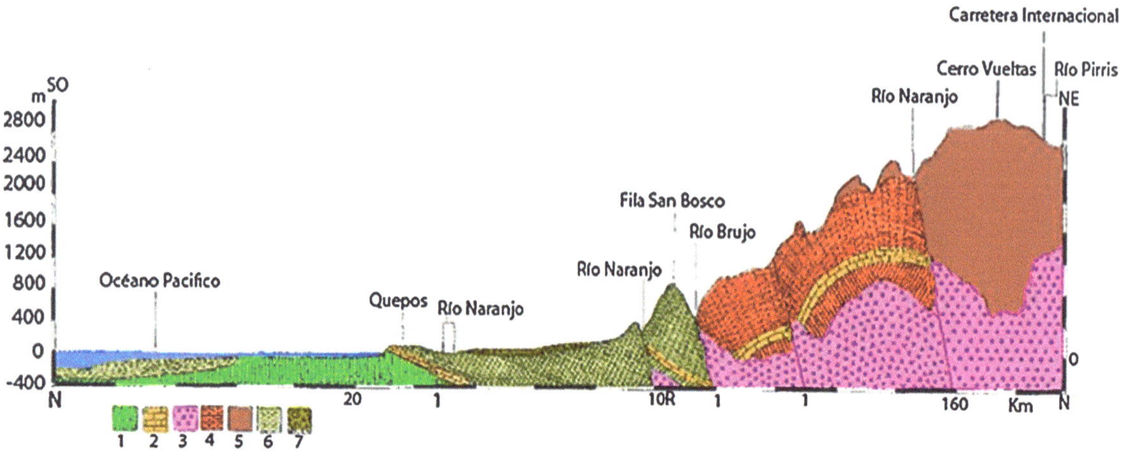

FIGURE 3.14 Geological profile from Cerro Vueltas Mountain to the Pacific Ocean (Bergoeing, 2007). (1) Cretaceous, ophiolitic Nicoya Complex. (2) Paleocene to Cretaceous, shale and sandstone from Rivas and Sabana Grande Formations. (3) Tertiary, intrusive. (4) Eocene to Miocene, lutites and sandstone from Brito. (5) Miocene, Talamanca's volcanism. (6) Quaternary, marine deposits. (7) Quaternary, piedmont deposits.

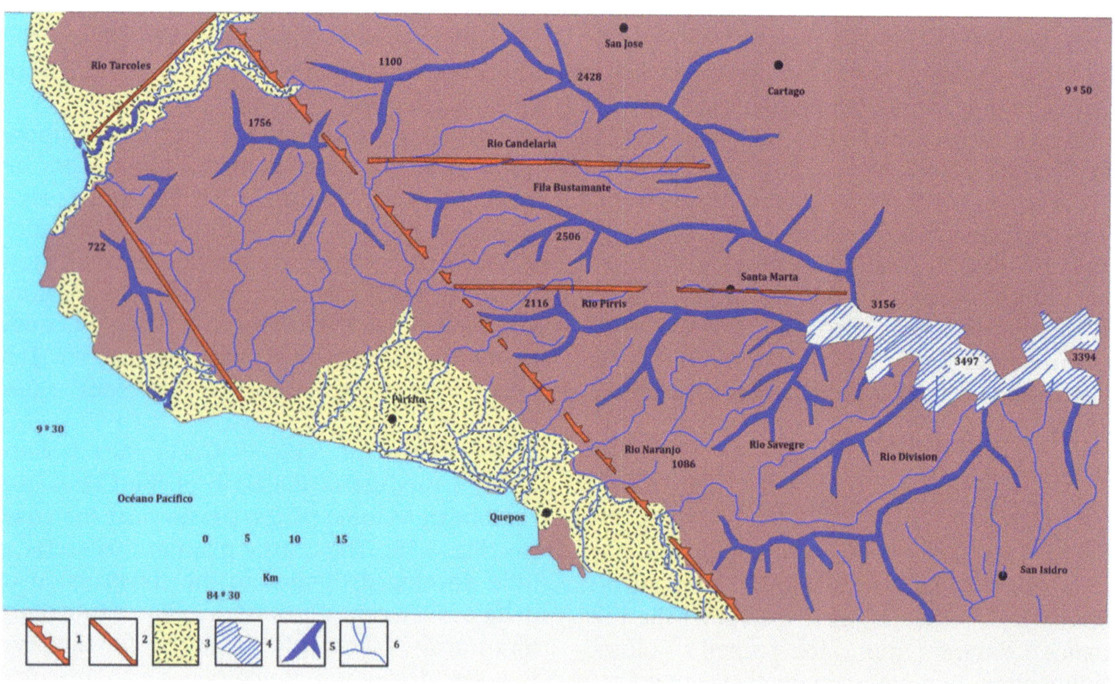

FIGURE 3.15 Talamanca Mountain Range's Pacific Ocean watershed (Bergoeing, 2012b). (1) Costa Rica Reverse Fault. (2) Normal tectonic faults. (3) Alluvial plain. (4) Paleonival and paleoglacier sectors. (5) Watershed continental division line. (6) Hydrographic network.

Pacific Talamanca watershed. Here the effects of the orogeny fold lutites of the Terraba Formation, forming a conform-relief where the anticlines are consistent to the topography. Vueltas Hill is the southernmost summit and serves as a boundary to the east Central Valley of volcanic origin, probably from the Miocene, and holds traces of the last Würm–Wisconsinian ice age. These deposits rest supposedly on the granodioritic batholiths of Talamanca, which emerge farther south of Talamanca (Weyl, 1955a,b).

Bustamante row, as well as the summits of Cerro Vueltas (11,535 ft) and Cerro de la Muerte (11,453 ft), presents typical volcanic forms, especially necks and ancient lava flows (outliers show a columnar dacite between Villa Mills and Cerro de la Muerte in the southern Inter-American Highway).

It is in this sector that the Candelaria River follows an important east–west fault alignment, with the same orientation as the major Telire River tectonic fault. This is at least what can be deduced from the observation of Landsat (7 MSS) imagery. If this is confirmed, this alignment would then be one of the most important of Costa Rica because it would separate the country into two sectors: one north and another south. A second tectonic fault of obvious regional importance, parallel to the Candelaria River, is as follows, to the south, the Pirris River. It is a fault, because a recent detachment presents true faulting mirrors near the Santa María de Dota sector. Starting from 9850 ft elevation the landscape changes dramatically. Indeed, the vegetation is greatly reduced, giving way to mounds of herbs, the dwarf bamboo *Chusquea*, and the presence of a few trees, the Lauriaceae. This is the moor. These few trees have long filaments that hang from the branches in the form of wool, which is testimony to the absorption by capillarity of the air humidity. Here dominates the cloudy climate, or "nebelwald," from the effects of the clash of moist ascending flows mostly from the Caribbean, but also from the Pacific.

The rocky ground is also much crushed. This is where the rock is nude, protected only by lichens. The Cerro de la Muerte summit is a place characterized by paleosnow forms, with small snow "niches" of accumulation in a leeward position (old snow bursts came from the northeast). The soil is the result of the snow paleoclimate composed by a layer of clays due to the melting snow, with corned elements due to cryoclastism (Beaudet et al., 1982a). Fig. 3.16 is a detail of a profile taken on the road between Cerro de la Muerte and the TV link station (Fig. 3.16).

Top: 2 ft of hydromorphic soil, linked to the thaw, with carried gelifracts 0.16 to 0.6 ft in diameter.

Middle: 0.10 ft of leached clay horizon, off-white, which is in the upper horizon of a second layer of clays, a similar thickness, but yellow.

Bottom: Altered bedrock outcrop (andesites). These volcanic rocks, exposed to the current bad weather, are under the influence of an intense biophysic activity, because there is cracking of the bedrock by daily and sudden changes of temperature. Lichens that have joined the rock, increasing the acidity of the rainwater that enters through joint and rock crevices, end up destroying healthy rocks by corrosion.

From the Cuerici summit (10,978 ft) and following up to Kamuk summit (11,644 ft), in the southeastern area of the country, it is possible through aerial photographs to observe the buildup of the paleonival surface. This surface of paleosnow accumulation must necessarily be the result of one or several wet periods in Costa Rica, which were translated by an upsurge of temperatures at altitude. However, it is only from 11,800 ft of altitude, going up from the Chirripó summit (10,978 ft), that we find the real effects of a recent ice age (Weyl, 1956; Bergoeing, 1978a), with cirques and glacier valleys whose small ice fronts decreased to give birth to lakes and moraines. This is clear in the Moraines Valley and the Conejos Valley (see Fig. 3.17)

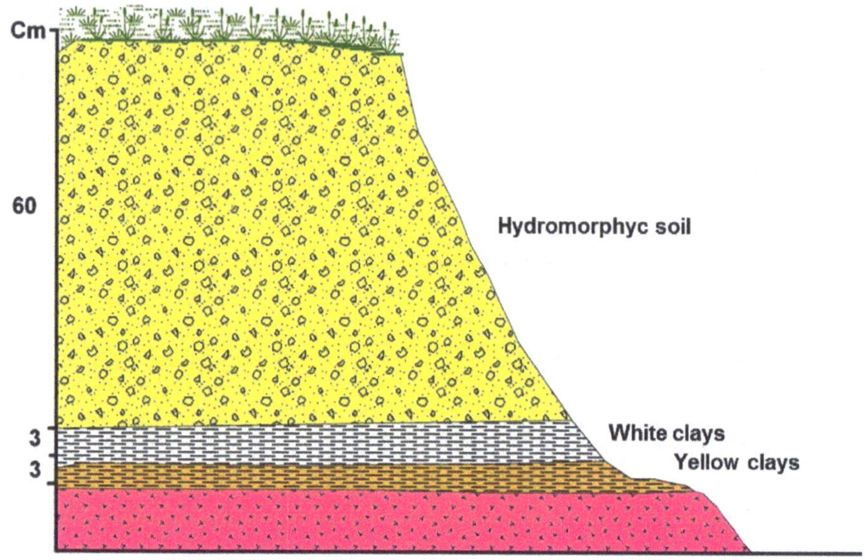

FIGURE 3.16 Cerro de la Muerte profile with paleonival Würmian–Wisconsinian deposits (Bergoeing, 2012b).

FIGURE 3.17 Chirripó Summit 12,539 ft, U-shaped moraine valley. *Photo Bergoeing and Malavassi (1981a).*

In this sector, it is common to find low temperatures, at dawn, of −9°C or 16°F. From Cartago City, the Inter-American Highway climbs up to Cerro de la Muerte summit, following closely the continental division of the water line. This road was built during the Second World War, with the help of the United States, eager to establish a ground loop with the Panama Canal. The frequent fogs in the area serve as excellent camouflage. The chosen path of this sector would provide difficult access to the mountain; a large number of bridges would need to built to follow a path along the coast. From the Cerro de la Muerte summit, the road descends rapidly toward the General Valley. In this way we can establish two important sectors for the Pacific watershed of the Talamanca Mountain Range:

- **The sector between the Candelaria and the Division rivers**. This is a compact high mountainous sector comprising a narrow coastal plain that boils down south to disappear down the river.

- **The sector southeast of the Division River**. This comprises the General Valley (swiveled faulty syncline) and the Diquis and Coto-Colorado coastal plains separated from the General Valley by the Coastal Mountain Range.

The first sector is occupied by major rivers following general tectonic orientations, as well as the paths of the Candelaria and Pirris rivers, going down which they converge to form the Parrita River. The Parrita River built the vast sector of coastal plain. The Naranjo, Savegre, and Division rivers have nearly parallel thalweg paths, and despite the steep slopes (more than 30 degrees), they failed to build an important coastal plain. This due to the brutal elevation of the Talamanca Mountain Range during the Quaternary period and because of the neotectonics conferred on this sector of the coast, marked by the tectonic Reverse Fault of Costa Rica, a subsiding characteristic between Quepos and Palmar.

The second sector is characterized on its high mountaintops by numerous intrusive outcrops, of the granodioritic batholiths above all, in the area close to the border with Panama (Cerro Echandi summit, 10,374 ft). This massive batholith is unique and in solidarity with the outliers of Cerro de la Muerte, Chirripó summit, and Santa Maria de Dota (radiometric dating gives a Miocene age; Tournon, in Saenz, 1981). The batholith was largely buried by the end of the Cretaceous and Tertiary volcanic deposits (Escalante, 1968a).

Volcano-plutonic rocks in this sector also suffer the effects of a very humid mountain climate, due to frequent orographic rainfall on both Talamanca watersheds, and that translates to a strong rock chemical alteration. Plutonic rocks wear out easily (becoming sandy, or the "arenitization" phenomenon), whereas volcanic rocks are clays of decomposition. It is true that vegetation at all altitudinal levels plays the role of a brake on erosion by runoff; hence the interest in preserving vegetation, above all on the summits and in the upper river beds.

The Talamanca Pacific watershed, at half slopes, where rivers empty into the General Valley, is still protected by the wild tropical forest. However, since 2005, humans have been diminishing this forest. The vegetation is the only brake on the erosion of these sheds. In the sectors of the Mosca and Ceibo rivers, human penetration translates, at the beginning, as a forest cleaning and, in a second phase, with the total exploitation of tropical hardwoods by logging companies, who disappear once the work has finished. They do not replant trees once they have done the logging work, although the law requires them to do so. Sectors of the Mosca and Ceibo rivers, upstream of Buenos Aires village, were colonized by peasants in small groups living autarkically for a time. Once roads of penetration were established, the owners of these small farms, through property speculation, sold the land that they had gained from the virgin forest to rich landowners, who exploited it for grazing or for citrus groves (Sánchez, 1979).

This activity has had an impact on the soil, depriving it of its natural protection, accentuating the effects of crawling and solifluction, and finally, by simple washing of rainwater, the material is dragged in suspension by rivers and then deposited on floodplains or, what is worse, into the sea. The grounds are soon furrowed by creeks that slowly become streams, and all this in the space of 1 or 2 years. In addition, it is that a period of 12 months, with more than 10 ft of rainfall, that works inexorably against a land devoid of plant protection. The soil of Costa Rica ends this way, slowly, in the sea, by the conjugated effects of humans and nature. This phenomenon is impressive in the satellite images of Landsat, in the spectral band MSS5, especially for the sectors described, as well as the Nicoya Gulf, Coronado Bay, and the southern Caribbean coast of Costa Rica.

3.2 Neotectonics and Structural Geomorphology of the Coastal Mountain Range

The Coastal Mountain Range is a morphotectonic feature that has developed in an area previously occupied by a forearc basin. According to Kolarski et al. (1995) in Alvarado (2000), the subduction of the marine Cocos ridge caused an inversion of the forearc and back-arc structural basins of Costa Rica. According to them, the surface subduction of this mountain range behaves as a rigid and nondeformable indenter, producing subhorizontal compressive efforts responsible for shortening of the forearc in front of the indenter. In other words, the penetration of such mountain range causes a compression in the Caribbean plate resulting in a cortical compressive deformation of the Coastal Mountain Range, by folding and producing a reverse faulting, perpendicular to the axis of the indenter. Owing to the compressive efforts, the rocky blocks tend to rise following the fault planes, raising the crust surface in areas formerly occupied by sedimentary basins. In this way the mountain belts of the Coastal Mountain Range and of Limon were built. According to the authors, this has entailed changes in the movement of faults that controlled those basins.

But not only have Kolarski et al. (1995) in Alvarado (2000) recognized the shortening and uplift of the Costa Rican anti-arc in the south of the country, but also Collins et al. (1995) in Alvarado (2000), Marshall et al. (2000) in Alvarado (2000), and Norabuena et al. (2004) in MacMillan et al. (2004). Collins said that lifting caused by the Cocos ridge subduction is continuous from the Pacific margin to the Caribbean. Marshall et al. (2000), in Alvarado (2000), suggested that in the southeast the indentation of the Cocos ridge directs the lifting and horizontal shortening within the Terraba belt (Coastal Mountain Range), and Norabuena et al. (2004), in MacMillan et al. (2004), reported a component of permanent shortening through the folding belt and push of the Coastal Mountain Range. In regard to the faulting, the Coastal Mountain Range is referred structurally to an overthrust pusher/belt formed by several reverse faults among the Terraba River, Palmar, Changuena, and Cajon and the great longitudinal fault (Denyer et al., 1993 in Denyer and Kusmaull, 2000), which begins in Panama and ends in Costa Rica, with the name of Candelaria Fault. All of these faults are heading northwest, parallel to one another and also parallel to the Mesoamerican trench. The great longitudinal fault is located at the foot of the Coastal Mountain Range, although Mora (1978b) interpreted it as part of the inverse Costa Rica fault. This active fault shows different morphological expressions and neotectonic movements along its segments, according to Montero (1999). For this author, the fault displays a prominent movement of landslide thrust in the segment of Quepos until the Panama border.

Despite the fact that the longitudinal fault had great activity in the Quaternary, the seismicity of the Coastal Mountain Range is relatively low. This is perhaps attributable to its location with respect to the rugged relief coming under the Caribbean plate. Much of the mountain range is across from a strip of the Cocos plate, having no sea mounts or very few; it seems to be influencing the generation of tectonic efforts. Seismicity is very low against the indicated strip and, perhaps due to that lack of bathymetric features, tectonic efforts are not so great against the Central Pacific and Burica. However, occasionally an earthquake of some importance takes place near Ciudad Neily.

The area comprises the piedmont sector of the Talamanca Pacific watershed, with flexures caused by the Quaternary orogeny, and the mountainous area of the Coastal Mountain Range characterized by east–northeast tilted blocks due to the same orogenic process. The lithology of the Coastal Mountain Range is predominantly of shale, sandstone, and limestone of the Terraba and Paso Real Formations in some sectors next to Palmar, forming "cuesta" structures in some cases, where the front is well defined, but in the majority of cases "pseudocuestas." The Coastal

Mountain Range, next to the General River sector, is also characterized by the presence of an early Pliocene volcanism (Mano de Tigre volcano; Bergoeing and Brenes, 1978b), as well as by intrusive Cretaceous and some outcrops of Jurassic limestone (Mora, 1979). The lithic material is therefore tilted, forming chevron folds in some sectors, and above all faulty, due to a NW–SE system of reverse faults. These are products of the collision of the Cocos and Caribbean plates, associated with perpendicular sliding faults forming insulated tectonic blocks, with a subsiding trend toward the northwest, which is demonstrated by the current Terraba River mouth. The Terraba River eroded its canyon in advance (antecedence) during the Lower Pleistocene, when the hydrographic network began to take shape and the relief of the Coastal Mountain Range rose gradually (see Fig. 3.18).

The General Valley is also undergoing neotectonics and proof of this is that the alluvial fans are folded, producing a very steep rising of the oldest alluvial fans (Lower Pleistocene) that are modeled by the General River. The folded axis is WNW–ESE. Already studied by Bergoeing in 1982 (Bergoeing et al., 1982b), this is a series of at least four or five alluvial fans that have been developing the Andean foothills of Talamanca and would have been produced by the orogeny of the Late Pliocene collision of the Coco's and Caribbean plates. Therefore the older cone (T4) is next to the General River and is found in an elevated position from the Lower Pleistocene and would be between 2 and 1.5 million years old. Boulders that comprise it are heavy gauge (0.10–1.4 ft in diameter), embedded in a red clay matrix, but completely altered, being "ghost" boulders. The water network in the T4 fan sector runs opposite the direction of the rivers coming down from Talamanca, because of the tilting of the blocks. These rivers eventually join large tributaries to the General River following the normal direction, having undermined deep canyons to reach the General collector.

FIGURE 3.18 Second tectonic level of the Terraba River watershed near the hamlet of Puerto Cortes with generalized soil solifluction. *Photo Bergoeing et al. (2010a).*

FIGURE 3.19 The General River alluvial fan seen from Paso Real. *Photography Bergoeing (1982a,b).*

3.3 The General Syncline, Alluvial Fans

The General T4 alluvial fan shape is a multiconvex exposed surface. The southern slopes of the T4 alluvial fan are formed of erosion talus where it is possible to find three or four fluvial–tectonic terrace levels, in some cases forming meandering shapes, like in Santa Cecilia, the lower Río Grande River, and south of Quebrada Pita River (see Fig. 3.19).

The T3 alluvial fan, overlapping chronologically the T4 fan, is logically more recent and it can be classified tentatively as from the Middle Pleistocene. Boulders that compose and identify it, though much altered, are recognizable and maintain a certain cohesion. They are also embedded in a clay red-brown matrix, but not as intensely as T4 (see Fig. 3.20). It is a cone that is more difficult to recognize in the field, being covered in part by the most modern T2, T1, and T0 alluvial fans, these last being put in the Upper Pleistocene. Of them, one of the most interesting is the T1 alluvial fan of the volcanic sector, inasmuch as it is lined with huge

FIGURE 3.20 Background: The General Valley alluvial fans seen from China Kicha Volcano. The General River is at the center of the photo. Foreground: Pliocene volcanic deposit, very altered. *Photo Bergoeing (2009c).*

FIGURE 3.21 Cañas River, Socorro sector: An erratic granodioritic glacier block, due to Chirripó summit's ice melting during the inter-Riss–Würm period or the post-Würm–Wisconsinian period? Note the rain erosion slots (pseudolapiaz). *Photo Bergoeing et al. (2010a).*

granodioritic blocks, highly eroded, forming fluvial erosion slots (pseudolapiaz) from the effects of rain erosion (see Fig. 3.21). These megablocks of 33 × 100 ft and 10 ft high are the consequence of the post-Würm–Wisconsinian melting ice of the Talamanca summits and made brutal and sudden landslides starting from the Chirripó summit due to the effects of the flood rains that occurred with the climate change of that period. It should finally be noted that Hs current deposits are Holocene and products of the Talamanca fluvial sedimentary accumulation from about 10,000 years ago (Bergoeing, 1977).

San Isidro de el General City, renamed as Perez Zeledon City, rests on a large alluvial fan built by the Pacific Chirripó River, which becomes in its middle course the General River. The observation of satellite images has allowed the identification of a series of four alluvial fan levels that have a parallel disposition. The effects of neotectonic lifting tilted the series to the west. They have a fine red clay matrix, with some big elements thickly rolled for the older elements, whereas an abundant fluvial rolled material of different lithology, resting on an outcrop over a dark brown matrix, characterizes the younger levels (see Fig. 3.21).

3.4 The General River

The upper and middle course of the General River runs in the north, encased, meandering among the alluvial fan talus of the Lower Pleistocene. In the south the steep slopes of the Coastal Mountain Range are characterized in this sector by a lithology of sedimentary rocks and pseudocuestas, with backs mostly very altered and

FIGURE 3.22 "El Brujo" sector. Left: T3 alluvial fan deposits resting on the T4 level. Right: Modern T3 and T2 deposits, in a deltaic intercrossed position. *Photos Bergoeing et al. (2010a).*

showing great landslides. Most of the tributary rivers are encased, forming canyons with fast streams and swift runways and presenting two climatic fluvial terrace levels that are a modern and recent expression of the General River. It was during the Pleistocene that the coastline receded, where the first incipient alluvial fans that would constitute the level T4 came to die, due to the rising Coastal Mountain Range monoclines (see Figs. 3.22–3.25).

FIGURE 3.23 Paso Real: the General and Coto Brus rivers' confluence area, which gives rise to the Grande de Terraba River. Note the large amount of rolled material contributing to numerous islands, which are covered during the annual river floods. *Photo Bergoeing et al. (2010a).*

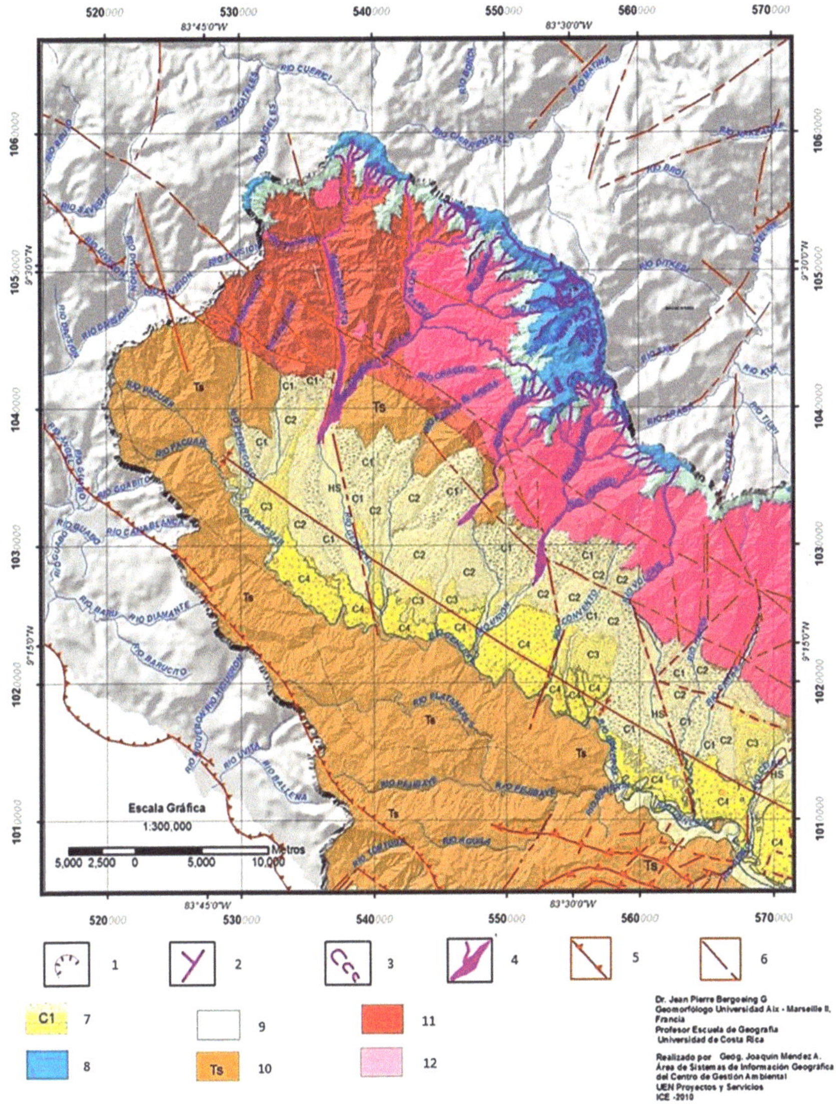

FIGURE 3.24 Talamanca Mountain Range southern piedmont alluvial fans bordering the General River. *C2* and *C3*, Middle Pleistocene; *CI*, Upper Pleistocene; *C4*, Lower Pleistocene; *HS*, Holocene deposits. In the north are paleoglacier and paleonival valleys of the Riss–Illinoian and Würm–Wisconsinian periods (Chirripó II and I). (1) Glacier circus. (2) Glacier peaks. (3) U valleys. (4) Paleoglacier valleys. (5) Reverse tectonic fault. (6) Tectonic alignment. (7) Pleistocene alluvial fans. (8) Paleoglacier areas. (9) Paleonival areas. (10) Tertiary sedimentary outcrops. (11) Tertiary volcanic outcrops. (12) Tertiary intrusive outcrops. *Photo, interpretation, and map Bergoeing (2013).*

FIGURE 3.25 The General River fluvial deposits T2 and T1 in the Paso Real area. *Photo Bergoeing et al. (2010a).*

In Quebrada El Tigre vs El Brujo, the lithology changes because we are in the presence of volcanic material that can placed between the Upper Miocene and the Pliocene and corresponds to an aborted volcanism of that period. The sector has been named as Mano de Tigre and studied by Bergoeing and Mora (1978b). The lithic material of this unit is also very altered on site, which translates to mass landslides and general crawling of some less consolidated slopes.

The Mano de Tigre sector presents at its top an ancient much-eroded crater and a smaller one where the hamlet of Volcancito is situated. The basalts of this sector have been dated by Gazel et al. (2011) in Alvarado (2000), providing an age of 4.23 ± 0.02 MY, which places it in the Pliocene and rejuvenates the basalt samples of Mano de Tigre Volcano, which were dated back to 14.10 ± 0.15 MY by MacMillan et al. (2004). Notably, at its base are very altered lahars. In fact, this volcanism is associated with a tectonic fissure going from the Mano de Tigre sector, comprising Ojo de Agua, Pueblo Nuevo, and China Kicha, and enabled the rise, during the Pliocene, of magmatic volcanic rocks ranging from basalt to the andesite. The sector was recognized also by Jean Tournon and reported in his 1995 geological map of 1:500,000 scale and it was transcribed in the geological map of Denyer and Alvarado (2007) in Alvarado (2009), at 1:400,000 scale.

3.5 Coastal Mountain Range

The Coastal Mountain Range is a morphotectonic unit, which developed in an area previously occupied by a forearc basin. According to Kolarski et al. (1995) in Alvarado (2000), the subduction of the Cocos ridge caused structural basins investment forearc and crust of Costa Rica. According to them, the surface subduction of this mountain range behaves as a rigid and deformable indenter produced by subhorizontal compressive efforts responsible for the shortening of the forearc in front of it (the indenter). In other words, the penetration of such mountain range causes a compression in the Caribbean plate resulting in a cortical compressive deformation in the coastal range, causing folding and inverse faulting perpendicular to the axis of the indenter. Through the compressive efforts the rocky blocks tend to rise by fault planes, causing rising from the surface of the crust in areas that were formerly occupied by sedimentary basins, thus building the mountain belts of the Coastal Mountain Range and of Limon. According to the authors, this has entailed changes in the movement of faults that controlled the basins.

But not only have Kolarski et al. (1995) in Alvarado (2000) recognized the shortening and uplift of the Costa Rican forearc in the south of the country. Collins et al. (1995) in Alvarado (2000), Marshall et al. (2000) in Alvarado (2000), and Norabuena et al. (2004) in MacMillan et al. (2004) have also done so. Collins et al. (1995), in Alvarado (2000), stated that the uprising caused by the subduction of the Cocos ridge is continuous from the Pacific margin to the Caribbean. Marshall et al. (2000), in Alvarado (2000), suggested that in southeast the indentation of the Cocos ridge directs the lifting and horizontal shortening within the Terraba belt (coastal mountains), and Norabuena et al. (2004), in MacMillan et al. (2004), reported a component of permanent shortening through the folding belt and thrust of the coastal row (coastal mountains).

In regard to the Coastal Mountain Range the faulting is structurally referred to as a pusher belt, formed by several reverse faults among the Terraba River, Palmar, Changuena, and Cajon and the great inverse longitudinal fault (Denyer et al., 1993), which begins in Panama and ends in Costa Rica, having the name of Candelaria Fault. All of these faults head northwest, parallel to one another and parallel to the Mesoamerican trench. The great inverse longitudinal fault is located at the foot of the Coastal Mountain Range, although Mora (1978b) interpreted it as part of the Coastal Mountain Range inverse fault set. This active fault shows different morphological expressions and neotectonic movements along its segments according to Montero (1999). According to this author, the fault displays a prominent movement of landslide thrust in the segment of Quepos to the border with Panama.

Despite the fact that the great inverse longitudinal fault had great activity in the Quaternary, the seismicity of the Coastal Mountain Range is relatively low. This is perhaps attributable to its location with respect to the rugged relief coming under the Caribbean plate. Much of the mountain range is across from a strip of the Cocos plate having no seamounts or very few; it seems to be influencing the generation of tectonic efforts. Seismicity is very low against the recommended strip and perhaps because of the lack of bathymetric features, tectonic efforts are not so great against the Central Pacific and Burica. However, occasionally some important seismicity takes place near Ciudad Neily.

This old system, dating back to the Cretaceous and Tertiary, and which has been very affected by Quaternary tectonics, therefore has been allocated the name of Coastal Mountain Range, replacing Costeña Row, which does not reflect the complexity of its formations, its size, and its extension.

The Coastal Mountain Range follows the general orientation of the country, NW–SE, and is characterized by the predominance of sedimentary rocks (shale, sandstone, limestone, and conglomerates) typical of the formation of a middle coastline that emerged at the end of the Tertiary by orogeny effects. The models that prevail in this sector are two: large fronts of slope or pseudocuestas (monoclines) situated to the north of the lower course of the Terraba River between Palmar Norte and Paso Real. Here a sedimentary series makes up vertical erosion slopes and the limestone of the base forms low cornices. These reliefs, held in parallel lines, are very easy to identify by the uplifted WNW–ESE scarp forming reverse faults. Softer dipping northeast slopes covered by a multiconvex modeling product of large landslides characterize the pseudocuesta back slopes.

The most surprising sector is probably the one going from La Virgen to Agua Buena in Filadelfia, where mudslides cover the surface of the ground completely in a vast extension product of a highly altered lutitic lithology. Big blocks are compartmentalized by a series of tectonic faults oriented NNE–SSE, exploited by a series of deep gorges that have fitted the rock crushed by plate tectonics. Also note the presence of the Cretaceous to Paleocene intrusive, composed of gabbros and dolerite, which corresponds to the

FIGURE 3.26 Maíz de Boruca intrusive dome, Upper Tertiary gabbros. *Photo Bergoeing (2011a).*

inner expression of volcanic activity occurring in the sector (see Fig. 3.26). These intrusives also are very disturbed and eroded by the actions of both rain and river, also giving rise to large mass landslides and erosion slopes.

4. ENDING TERTIARY VOLCANISM

From the Paso Real sector to China Kicha, the Coastal Mountain Range, on its border with the General River, is characterized by a volcanic modeling, where is a series of structures very altered by erosion. The sector, known geologically as the Paso Real Formation, composed of volcanic conglomerates, is actually a volcanic complex of the Pliocene. Bergoeing et al. (1978a) had already recognized the structure of the Mano de Tigre Volcano, which is situated in El Brujo sector. From this point, there is a diversity of volcanic forms, remains of craters, calderas, and planezes, associated with lahars in much altered state. Farther north is a particular volcanic complex that corresponds to the volcanic cone of China Kicha, where it is possible to observe a crater in the sector of the hamlet and two structures inside (see Fig. 3.27).

In this last sector, we have found a predominance of basalt in advanced alteration condition as well as very well preserved andesitic rocks. However, the China Kicha complex surface is coated with a profound alteration on site that translates into a significant thickness of weathering that has given way to processes of mobilization of material altered by the effects of the gravity associated with rain moisture and ranging from terraces to mass landslides. This phenomenon is general throughout the area of the Coastal Mountain Range, due to the prevailing very humid tropical climate in the sector, which is under anthropogenic deforestation.

These volcanic shapes are currently subject to tectonic thrusts by the muffling of the Caribbean and Cocos plates and where the reverse faults primarily affect the sedimentary structures of the Terraba Formation, creating a tilted series of monoclines that press more to the north of the Upper Tertiary volcanic complex and thus to the large alluvial fans of the Talamanca piedmont. Because of this effect of tectonic thrust, older alluvial fans (T4) of the Lower Quaternary present themselves in contrary position to the initial runoff that created them as well as to the current general runoff of Low Talamanca.

The drilling of wells and tunnels in the Mano de Tigre Volcano—El Brujo sector by the ICE following the Diquis hydroelectric project dam site has made it possible to confirm the volcanic structure of Mano de Tigre Volcano. It is dominated by tuffs, breccias, and above all volcanic

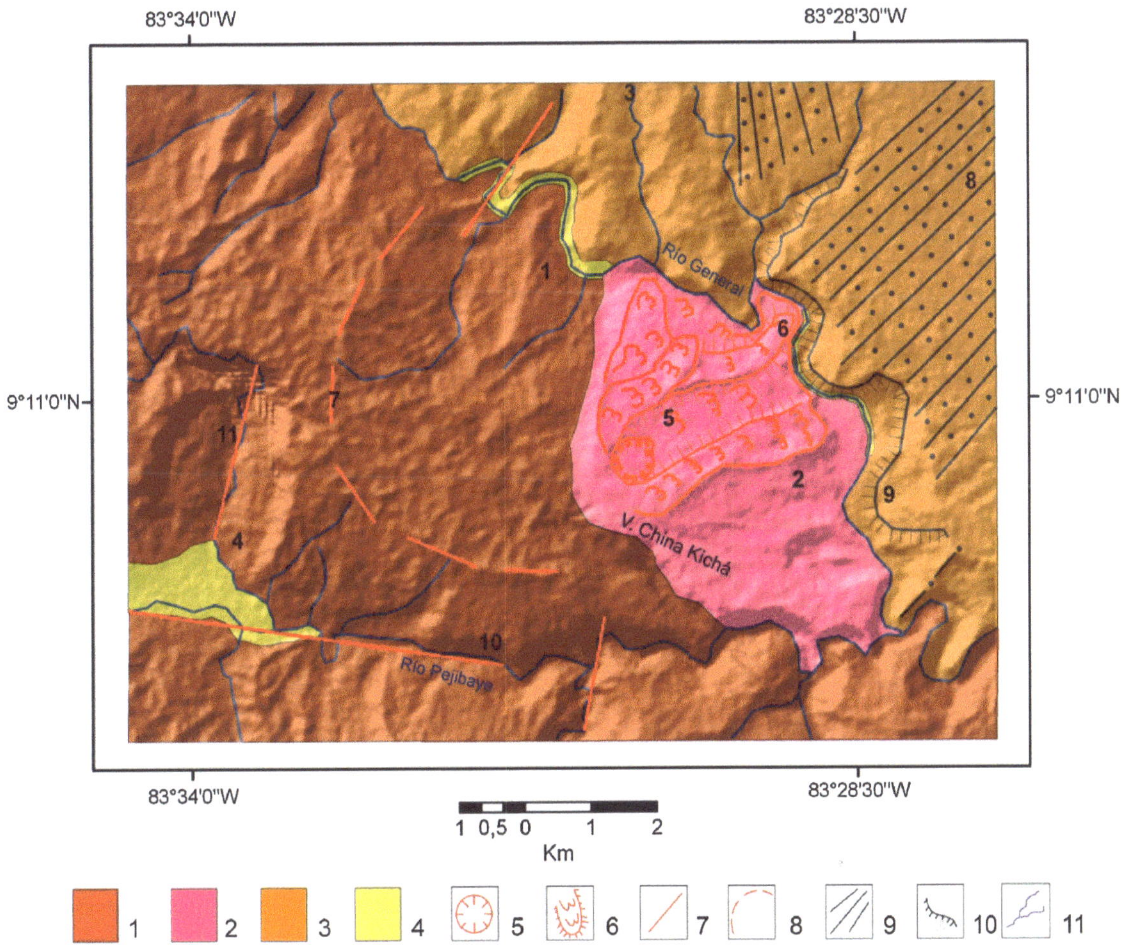

FIGURE 3.27 Satellite radar image of China Kicha Volcano where it is possible to observe a circular structure and a central crater with lava flows (Bergoeing, 2012b). (1) Tertiary sediments of Terraba Formation. (2) Miocene–Pliocene volcanism. (3) Lower Pleistocene sediments. (4) Holocene fluvial sediments. (5) Volcanic crater. (6) Lava flows. (7) Tectonic faults. (8) Circular tectonic structure. (9) The General Valley T4 alluvial fans. (10) Erosion talus. (11) Fluvial network.

conglomerates that have already been observed by Bergoeing in 1978a. While waiting for K/Ar dating that would confirm that it is a volcanic fissure of the end of the Tertiary, we can anticipate that the Paso Real Formation comprises a series of small andesitic–basaltic volcanic structures built in the middle of coastal marshland, following a NW–SE direction, products of the first clash of the Cocos and Caribbean tectonic plates. In addition, in the second instance, the tilting of the monocline sedimentary volcanic structures of the Terraba Formation (Oligocene–Miocene) pressured the set through a series of reverse faults out of which emerged the Coastal Mountain Range.

Selected samples in the Coastal Mountain Range, particularly in the following sectors, have given these results (according to a study conducted by Luis Guillermo Obando A., MSc, Central American School of Geology at the University of Costa Rica.

Sample A2, Daboncragua sector: andesite with augite (this sector is a few miles northwest of Mano de Tigre Volcano)
Components: 32% phenocrystals
Plagioclase: 20% twinned, hypidiomorphic with clayed zoning
Augite: 7% with xenomorphic zoning
Brown hornblende: 2% xenomorphic with opacity
Opaque: 3% magnetite
Matrix: 68% contains plagioclase microlite, opaque and weathered glass

Sample A3, Bijagual sector: basaltic andesite (following to the northwest of Daboncragua is a sector that is characterized by a series of crater and caldera structures)
Components: 40% phenocrystals
Plagioclase: 25% twinned, hypidiomorphic, with clayed zoning.
Augite: 8% with xenomorphic, locally twinned zoning
Olivine: 2%, in ghosts, completely altered to phyllosilicates
Opaque: 5% magnetite
Matrix: 60% contains plagioclase microlite, olivine and opaque
Other features: sample weathered with impregnation of iron oxides

Sample A4, sector of China Kicha: leucomonzogabbro (upper sector is massif)
Components:
Plagioclase: 72% twinned, hypidiomorphic, with clayed zoning
Quartz: 6%, xenomorphic, micrographic texture
Orthoclase: 12% shows clayed, xenomorphic
Opaque: 4% magnetite
Hornblende: 6% green, altered to chlorite and opaque

All these samples indicate the eminently igneous nature of this sector of the coastal mountains in the NW–SE alignment.

The last sample of a leucomonzogabbro of the China Kicha volcanic sector is particularly interesting because rocks originating in the Pliocene are an oceanic crust vector to a continental medium by the oceanic closure due to the collision of plates. The studied sector has a history that begins at the end of the Cretaceous period with the emergence of a string of volcanic islands in the middle sea. Tertiary volcanic structure erosion by the sea during this period produced significant accumulations of fine sediments (shale, sandstone) deposited in a marine coastline where coral formations were transformed later into limestone. At the end of the Tertiary the Cocos and Caribbean tectonic plates collided, resulting in the beginning of the orogeny, which continued throughout the Quaternary period. The emergence of a first relief that originated Talamanca's mountain range created more ancient alluvial deposits designated by the T4 alluvial fans and the small volcanic row Mano de Tigre–China Kicha. The continuous pressure of the Cocos and Caribbean plates produced a rise, increasing both the Talamanca Mountain Range and the Coastal Mountain Range, by tilting and in some cases folding the Terraba Formation sedimentary series. It was only during the Upper Pleistocene that Talamanca reached altitudes higher than 9800 ft.

This would enable at least two ice age stands in the high peaks and deposits of snow around 8500–9800 ft above sea level (Bergoeing, 1987b,c). The post-Würm–Wisconsinian melting built powerful alluvial cones (T1) at the foot of the Talamanca Mountain Range's Pacific slopes. Those alluvial fans are superimposed on the old cones, taking the appearance of a chaotic mass several hundred feet thick, with blocks of more than 17,600 ft^3 uprooted from the Talamanca Mountain Range, where prevailed the trending of the summit of Chirripó. The landscape thus constituted is again modeled by the current interglacial, which has shaped runways, sometimes deep, through the system of alluvial fans, in particular at level T4, which is tilted by the neotectonics of the Cocos plate push product, and in this sector T4 the fans draw the opposite river direction, plotting a general NNE–SSW course.

5. LATE TERTIARY VOLCANIC CONE EVIDENCE ON TALAMANCA'S SOUTHWEST SLOPES

Talamanca's southwest slopes, between the Canasta and Coton river headwaters, is characterized by a series of isolated reliefs from an average altitude of about 4900 ft, covered by rainforest, contrasting with some of the same Talamanca slopes. We speak of the hills Frantzius, Cedro, Irkibi, and Bellavista, to cite the most eminent. Among them are rivers that have dug deep channels, such as the Canasta, Las Gemelas, Cedro, Bella Vista, Cotito, Sura, and Coton rivers, which leave surface bedrock outcrops and already give an index of the volcanic composition of the sector.

5.1 Geological and Geomorphological History

The geological maps of Tournon and Alvarado (1995) and Denyer (2007) in Denyer and Gazel (2009) present this sector as eminently volcanic. For Tournon the ages range between Upper Miocene and Pliocene, whereas Denyer places them in the Miocene. Bergoeing (1977) describes for the first time the presence of the Mano de Tigre volcano in the Coastal Mountain Range near Paso Real. It is a pyroclastic volcano, probably Pliocene. Research conducted in 2010 in the Mano de Tigre field (Bergoeing et al., 2010a) allowed the observation of a volcanic set ranging from Mano de Tigre to China Kicha. It has a volcanic shape, much eroded, but in which structures such as craters, calderas, lava flows, and pyroclastic deposits are recognized. All this leads to confirm that during the Pliocene, a NW–SE fissure occurred from which magma extruded, ranging from basalt to andesite. The presence of volcanic cones in the Canasta–Coton rivers sector suggests the same phenomenon of magma extrusion products, in the same fissure prolongation. This is due to the collision of the Cocos and Caribbean tectonic plates, giving a home to the orogenesis of the Talamanca Mountain Range as well as the monocline sedimentary series tilting of the Coastal Mountain Range, reorganizing by advance the fledgling river system that was created.

5.1.1 Frantzius Volcano

Located between the Platanillal and Canasta rivers, Mount Frantzius bears the name of Dr. Alexander Von Frantzius, a German biologist who studied the sector in the mid-19th century. Here is La Amistad National Park. Frantzius Summit rises to 7001 ft above sea level and is characterized by two peaks: Biolley (5794 ft) and Cerro Tonduz (6332 ft) (see Fig. 3.28)

Between the two summits is described a topographical curve that can be associated with an ancient very upset erosion crater. The Frantzius summit, a perfect volcanic cone, dominates the northeast. The core of the complex is characterized by a series of lava flows, choppy, very altered by rivers, and tributary streams of the Coton River, which here joins the Coto Brus River.

Materials collected at the foot of Frantzius Volcano provided two volcanic samples for analysis:

Sample 1: Cerro Franzius–Pittier (La Amistad Park)
Latitude 9°01.744′ north, longitude 83°00.485′ west, altitude 4564 ft

FIGURE 3.28 Frantzius Volcano covered by a tropical dense forest. In the foreground can be seen volcanic lava fitted by the fluvial system. *Photo Bergoeing (2011a).*

FIGURE 3.29 Hacienda Alturas Quarry, where volcanic flows of the Chai–Irkibi Complex are exposed. *Photo Bergoeing (2011a).*

Hydrothermal propylitic altered gabbro
Minerals observed: sericite in the plagioclase, chlorite, calcite, epidote, and pyrite

Sample 2: Cerro Franzius (quarry)
Latitude 9°01.529′ north, longitude 83°00.073′ west, altitude 4239 ft
Andesite: plagioclase; both phenocrysts and microlites, relatively unaltered, twinned with zoning; they can present a glomeroporphyritic texture.
Green hornblende: both fine microlites and phenocrystals. They may present a glomeroporphyritic texture. There is scarce augite.
Opaque: magnetite and hematite. Due to the characteristics of the presentation of the hornblende, it is not clear if the microlites of hornblende are primary or secondary by hydrothermal alteration (Fig. 3.29).

5.1.2 Irkibi Volcano

It is possible to access Irkibi Volcano through the Las Alturas estate. The Irkibi volcanic cone lays east of Frantzius Volcano following the magma extrusion alignment as already indicated. It is composed of several summits that leave one to assume many other craters where lava flowed. The Mount Chai cone (6890 ft) is with the Irkibi cone (7251 ft), and further east, Cerro Bellavista (6719 ft) is a volcanic package fitted only by the Cotilo River. At the foot of Mount Chai abound dacite lava flows according to the Central American School of Geology analysis. West of the first volcanic set, separated by the Bellavista River, is a new volcanic system. It is the Cedro row of 6995 ft elevation, whose summit describes an old explosion caldera. All these volcanic sources, limited to the south by the Coton River, are probably Pliocene. Among them are granodiorite outcrops that indicate a Pliocene volcanic extrusion through the Cretaceous–Miocene Talamanca batholith and are connected northwest with the Mano de Tigre. The Doboncragua volcanic complex was dated by K/Ar by Kessel in 1983 (Alvarado, 2000) as Pliocene, from 4–5 million years ago.

Sample 3: Mount Irkibi
Latitude 8°57.161′ north, longitude 82°50.972′ west, altitude 5013 ft
Lava or volcaniclastic breccia: rock with heavy hydrothermal propylitic alteration
Observed minerals: chlorite, epidote, quartz, and clay
Epidote crystals are in nodules or rounded tonsils
Locally one can see remains of twinned plagioclase
Propylitic alteration impeach to determine the exact type of rock.

Sample 4: Mount Irkibi
Latitude 8°57.026′ north, longitude 82°50.356′ west, altitude 4790 ft
Dacite, with heavy hydrothermal propylitic alteration
Observed minerals: chlorite, epidote, quartz, and clay in the plagioclase

CHAPTER

4

The Central Volcanic Mountain Range

OUTLINE

The Central Volcanic Mountain Range was formed by a series of WNW–ESE aligned modern volcanic edifices. To the west, its natural limit is the Balsa River, which separates it from the Tilaran Mountain Range. To the north and east it is bordered by the vast Quaternary northern flooded plain. To the south, the Central Valley constitutes its limit (see Fig. 4.1).

The Central Volcanic Mountain Range is composed of several volcanic structures that house many forms, adventitious cones, craters, lava flows, etc., giving them their own characteristics. These structures are, from west to east, the Platanar–Porvenir, Poas, Barva, Irazu, and Turrialba volcanoes. The first three constitute the northern boundary of the western Central Valley. They are individualized and separated by important and profound thalwegs. In this way Platanar Volcano (Tournon, 1983) is separated from Poas Volcano by the Toro River, which fits in a great NNE–SSW tectonic fault line. Poas Volcano is separated from Barva volcano by the Sarapiquí River. (On the left bank of the Sarapiquí River is traced the road from San José to Sarapiquí in the Great Northern Plain.) Finally, Barva Volcano is separated from the Irazu–Turrialba volcanic complex by a vast depression known as La Palma Pass (in contact

Geomorphology and Volcanology of Costa Rica
http://dx.doi.org/10.1016/B978-0-12-812067-5.00004-6

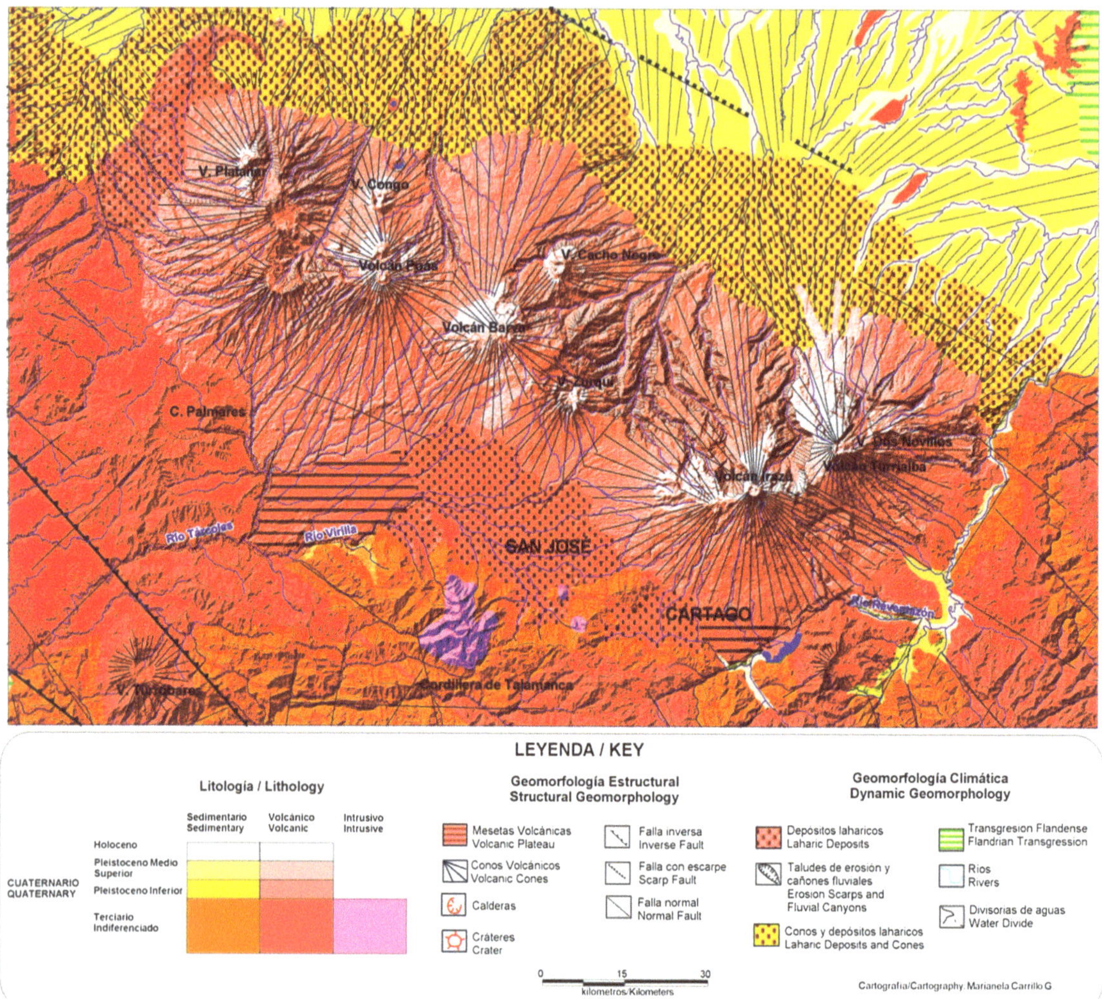

FIGURE 4.1 Central Volcanic Mountain Range geomorphology. From left to right: Platanar, Porvenir, Poas, Barva–Congo–Laguna de Hule, Barva, Cacho Negro, Irazu, and Turrialba–Dos Novillos volcanoes (Bergoeing, 2009b).

with the fundamental area of the Central Valley and through which the northeast trade winds preferentially penetrate), and also by the name Bajo de la Hondura. This sector forms many river thalwegs, such as the Patria, Hondura, and Sucio rivers, among others, which converge to make contact with the main river flood, and therefore plain, and have built a huge alluvial fan by alluvial load-break. The river formed in this way is the Chirripó River (which should not be confused with the river of the same name in the Talamanca Mountain Range Caribbean watershed).

1. PLATANAR–PORVENIR VOLCANIC COMPLEX

As we have just shown, this is the westernmost volcanic system of the Central Volcanic

FIGURE 4.2 Platanar Volcano upper cone and crater. *Photo courtesy of Francisco Solano (2008).*

Mountain Range. It has many craters, of which the most important are Platanar (7162 ft), Porvenir (7438 ft), El Viejo (6759 ft), and, to the south, Cerro Avion (5640 ft). La Picada row, destroyed by erosion, is witness to ancient craters, of which the Chayote crater (6890 ft) is the only one that remains partly intact (see Fig. 4.2). In addition to the craters, there are numerous explosion caldera forms (geomorphology map, see Fig. 4.1). This volcanic complex covers an area of 174 square miles. The eastern sector is cut by the large Chocosuela scarp with more than 2297 ft of free fall, in the background of which runs the Chocosuela River. According to Tournon (1983), this scarp is the remains of an old collapsed caldera rim. Because of its size, it is one of most important calderas in the country, comparable to only the Guayabo and Palmares collapsed calderas.

1.1 Porvenir Volcano Crater

This is practically the twin of Platanar. The Porvenir cone, not very high, overlooks a small depression (Lake Congo), is situated to the west and marks the location of a former caldera (probably of an explosion). The Chocosuela scarp structure, already mentioned, which corresponds to an ancient collapsed caldera, covers the two volcanic cones, Platanar and Porvenir, and touches the western part of the Viejo volcano crater (Carr et al., 2007, in Alvarado and Gans, 2012). In the Toro Machines House Road sector, Carr dated andesite lava by the $^{40}Ar/^{39}Ar$ method, which gave an age of $0.554 \pm 0.0.20$ MY, which is Upper Pleistocene.

South, Porvenir Volcano is deeply notched by a canyon, modeled by La Vieja River, whose cut can easily reach 656 ft. The eastern Porvenir cone is deeply cut by a fault scarp oriented north–south and that extends over the Platanar cone (see Fig. 4.3).

1.2 Pelon, Viejo, and Avion Craters

These are aligned following an ENE–WSW direction. The westernmost is the Pelon crater, precisely where La Vieja River is born. This

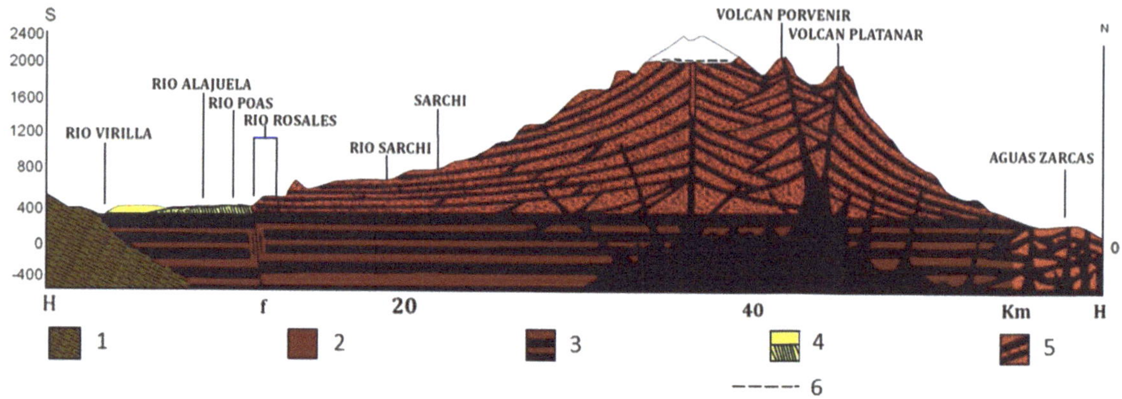

FIGURE 4.3 Platanar and Porvenir volcanoes. (1) Miocene–Pliocene volcanism; Aguacate Formation. (2) Lower Pleistocene; Aguas Zarcas volcanism. (3) Lower Pleistocene; basal intercanyon lavas. (4) Middle Quaternary avalanche deposits (ignimbrites). (5) Upper Quaternary; interstratified lavas (stratovolcano). *Geological profile from the Central Valley to Aguas Zarcas Bergoeing (2012a).*

sector of the first volcanic cone is a flat land, which ends in a circular escarpment for more than 984 ft and which is the "rim" of an old collapsed caldera. The Viejo cone touches this volcanic structure in the south, as well as that of Chocosuela in the north. The Viejo and Avion volcanoes rest on a deeply notched sector of gorges and canyons. Dating in basaltic andesite of the sector gives ages ranging from 0.55 to 035 MY (Alvarado and Gans, 2012), making it a young Upper Pleistocene volcanism.

The east end of the structure is crossed by a fault that has been used by the Toro River to dig its bed. The upper part of its course is a valley with a flat bottom, where Bajos del Toro village is located and which corresponds to triangular basaltic plateau elements. Following this tectonic faulting alignment upstream, it is possible to observe the diminished forms of the ancient Cerro Chayote volcanic focus that allowed the creation of a part of the north piedmont of the western Central Valley.

From La Vieja River, the west piedmont consists of large lahar deposits that are exposed along the Zarcero village road. No wonder, then, that these rivers form deep canyons, given the friable fluvial material. On the surface, the landscape presents as a series of rounded hills (multiconvex relief), where a stepped land (first stage of the soil erosion) is present more or less everywhere, due to the effects of cattle trampling. The cattle use the grasslands, which sometimes slope more than 32 degrees and on which a thick grass blanket, soaked in water, which is easily disturbed by the cattle hooves, covers the clay soil.

Near Zarcero, Tournon [Opus citate (op. cit.)] describes lava flows, quite old, important gaps in formations associated with lahars and pyroclastic flows (nuées ardentes, or burning clouds). Here, where this material is well exposed, a quarry exists that has been used by the Public Works and Transports Ministry to remove it to tamp down the road. According to Tournon (1983), the volcanic sequences that have occurred are the following:

- Emergence of a great volcanic edifice
- Volcanic collapse of the northwest sector and formation of the collapsed Chocosuela caldera

- Strong relief erosion and accumulation of the material to the north of Quesada City
- New volcanic activity in the caldera and formation of the current Platanar and Porvenir cones

A series of small hills, which serve as a transition, is located to the north of the Platanar volcanic edifice, between the volcanic piedmont and the floodplain. This is the sector of Aguas Zarcas, where a series of small aligned Strombolian cones runs for 6 miles in the north–south direction. These small cones, such as the Loma Morena (2037 ft), Buenos Aires (1512 ft), Loma Juan Murillo (1700 ft), Loma Barrantes (951 ft), Vuelta de Kopper (950 ft), Valle Hermoso (705 ft), and Los Chiles (820 ft), are the products of a recent magmatic activity (Upper Pleistocene). Large amounts of volcanic materials, such lapilli, basalts, and andesites, are present, but here we find mostly alkaline basalt (Tournon, 1980) (see Fig. 4.4). Of the volcanic cones of Altamira, Alvarado and Gans (2012) obtained a radiometric dating by $^{40}Ar/^{39}Ar$ in alkali basalts, which showed respective ages of 0.3 and 0.295 MY.

1.3 Platanar Volcano Concentric Calderas

Volcanic structures formed by concentric calderas exist in some areas of the planet. Landsat and Spot satellite images have already been identified and described that allow us to obtain a global vision of the areas concerned. In Costa Rica, Platanar Volcano, located northwest of the Central Volcanic Mountain Range, is formed by a stratovolcano compound, which we can date tentatively as from the Upper Pleistocene, constituted by the Platanar cone itself and, a little farther to the south, the Porvenir cone, separated by a distance of 2 miles from each other. The volcanic complex covers more than 44 square miles. Platanar crater is located at 7162 ft above sea level, whereas the Porvenir crater reaches an altitude of 7438 ft, showing an important and fast buildup since the Upper Pleistocene.

On the northern flank of the Platanar, overlooking the Great Plain, is drawn the collapsed Palmera caldera, probably from the Lower Pleistocene, filled largely by lahars arranged in

FIGURE 4.4 Aguas Zarcas Holocene crater and Palmirita Strombolian volcanic cone. *Photo Bergoeing (2008).*

large cones coming from the Platanar slopes (see Fig. 4.5). This filling can be reasonably considered as a constant fluid that filled, during the Quaternary period, when the depression collapsed. Farther east are nine Holocene cinder cones, aligned north–south and known as the Aguas Zarcas volcanism, presented earlier. They reach elevations not over 328–525 ft, but they have well-preserved structures and some even hint at craters. They are the manifestation of the latest magma ascension of the sector. The Platanar and Porvenir cones are a modern construction (Upper Pleistocene), rebuilt on older structures, notably, the east Chocosuela caldera (Alvarado and Carr, 1993), which forms a deep canyon that can reach 1640–3280 ft of free fall and where runs the Aguas Zarcas River in the thalweg. If we consider it, this whole structure rests on a bigger caldera structure. It would go from El Toro River, east, to the San Lorenzo River, west. Thus the area takes on another dimension and the Platanar–Provenir assembly fits within the framework of the large world calderas, because the diameter of this structure would be about 22 miles, which would give approximately 386 square miles of surface area.

The Platanar–Porvenir volcanic complex is characterized by magmatic calc-alkaline, alkaline, and magmatic basalt–andesitic series and also dacitic and rhyolitic series. The basalt is mafic and transitionally alkaline. The negative anomalies present in the alkaline magmatic series make it a somewhat unusual volcanic complex within the Central American volcanic activity (Alvarado and Carr, 1993). The Chocosuela caldera would be an episode of collapse during the Middle Pleistocene and the former old stratovolcano that disintegrated left pumice and rhyolitic deposits.

Satellite images (Landsat satellite image mosaic using multiband system, year 2000, raster clipper; NASA website) have allowed us to highlight at least three structures of concentric calderas in this area (see Figs. 4.5 and 4.6), located to the west of the Platanar–Porvenir complex cones. These structures, studied in the field, present as concentric depressions surrounded by slopes (rim), whose elevation does not exceed 300–400 ft, formed by volcanic material (basalt to andesitic lavas, slag, and ash, all very altered, edaphologically speaking). One of the most interesting points is located in Buena Vista and concerns populated rural areas, such as La Quina and Santa Elena, where stands old collapsed calderas filled with important thicknesses of altered sediments (brown clays).

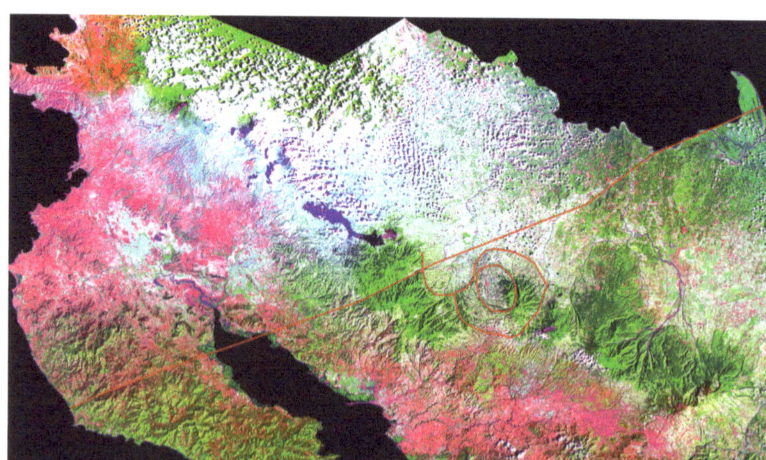

FIGURE 4.5 We see in this image the tectonic ENE–WSW accident starting in the lower course of the San Juan River (extension of the Hess escarpment) cutting the Nicaragua graben, the Tilaran Mountain Range, and the Nicoya Peninsula, crossing north of the Platanar concentric calderas as well as the San Lorenzo paleocaldera. Satellite 3 Landsat 2000, multiband raster clipper mosaic image.

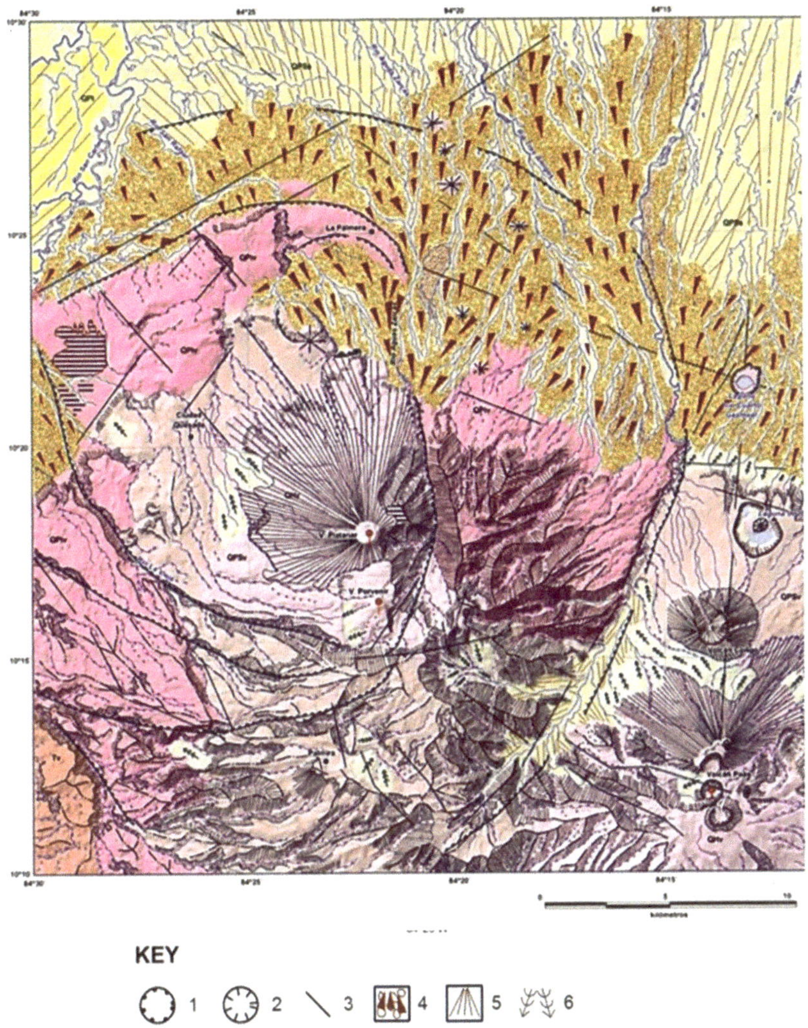

FIGURE 4.6 Structural map of the Platanar–Porvenir complex (Bergoeing, 2007).

Indeed, the sector presents a semicircular volcanic structure with weathered red-brown clays revealing a multiconvex shape, where, however, emerge andesitic blocks. The drainage system has deep eroded canyons through which the Santa Clara River waters evacuate. The west, with the San Lorenzo river banks and Santa Ritas row that mark the external limits of the concentric calderas, limits the volcanic structures. It is likely that a megavolcanic structure has existed there, limited to the east by the Toro River and to the west by the Balsa River, comprising all the Platanar volcanic system to the Chayote Hill volcanic system. If this hypothesis is confirmed, we would be in the presence of one of the largest and most active caldera structures remaining in Costa Rica. Next to this volcanic caldera, it is necessary to mention a second circular structure of imposing dimensions, which is situated to the west of the first and could be called the San Lorenzo caldera. It would be the relict of a collapsed volcanic manifestation whose magma migrated eastward from the displacement of tectonic plates, giving birth to the Platanar–Porvenir complex.

Another feature of the sector, which the satellite image permits us to observe, is the great tectonic ENE–WSW accident that runs north of the volcanic Platanar–Porvenir complex (Fig. 4.5). It could correspond to the inclination of the Caribbean plate subduction zone boundary, with the Cocos plate: "We have found that under the border between Nicaragua and Costa Rica the Wadati–Benioff contortion zone, is very inclined in Nicaragua unless[sic] inclined in Costa Rica, but does not show evidence of faulting as it was postulated by other authors" (Protti et al., 1994). Farther south, the subduction zone shows a ripping, named the "Contorsión Brusca de Quesada" (Quesadas abrupt contortion), at intermediate depths greater than 229,659 ft. In general, the angle of the Wadati–Benioff zone decreases from 84 degrees under Nicaragua to 60degrees under the central part of Costa Rica. The maximum depth of subduction earthquakes also decreases, from about 656,168 ft under Nicaragua to 410,105 ft under the Central Volcanic Mountain Range. In the south of Costa Rica to the east of the meridian at 83°55′, we found no evidence of the Wadati–Benioff zone below 164,042 ft deep (M. Protti op. cit.). This would indicate to us that the volcanism north of the aforementioned fault, beginning at Arenal Volcano, would be of more intense activity, and consequently the magma ascent would be more important and more acidic, whereas south of the tectonic accident, the magmatic ascension would be less active.

The Platanar–Porvenir volcanic complex is an active stratovolcano system that was rebuilt during the Quaternary. Witnesses to this are the concentric caldera remnants of the previous structures and the old megacaldera structure of San Lorenzo to the west. The characteristics of that caldera will be the subject of another research; it indicates a shift from west to east of the magma ascent of the volcanic edifices from the advance and collision of the Caribbean plate westward with Cocos plate. The most spectacular caldera rim certainly is that of Chocosuela. The volcanic Platanar paleocones are of the explosion type rather than the collapse, except perhaps in the case of Palmera's collapsed caldera structure, located to the north and associated with the volcanic complex, which presents as a very old collapsed structure because of the highly altered lavas that compose its rim (Lower Pleistocene?). The great volcanic activity of the Platanar–Porvenir complex is demonstrated by the diversity of the lava flows and lahars that cover its flanks going toward the Northern Plain.

The Aguas Zarcas cinder cones are another example of the strong zonal volcanic activity which was initiated during the Holocene and are therefore very recent structures, associated with magmatic ascents through the Nicaragua graben (Cocos and Caribbean plates subduction zone). The satellite image allows us also to perceive that the volcanic Platanar–Porvenir complex fits within one of the larger collapsed calderas of Costa Rica. Inside this megacaldera, Quaternary volcanic activity has left traces of a series of smaller calderas and of the present stratovolcano, so we could categorize this structure as a collapsed caldera of the first group type, Katmai or Valles, following the Williams classification.

2. POAS VOLCANIC STRUCTURE

This is the second volcanic structure of the Central Volcanic Mountain Range western sector. It consists of two well-formed craters situated at the top (see Fig. 4.7). The first is inactive, but well preserved, and covered by vegetation. It is the culminating point of this structure of 8871 ft altitude. A lake of cold freshwater of pluvial origin occupies the bottom of the crater. To the south of it is the Poas active crater, a circular structure of 0.6 mile diameter, devoid of vegetation, with a formation of volcanic terraces attached to the inner walls. In the crater bottom is an acidic lake of slowly boiling water, sometimes with geyser-type eruptions, whereas from its southern wall sulfur fumaroles emerge (Bullard, 1956).

The Poas cone summit is occupied by other eroded volcanic cones also devoid of vegetation, especially the western sector, because of the strength of the northeastern trade winds blowing violently at times, taking to the southwest the ashes and other volcanic elements in times of volcanic eruptions. The western sector, devoid of vegetation, is crossed by two major rivers, the Anonos and Desague rivers, which are tributaries of the powerful Toro River, whose thalweg separates Poas from Platanar Volcano (see Figs. 4.8–4.10).

From this structure begins a north–south alignment of cones and volcanic craters, similar to the Aguas Zarcas volcanism, but of much more important dimensions, because it stretches 13 miles in a straight line. On this alignment, we also find the two craters of Poas Volcano, the semidestroyed cone of Congo Volcano, the Laguna de Hule collapsed caldera with its small pyroclastic cone, and the Laguna de Río Cuarto gasmaar crater, occupied by a cold freshwater lake (Bergoeing and Brenes, 1977). "Poas volcano and the three devices aligned North–South and are framed by two deep straight valleys, equally oriented North–South: Bajos del Toro and Sarapiqui Rivers. These two valleys seem to have followed the Northern–South tectonic faults. The Poas would then be housed inside a graben" (Tournon, 1984).

This volcanic edifice displays a sequence similar to that of the Platanar–Porvenir complex, with basal basaltic lavas and ignimbritic series in the upper part of the central plateau, covered in part by powerful lahars that

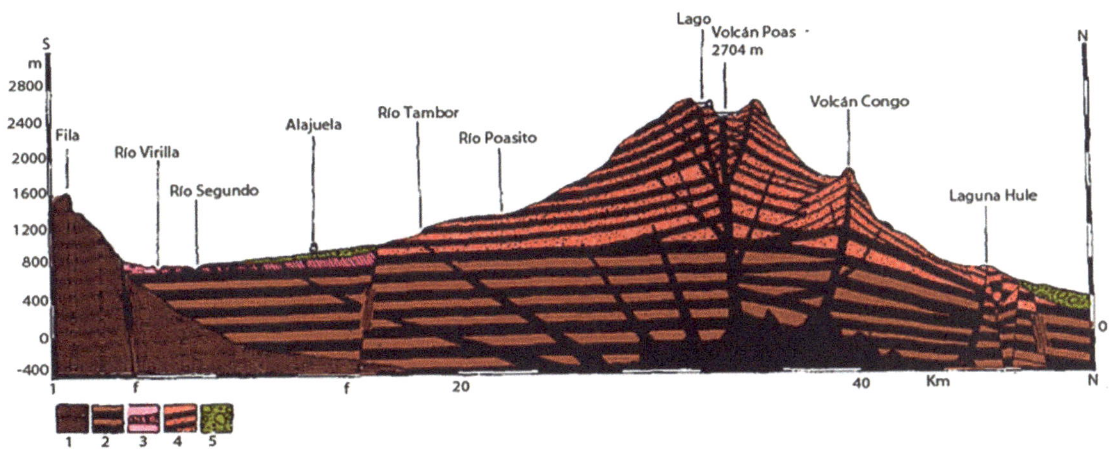

FIGURE 4.7 North–south geological profile of Poas Volcano between the Central Valley and Laguna de Hule (Bergoeing et al., 2010e). (1) Santa Ana Miocene volcanism. (2) Lower Quaternary basal intracanyon basaltic lavas. (3) Upper Quaternary ignimbrite avalanche deposits. (4) Middle Quaternary intracanyon lavas. (5) Middle to Upper Quaternary piedmont deposits.

FIGURE 4.8 Poas Volcano's active and explosive calderic crater with a sulfurous lake. In the foreground is the Botos pluvial crater lake. *Oblique aerial photo Bergoeing (2012a).*

FIGURE 4.9 Inactive Botos crater is just 300 ft from the main crater occupied by a stormwater pond and surrounded by high-altitude tropical vegetation, where *Gunnera talamanquensis* thrives in front. *Photo Bergoeing et al. (2010e).*

sometimes arrive up to the Virilla River, and finally the construction of the modern stratovolcano. Poas is an active volcano with an explosion caldera that contains in its interior a pluvial sulfuric acid lake due to its gaseous emanations (geyser activity) (see Fig. 4.7).

The volcanic–tectonic alignment emerges more northward with the modern Congo Volcano, with a much-eroded cone, and the Laguna de Hule collapsed caldera that built a postcollapse pyroclastic cone. This sequence can be attributed to the Holocene and ends with the Río Cuarto gasmaar. The base of the current cone of Poas Volcano, according to radiometric dating in basalt and andesite carried out by Alvarado and Gans (2012), is located in a period ranging from 0.6 to 0.56 MY and continues to the present day.

FIGURE 4.10 Northern very eroded parasitic cone of the active Poas crater at 8658 ft above sea level. *Aerial oblique photo Bergoeing (2012a).*

3. CONGO VOLCANO

Congo Volcano (6608 ft) is a well-formed cone, whose summit is devoid of a crater because of erosion. Where the crater was previously, there is a deep gorge formed by the Seco River (see Fig. 4.11). The cone is covered by the tropical mountain rainforest. Other rivers that are born on these sides are launched as tributaries of the Laguna de Hule caldera, such as the Hule River. The Laguna de Hule caldera has been the subject of several investigations (Bergoeing et al. op. cit.). The study shows that it is a collapsed caldera, probably from the Upper Pleistocene, which after its collapse issued pyroclastic material for some time, which allowed the construction of a small secondary cone right in the center of the caldera. Its volcanism is dominated by andesite. Once the paroxysmal activity stopped, rainwater filled the depression, thus forming a large lake, which today has declined and so has divided into two small lakes, the eastern one being the larger (Lake Hule) and the western the smaller (Lake Congo) (see Fig. 4.12). This northern part of the Poas watershed is fitted by rivers near the top due to a soil formed by lava flows poorly resistant to erosion, especially of the fluvial type. The Salto de la Paz waterfall (on the road to Sarapiquí) is a good example, because it is formed by welded tuffs, which allowed the construction of this 65-ft waterfall, part of the Angel River, which is born in the northeastern slope of Poas near its active crater. The sector lies within the volcanic structures of Poas and has an age of 0.527 MY (basaltic andesite dating by Alvarado and Gans, 2012).

Downstream, the piedmont turns, giving more rounded and monotonous forms, out of the volcanic accident of Congo Volcano, the Laguna de Hule caldera, and the Río Cuarto gasmaar (the last dated to 0.21 MY by Carr et al., 2007). Finally we reach the Northern Plain by means of alluvial coalescent fans, consisting of a material that is certainly in part altered, but is predominantly volcanic.

FIGURE 4.11 Congo has an eroded volcanic cone to the west. *Aerial photo, courtesy of Andre Quiros Tacsan (2012).*

FIGURE 4.12 The Holocene Laguna de Hule collapsed caldera. In the center is the postcollapse cone bordered by three lakes. *Aerial photo courtesy of Andre Quiros Tacsan (2012).*

The southern slope of Poas, which looks toward the western Central Valley, much less rugged, however, is fitted by rivers forming a subparallel network, whose direction is ENE–WSW. Humans have occupied this sector since colonial times and exploited it for pastoral purposes.

4. BARVA VOLCANO

Barva Volcano is the third major volcano in this mountain range and the last structure of the western Central Valley sector. At its peak open small craters containing lakes forming from rainwater. The main summit reaches 9534 ft above sea level and is where the first crater lake remains: three volcanic domes and a caldera rim that is oriented toward the east surround Lake Barva. Northward 1 mile as the crow flies, the second set consists of the Gongolona cone, which houses a rainwater lake inside the main crater, surrounded by three volcanic domes, and where researchers from the National University of Heredia described the presence of an important thermal water source (E. Malavassi, oral communication). The current adventitious cones and described craters are of recent formation (Holocene). They come from an explosion caldera structure (at the base of the current small cones that are at the top). Whereas historically Barva has no records of volcanic eruptions, it has a very recent volcanic activity, given the conservation of forms and fumarole activity (Boudon, 2003a,b). Eastward, this complex is separated from the Cacho Negro volcano cone by a structure known as the Molejon caldera, being crossed by the river of the same name. This structure from the Upper Pleistocene is partly buried by lava flows from the Gongolona crater and the Cacho Negro cone such that the rim of this caldera flourishes only in the north and south. It is just east of the summit of the Gongolona, where is drawn the most beautiful structures of the caldera. It is a semicircular (rim) escarpment with about 300 ft of domain and where is born the General River, a tributary of the Chirripó River. All this is deeply notched by the rivers that have eroded deep canyons (see Fig. 4.13).

This is the third major volcanic edifice of the western Central Valley. Like the Poas, this volcanic complex has been dated by radiometric analysis of its andesitic lavas. From analyses of the andesite and ignimbrite of the Braulio Carrillo sector done by Alvarado and Gans (2012), the oldest are in the range of about 1.2 MY and the rest range from 0.57 to 0.43 MY. To the south of this profile emerges the Escazú granite intrusive, which dates back to the Miocene (Bergoeing et al., 2010e) and may be associated with Talamanca's intrusions. It is fractured, and monzodiorite outcrops emerge. Pacacua Formation sandstone affected by contact metamorphism (cornubianite) surrounds the intrusive. The Pacacua Formation sediments are discordant and faulted on the folded strata resting on the Terraba Formation (Oligocene shale). Staggered tectonic faults south of the western Central Valley are partly covered by Upper Tertiary and Quaternary volcanic deposits. Barva Volcano is a modern stratovolcano composed of several craters on top hiding some structures

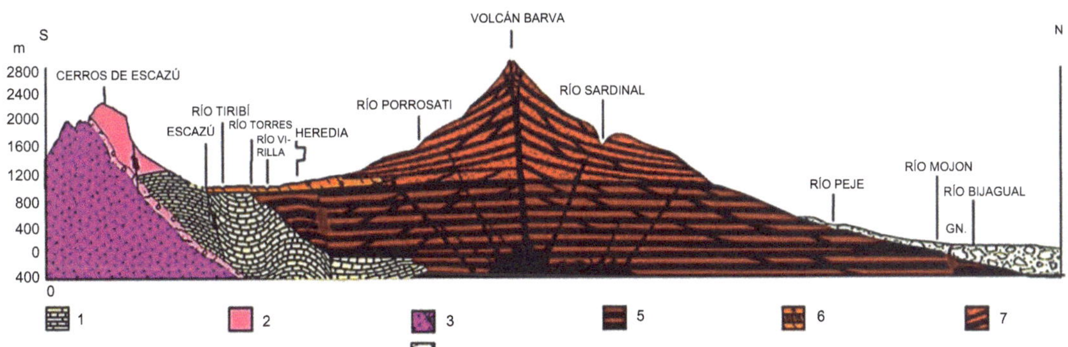

FIGURE 4.13 Barva Volcano's south–north geological profile from Escazú Mountain to the Bijagual River (Bergoeing, 2012a). (1) Oligocene–Miocene lutites, sandstone, and conglomerates; Terraba Formation. (2) Miocene cornubianite; Pacacua Formation. (3) Miocene Escazú intrusive. (4) Lower Quaternary basal lava flows. (5) Middle Quaternary ignimbrite avalanche flows. (6) Upper Quaternary interstratified lava. (7) Quaternary piedmont deposits.

FIGURE 4.14 Central Volcanic Mountain Range observed from San José. From left to right: Poas, Guarari Volcano summit, and Barva, with three domes. *Aerial photo Bergoeing (2012a).*

inside, such as the Molejon caldera. Northernmost sits the Holocene Cacho Negro Volcano, whose crater is open westward from the effects of erosion. At its feet, farther north, it is possible to observe the relicts of the Santa Clara caldera that mark the boundary of the volcanic system with the vast Northern Plain (see Fig. 4.14).

Barva's watershed is different from the others, because it is directly exposed to the northeast, where the humid flows from the Caribbean penetrate and often precipitate in the form of torrential rains, before reaching the volcanic peaks or the Central Valley, which makes this aspect more irrigated than the others elsewhere. It is logical that the tropical mountain forest is impenetrable in this sector.

5. MOLEJON COLLAPSED CALDERA

Located between the Mount Gongolona cone (8399 ft) and the Cacho Negro cone (7598 ft) the volcanic depression reaches 4921 ft above sea level, and at its center is a lake of pluvial origin. Molejon caldera is sectioned through its center by the Sardinal and Molejon rivers, which are the main tributaries of the Puerto Viejo River, whose origin is south of the volcanic depression. The caldera is equally deformed by a NNE–SSW tectonic normal fault where the Puerto Viejo River begins. It can be said that this volcanic structure predates the episodes that built the Cacho Negro and Gongolona cones because lava flows from these two cones bury much of the caldera structure; however, testimonies that mark its evidence are formed by the northern rim, where the Sardinal River starts, as well as the southern rim, fitted by the General River (see Fig. 4.15). It is very likely that the Molejon caldera emerged in the Upper Pleistocene, and the large pyroclastic flow that covered all the western Central Valley to the confluence with the Grande del Virilla River came from here.

6. CACHO NEGRO VOLCANO

Cacho Negro Volcano (7598 ft) is an isolated cone that dominates the northern slope of Barva volcanic complex. It is not active and, like the Congo Volcano, its crater is half-destroyed by erosion. In addition, it is covered by a thick blanket of tropical mountain vegetation (see Fig. 4.15). From this point, Barva's watershed descends gradually into the northern floodplain, forming a enormous piedmont, with megalandslides and secondary landslides, which hide an important part of a tectonic alignment (see Fig. 4.16). This alignment must be in close relationship with the southern edge of the Nicaragua graben. It is also in this sector where the Santa Clara collapsed caldera sits (see Fig. 4.17).

The eastern Barva structures, just on the edge of the Central Valley, are present as small inactive volcanic focuses. They are much eroded, but they retain their typical forms that are easy to recognize. They are the volcanic cones of Delicias, 7411 ft; Tibas, 6955 ft; Turú, Zurquí, 6562 ft;

FIGURE 4.15 Molejon caldera, northern slope of Barva Volcano. In the foreground lies part of the open-mouthed Cacho Negro Volcano cone. *Aerial oblique photo Bergoeing (2012a).*

FIGURE 4.16 Megalandslide on the northeast Barva Volcano flank. *Aerial oblique photo Bergoeing (2012a).*

and Hondura, 6758 ft, which operate in a much faulted and deep terrain. These forms are very young, because Tournon (in 1981 Sáenz) carried out K/Ar dating in an andesite, which gave an absolute age of 0.30 MY (Upper Pleistocene). Tectonic faults here are oriented preferably along a WNW–ESE axis, and this orientation is also found in the sector of Bajo de la Hondura, as well as on the west slopes of Irazu Volcano. It is a tectonic zone par excellence, and if the tremors that have occurred have not given victims so far, it is because it is an uninhabited

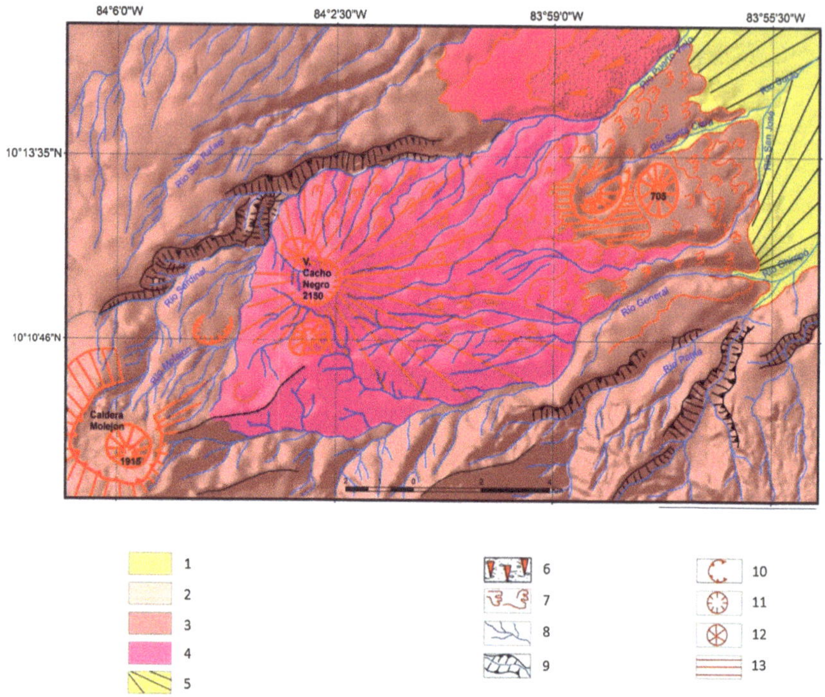

FIGURE 4.17 Geomorphology of the Molejon caldera area, the open crater of Cacho Negro Volcano, and the Santa Clara caldera complex. (1) Lower Pleistocene, 2–0.7 MY. (2) Middle Pleistocene, 0.7–0.2 MY. (3) Upper Pleistocene, 0.2–0.012 MY. (4) Holocene. (5) Alluvial fans. (6) Lahars. (7) Lava flows. (8) Fluvial network. (9) Fluvial canyons. (10) Collapsed caldera rim. (11) Crater. (12) Volcanic dome. (13) Structural volcanic plateau. *Based on a satellite radar image Bergoeing (2012a).*

region where the virgin tropical mountain forest reigns, despite the strong erosion of soils, in a low rain-erosion-resistant volcanic material.

It is precisely on the watershed dividing line of the eastern Barva Volcano flank, in its most depressed sector (Bajo de la Hondura), where the new road to Puerto Limon was built, but not without difficulties. It was necessary to dig a tunnel in volcanic soils, which lava flows will overwrite with a softer volcanic material (stratovolcanoes). Sunny days are rare, and there are strong condensation and daily rainfall, related to the northeast trade winds. This road joins San José, the capital city, to Guapiles City (the heart of the banana region in the Caribbean plain), a 1-h route today that formerly took 8 h by the eastern Central Valley. This layout is part of an old railway line built in the 1880s and abandoned at the end of the 20th century.

The Barva thermal activity occurs at various points. The Quebrada la Bomba and Quebrada el Cabro rivers are main points. Here the waters reach 140°F at the surface, but the most important place is at the Gongolona summit. It is a vast depression over the northeastern edge of the Camaquire, where there are sulfur dioxide fumaroles. The temperature here reaches 185°F (Yehudi Monestel, 1980).

7. SANTA CLARA CALDERA

Located at approximately 10°13′ north latitude and 83°58′ west longitude, about 3.7 miles north of Cacho Negro cone, a structural volcanic plateau presents itself at about 2297 ft above sea level, within which are the vestiges of two volcanic cones, probably postcollapse, very eroded, and in whose center is born the Santa Clara River, which forms a deep canyon. The old volcanic relict should go back to the Middle to Upper Pleistocene. Farms are located in the northern sector, and between the Santa Clara and the San José rivers mainly dacite rocks cover a series of craters. This old structure of Santa Clara marks the beginning of the elevation of the Central Volcanic Mountain Range and is located in a northern volcanic alignment, in which is also inscribed the western Flores caldera. Also in this alignment is situated the Laguna de Hule caldera, the Río Cuarto gas-maar, and finally the Aguas Zarcas small volcanic cones, placed by what might be a neovolcanic phase.

8. IRAZU VOLCANO

As already stated, this volcano shares the same structural basis with Turrialba Volcano. However, they are two individual cones, for which reason we will study them separately. The Irazu crater is 11,329 ft above sea level and, as a result, is the highest volcanic peak of Costa Rica. From the La Palma depression in the west piedmont, the cone slowly rises toward its volcanic summit. Because of this, the high part of Irazu dominates both the cities of San José and Cartago (former colonial capital) (see Fig. 4.18).

9. LAS NUBES VOLCANO, PART OF THE IRAZU COMPLEX

On Irazu's west side the only accident that disrupts the almost perfect ascent is an ancient crater, located near the hamlet of Las Nubes and given this name by the author (Bergoeing, 1979) (see Fig. 4.19). La Nubes is a crater that should correspond to a side activity of Irazu Volcano. It is situated in the Upper Pleistocene, according to dating and little alteration of the lava (andesite from 0.544 ±0.03 MY ago by Ar/Ar in Soto and Alvarado, 2006, and Carr et al., 2007), and that should compare at the same time, perhaps, with the Zurquí volcanic activity.

The crater interior was destroyed by erosion, forming a circle with a diameter of 0.62 mile, inside of which is a dairy farm, very well exploited. Also in this place begins the Virilla River, the main river of the western Central Valley. Following the rising line of the water division, there are other important summits. They are Irazu's small adventitious cones, especially Cabeza de Vaca (10,039 ft); Cerro Retes (10,371 ft), which still holds traces of the crater; and Sapper (11,154 ft), finally reaching the current Irazu craters.

Irazu has two well-formed craters, which communicate with each other. The western one is the main one (Calvert, 1918), whereas the eastern is less deep. After the 1963 ash eruption the manifestation of fumaroles stopped at the bottom of the main crater, which is filled with water during the rainy season. A third crater, completely filled, is located about 0.20 mile southeast of the first two, following the path that leads to the lookout, or "mirador."

Kruschensky (1972) noted, between the last point and the cone of the Turrialba volcano, two semicircular escarpments, open to the north from rain erosion effects, which must relate to possible caldera edges. These semicircular edges would be the only witnesses to the southern sector of the calderas. Very close to the current resting crater (Gutierrez, 1963a), there are important adventitious cones, situated to the south and the north. The large, high summit Alto Grande (9196 ft), which saves an element of triangular volcanic plateau, is oriented to the north. To the south are much more numerous adventitious cones; very close to the main crater is Mount Noche Buena (10,499 ft), and farther to the south

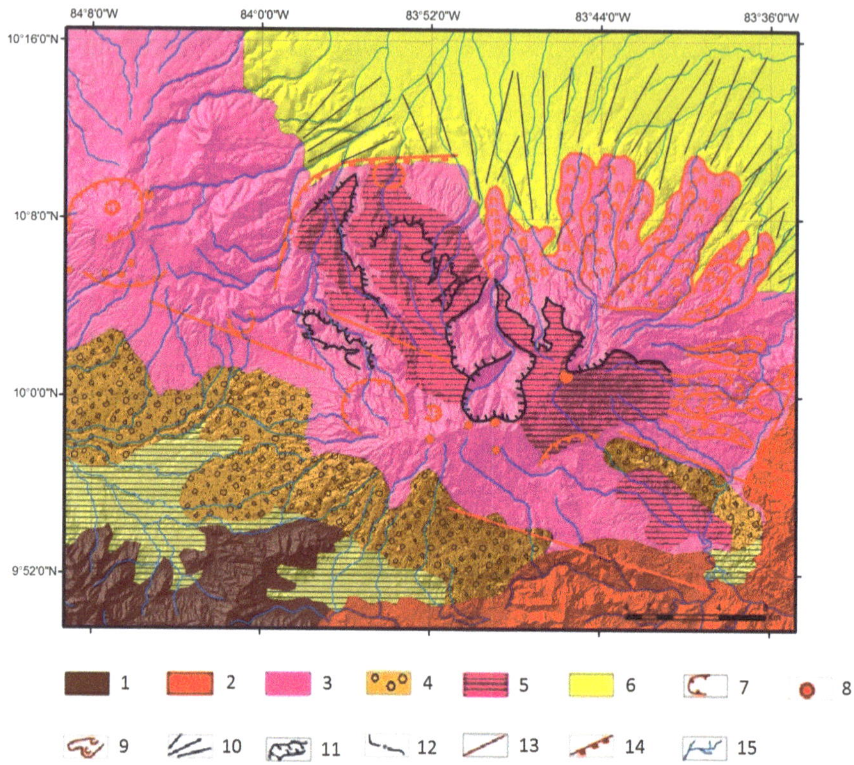

FIGURE 4.18 Irazu–Turrialba volcanic complex geomorphology as well as that of the Barva Cacho–Negro massif (Bergoeing, 2012a). (1) Tertiary sedimentary. (2) Tertiary volcanic. (3) Quaternary volcanic. (4) Lahars. (5) Planezes. (6) Topographic plateaus. (7) Calderas. (8) Volcanic craters. (9) Lava flows. (10) Alluvial fans. (11) Erosion scarps. (12) Water dividing line. (13) Tectonic faults. (14) Reverse fault. (15) Fluvial network.

is a series of three aligned small adventitious cones, whose main peak is Mount Gurdian (10,059 ft). Southeast of these three cones is Mount Pasqui (8379 ft), an important cone from where it is presumed the Cervantes lava flow originated (23,000 years BP) by Madrigal (1977) (see Fig. 4.20).

The various geological works in this region allow one to posit the hypothesis that in the place where the town of Paraiso stands, the lava surface that supports the city (topographic plateau formed of basaltic–andesitic lava flows exhibited at more than 656 ft) corresponds to the basal lava system from the initial volcanism (Lower Pleistocene). Here, in the eastern sector of the Central Valley, is a lava plateau known as Lavas de Paraiso ("Paraiso lavas"). They are basaltic–andesitic lavas of several hundred feet, very well preserved, with tuffs and volcanic conglomerate intercalations visible on the Turrialba City road slopes. The Cervantes lava flow is undoubtedly the most visible of the southern sector of the Irazu volcano phenomenon, because it covers an area of approximately 16 square miles. According to Kruschensky et al. (1976), this lava flow is composed of augitic and olivinic andesites with tuffs and lapillus deposits. It is the most recent lava flow from this

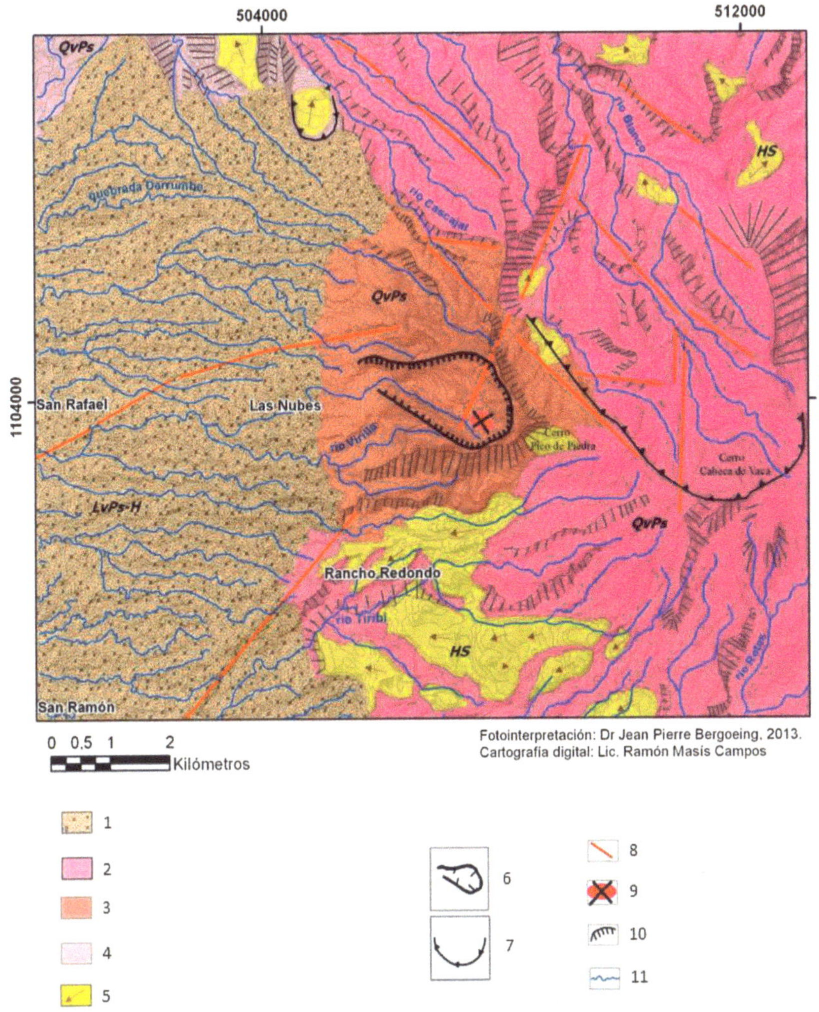

FIGURE 4.19 Las Nubes Volcano geomorphology (Bergoeing, 2012a). (1) Upper Pleistocene–Holocene lahar area *(LvPs-H)*. (2) Irazu volcanic area *(QvPs)*. (3) Las Nubes volcanic area (0.5 MY) *(QvPs)*. (4) Zurquí volcanic area *(QvPs)*. (5) Holocene mass landslides. (6) Crater. (7) Volcanic calderas. (8) Tectonic faults. (9) Volcanic dome. (10) Erosion talus. (11) Fluvial network.

sector, and it descended to the Reventazon River thalweg. This author recognizes other lava flows near the top of the Upper Pleistocene to which he gives the names of Birrís, Sapper, and Reventazon Formations. The last corresponds to the Lower Pleistocene and is related to the Paraiso lava flow, thus constituting the base of the stratigraphic column.

The east Central Valley, characterized by a northern slope, is formed by a volcanic edifice housing two main modern volcanic cones. They are the stratovolcanoes Irazu and Turrialba, both active. The geological section begins with the last foothills of the northern Talamanca Mountain Range, characterized by Tertiary volcanic deposits (Pliocene), which are discordant

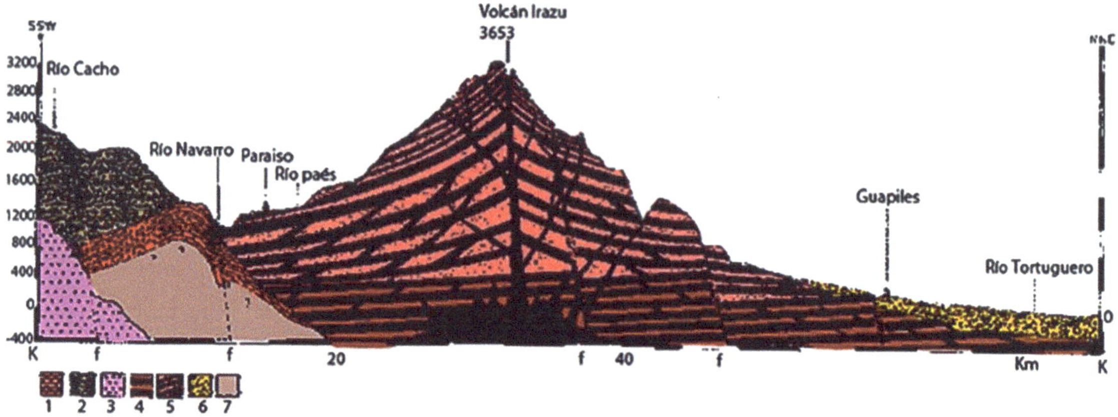

FIGURE 4.20 Irazu Volcano geological profile (Bergoeing, 2007). (1) Miocene sandstone, shale, and lutites; Coris and San Miguel Formations. (2) Pliocene volcanism; Aguacate Formation. (3) Talamanca's Tertiary intrusive. (4) Basal Quaternary volcanism. (5) Middle to superior volcanism. (6) Quaternary piedmont alluvial fans constituted by lahars. (7) Tertiary volcanism.

about the underlying sedimentary Miocene series. Sedimentary formations are limestone, with *Pectens* fossils in the San Miguel Formation and siliceous sandstone in the Coris Formation. A stratification strongly folded by the effects of the Plio–Quaternary orogeny provides anticlines and synclines in the south of La Carpintera row. The lavas of Irazu reached southerly to the Navarro River banks and entered its bed following a tectonic WSW–ENE fault line.

The town of Paraiso is seated on a basaltic–andesitic plateau (Paraiso lavas) ranging from the Lower to the Middle Quaternary and which has a thickness of several hundred feet. Near Paes River, piedmont deposits intermingle and intersect with lahars.

Irazu is an active volcano whose last eruption dates back to 1963. Volcanologist Phillip Ruprecht of Columbia University, New York, studying the 1963–65 eruption lavas of Irazu, classifies it as a supervolcano because its magma chamber is connected directly to the mantle and he therefore has dubbed it "the highway from hell." He came to this conclusion after checking the presence of trace amounts of nickel in the olivine of the lavas that should have melted before being ejected by the eruption, which is proof of the rapid magma ascent from the mantle. Ruprecht said that Irazu is the most important stratovolcano of the Central American arc, because, during the eruption of 1963–65 it ejected basaltic–andesitic lava containing a variety of olivine crystals that have preserved original chemical traces of the mantle as nickel and forsterite, as measured by inductive laser ablation coupled to a plasma mass spectrometer (Ruprecht and Plank, 2013).

Irazu's northern slope, exposed directly to the moist Caribbean trade winds, has canyons and planezes in relief covered by a thick mountain rainforest. Mount Alto Grande is a parasitic cone of Irazu, which marks the progressive decrease in the lava flows toward the Northern Plain (Fig. 4.21). The edge of this plain is marked by a series of volcanic structures of the Upper Pleistocene such as the Flores caldera.

In the sector of Coliblanco, traces of an old lake of volcanic origin, which today is filled and presents a flat surface, are the remains of an old explosion caldera that corresponded to the top of Turrialba Volcano during the Middle Pleistocene. It would have to be related it to a

FIGURE 4.21 Adventitious Alto Grande cone (9196 ft), Irazu Volcano northern watershed. *Oblique aerial photo Bergoeing (2012a).*

former level of Turrialba's volcanic summit, which at that time should have been at an altitude of 5742 ft. The current Turrialba River would drain this lake, later, and perhaps due to volcanic eruptions.

Irazu's southern watershed is divided into two different areas:

Between the hamlet of Cot and the summit, the area is formed by a series of lahar layers, with 0.30 ft of ash-coated thickness. On this slope, the Reventado River is the combined result of fluvial erosion and volcanic eruptions, because by this natural leak descended large chaotic lahars and torrential lava flows that built the alluvial cone on which Cartago City has been established since the 16th century.

The lower sector of the southern slope, composed of lahars, extends westward, and descends also toward the western Central Valley. Important lahar deposits cover this sector but are re-covered in large part by a thick blanket of volcanic ash; this is due to the exposure of this sector to the northeast winds that cross the top of Irazu. This is evident especially in the month of December, when the winds are the coldest of the year and lift dust from the recently ploughed lands of the southern slopes of the volcano, leaving one to think that it is volcanic activity because of the darkness of the air.

The northern Irazu watershed, as we saw in the case of the northeast slope of Barva, is subjected to an intense pluvial and river erosion, the latter aided by very numerous southeast tectonic alignments in the sector. Two major rivers have undermined deep canyons and ravines on this aspect: they are the Sucio and Toro Amarillo rivers. Parts not exposed to the fluvial action of the slope are under the influence of widespread solifluction or creeping soils.

At the foot of this slope register important coalescent alluvial fans before reaching the northern floodplain. These alluvial fans are very active, and they have tested the railway road from the banana plantations during an exceptional flood of the Toro Amarillo River. During the rainy season of 1980, the iron beams of the bridge that crosses the river were spiraling as if they were lead. The force of the storm is not exercised, but only in times of strong floods, when the phenomenon of resumes of fluvial material rupture load, deposited in the period of decline.

This means that frequently river beds also change with time. This is recorded in the topographic maps of the Instituto Geografico Nacional (IGN) of Costa Rica, in which from one edition to another we can observe successive changes (see Fig. 4.22). Also at the foot of Irazu Volcano, already in contact with the Northern Plain, are the relicts of a collapsed caldera that we call the Flores caldera complex and which probably correspond to a volcanic activity of the Upper Pleistocene in this sector.

10. FLORES VOLCANIC COMPLEX

The Flores volcanic structure, predominantly andesitic, consists of a much-eroded caldera where the Flores River is born. To the north it is bordered by the Siquirres–Matina reverse fault that has lifted up the outer edge of the rim. Inside of the caldera are much-eroded remnants of two postcollapse cones whose altitudes do not exceed 1194 ft. The caldera, made by a multiconvex relief, denotes a profound alteration in situ of andesite made by landslides, which are arranged concentrically. To the west of the caldera and separated by the Corinto River is other volcanic structure, also very eroded, with remains of planezes and whose center is crossed by the Molinete River, forming superimposed or deep-boxed meanders, exploiting the weakest volcanic structure areas. This structure is separated from the collapsed Flores caldera not only by the Corinto River but also by two tectonic faults, one of heading NNW–SSE and another NNE–SSW, exerting a remarkable structural control on the Corinto River. To the south, the caldera is bordered by the latest Irazu complex lava flows and on the east by a series of laharic alluvial fans, where the Costa Rica River is born (see Fig. 4.22).

11. TURRIALBA VOLCANO

From the top of its 10,892 ft, this last volcano in the Central Volcanic Mountain Range dominates the Northern Plain, the Limon basin, and the eastern Central Valley, where is drawn the lower course of the Reventazon River. Turrialba's cone dates from the Upper Quaternary, and it is very well preserved because its forms resist pluvial and fluvial erosion. The current great wide cone houses four small craters, which are located in the southwest. Since 2010, Turrialba Volcano has presented intense fumarole activity, and in 2016 it began large ash eruptions (Fig. 4.23).

The oldest historical eruption of this volcano was recorded in the 18th century; another occurred in 1852–53, the ashes of which reached the northern Nicaraguan port of Corinto; another eruption occurred at the beginning of the 20th century (Miyamura, 1980) and the most recent began in 2010 and continues today. Turrialba lavas have been dated and placed in a margin ranging from 0.983 MY (basaltic andesite on the Siquirres River) to 0.251 MY (aged andesitic–basaltic lava of Finca Convenio Farm) to 0.090 MY (andesitic–basaltic lava flow of Hacienda Bonilla) and to the recent time of 0.0003 MY ago (dacitic lava from La Silvia Farm). This shows the effusive continued activity of this volcanic massif (Ruiz et al. 2010 in Alvarado and Gans, 2012).

Near the main cone, small adventitious cones can also be discovered, above all that of Mount Tiendilla (9157 ft); a second cone close to the first one, anonymous, but whose elevation is 9022 ft; and then a third, halfway to Irazu Volcano, whose elevation is 10,085 ft. This last cone divides two semicircular structures separating the two large volcanic cones. Finally, a parasitic cone can be observed to the northeast, where

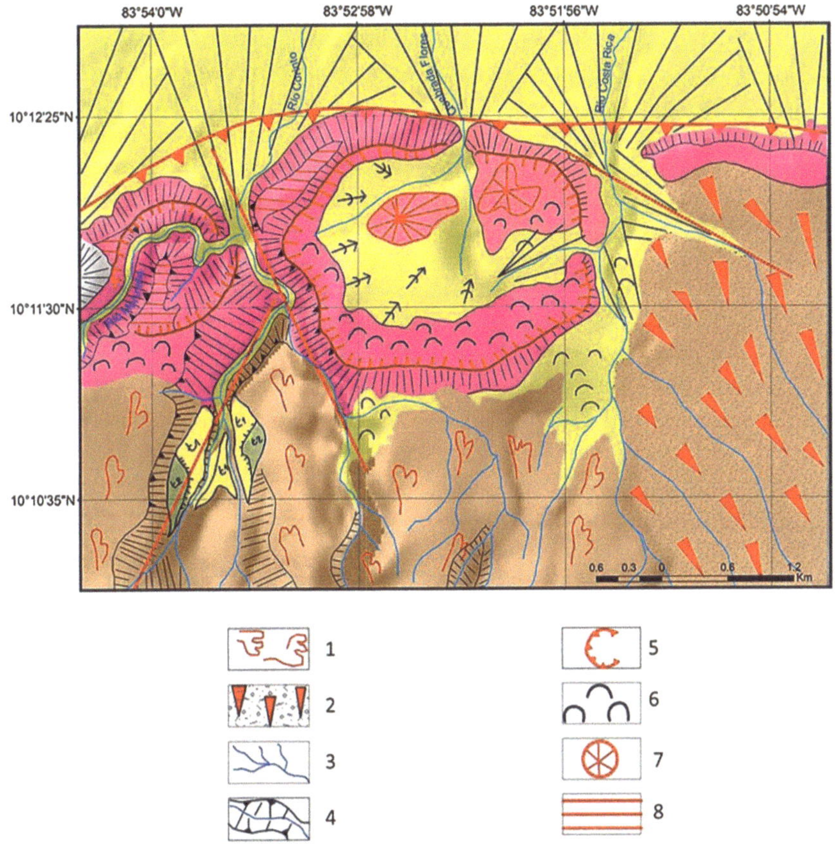

FIGURE 4.22 Middle to Upper Pleistocene Flores collapsed caldera and volcanic complex geomorphology (Bergoeing, 2012a). (1) Lava flows. (2) Lahar fans. (3) Fluvial network. (4) Fluvial canyons. (5) Collapsed caldera rim. (6) Multiconvex shaped area. (7) Volcanic dome. (8) Structural volcanic plateau.

FIGURE 4.23 Turrialba Volcano craters and the beginning of an eruption. *Aerial photo Bergoeing (2012a).*

the Guacimo, Parismina, and Dos Novillos rivers are born. Turrialba has not had recent lava flows similar to the Cervantes lava flow of Irazu. However, major lava flows are known during the Quaternary on the southern volcanic slopes, such as the Aquiares flow, descending on Turrialba City, or the Pascua lava flow (Tournon, 1980), as well as the Peralta lava flow, which gave rise to the natural dam of the lower course of the Reventazon River, which resulted in the formation of the small Bonilla and Lancaster lakes. On the northeast slope, a beautiful lava flow is drawn, between the Dos Novillos and Pascua rivers. It probably comes from the parasitic northeast cone, whose altitude is 7218 ft, and covers a distance of at least 5 miles and disappears under the Piedmont deposits approximately at 2300 ft above sea level.

The 7316-ft summit of the adventitious Turrialba cone volcano marks the starting point of the descent of the northern watershed, whose transition to the tectonic northern depression was fed by volcanic lava flows from the Upper Pleistocene and Holocene resting on ancient lahar deposits (near Guapiles sector) (see Fig. 4.24). The Northern Plain is covered by significant thicknesses of volcanic deposits that are equally powerful alluvial fans in the contact area. The Lomas de Sierpe volcanic row emerges north of the geological section and corresponds to a Pliocene–Quaternary calc-alkaline volcanic series surrounded by the coastal sediments of the Holocene. This is where the little Tortuguero volcano cone emerges as an aborted attempt of littoral volcanism in the middle of the Flandrian parallel coastal cords that gave rise to the Tortuguero canals.

The northern Turrialba Volcano piedmont in general is less dissected like the Irazu Volcano watershed. The ravines are present only at the very birth of the Guacimo and Dos Novillos Rivers, in the place where the already described parasite cone of 7218 ft above sea level lies. On the other hand, the southeast piedmont, which descends toward the limits of the east Central Valley, is much more dissected, exposing powerful triangular basaltic plateaus (planezes), products of deep lava deposits. Here the alteration mantle of the northern side that plays a protective role is practically nonexistent, and lavas are directly attacked by erosion. The Pascua River runs through a deep canyon of more than 1300 ft, with vertical, narrow walls. This river follows a straight path, which should relate to the southeast tectonic alignments.

The most important tectonic alignment, and marking a local neotectonic phase, is the Tuis River fault. It extends in a direct line, goes up

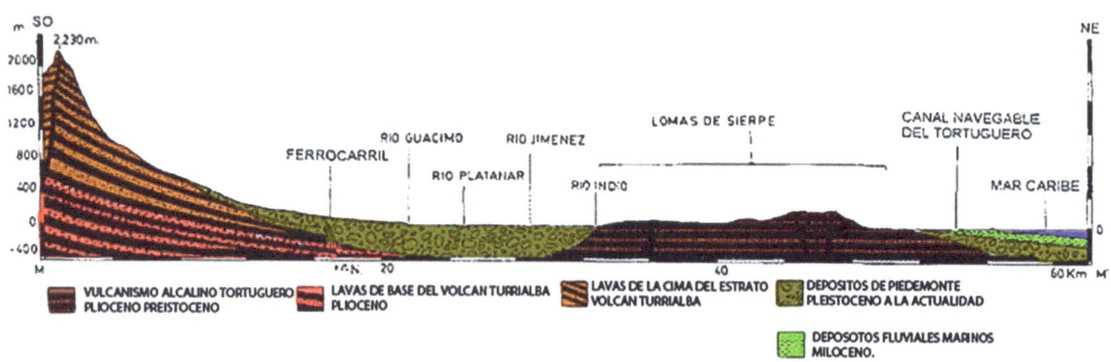

FIGURE 4.24 Geological profile from Turrialba's northern slopes to the tectonic northern Costa Rica depression going through Sierpes volcanic row and Tortuguero channels near the Caribbean sea (Bergoeing, 2007). (1) Pliocene–Pleistocene, Tortuguero alkaline volcanism. (2) Pliocene, Turrialba's basal lavas. (3) Turrialba's summit stratovolcano lavas. (4) Quaternary piedmont deposits. (5) Miocene marine and fluvial deposits.

FIGURE 4.25 Turrialba's cone and planeze with vegetation burned by sulfurous volcanic emanations. *Oblique aerial Photo Bergoeing (2012a).*

the Turrialba River, and then crosses the upper part of the volcanic cone near the current crater to continue on the northern slope in the Bajo de la Hondura sector. We think that this tectonic fault is closely related to the 1952 Patillos earthquake in the northern Irazu slope, which claimed victims among the members of the agricultural colony recently implanted in this sector (Miyamura, op. cit.). In some areas, the ground sunk more than 16 ft, swallowing household objects. The effects of this earthquake are still visible in part.

It is important to point out, on Turrialba Volcano's southern slopes, at the base of the volcano, sedimentary outliers, none dated, which must necessarily be prior to the Quaternary and older than the calcareous deposits of the Las Animas Formation (Miocene). Limestone is partly covered by volcanic Quaternary deposits. It is very likely that calcareous deposits were placed on the volcanic base of the Nicaragua graben and are the ancestors of the current volcano structure. All have suffered the effects of the Pliocene–Quaternary uplift (orogeny).

In conclusion we can say that the Central Volcanic Range activity, during the Pliocene and the Lower Pleistocene, must have been very important, allowing the construction of the solid bases that we currently appreciate and that enabled it to later reach its current dimensions. Lava flow outcrops on the sides of these volcanic structures, like those of Florida (Platanar volcano), Los Angeles (Barva's southern slope), Cervantes (Irazu's southern slope), and Aquiares and Pascua (Turrialba's southern slopes), must be located in the Upper Pleistocene. According to Tournon: "The great stratavolcanoes construction of the Central Mountain Range appears to have started in the Upper Pleistocene. The volcanoes layer overlaps Aguacate's Pliocene series to the West and to the East, the alkaline Pliocene lavas" (Tournon, 1978 in Bellon and Tourno, 1978).

As we see it then, in the Pliocene and the Pleistocene, volcanic activity had no paragon, and concerning rest periods, such as those currently living volcanoes in the Central Volcanic Mountain Range. Outside the small fumarole effusions either volcanic eruptions such as Irazu in 1963 and Turrialba (2010–present) are not but a phase of relative calm, in the course of the past 500 years (known historical period), which does not exclude that these volcanoes can emit powerful lava flows (Fig. 4.25).

CHAPTER

5

Central Valley Geomorphology

OUTLINE

The Central Valley of Costa Rica covers a total of 1253 square miles and follows an east–west depression that separates the northern Central Volcanic Mountain Range from the southern Talamanca Mountain Range, becoming the natural northern limit of the latter. According to Jean Tournon (oral communication), the Central Valley and the Central Volcanic Mountain Range form part of the Nicaragua graben, being the southern boundary of this tectonic depression until reaching Matama row (Talamanca's Caribbean watershed). This would explain the Quaternary volcanic forms of Matama described earlier (see Fig. 5.1).

For Enrique Malavassi V., with which I undertook the geomorphology mapping work of the Central Valley, this valley would be the southern limit of the Nicaragua graben. Therefore, it defines two different areas.

- **Perigraben sector**: This is where the Central Valley and the mountain ranges are located.
- **Intergraben sector**: This begins at the foot of the northern slope of the Central Volcanic Range and extends northward to Tortuguero's volcanic calc-alkaline outliers.

On the other hand, according to Landsat satellite images, it is possible to observe, at the foot of the volcanic mountains' northern slopes, a large tectonic accident, or southeast tectonic alignment, that is interrupted or hidden by modern piedmont deposits on the northern Barva–Cacho Negro volcano system slopes (Bergoeing, 1981).

http://dx.doi.org/10.1016/B978-0-12-812067-5.00005-8

FIGURE 5.1 Costa Rica's western Central Valley bordered by the northern volcanic mountain range and the southern Talamanca Mountain Range and drained by the Virilla-Tarcoles River. *Geomorphology Atlas of Costa Rica, 1:350,000 scale, by Jean Pierre Bergoeing et al. Instituto Costarricense de Electricidad, 2010.*

This piedmont then hides this fault alignment, interrupting its path, as can be seen on both sides of this enormous piedmont megalandslide cone.

The Central Valley's location can be defined as a tectonic depression, consisting of a series of stepping faults of the Talamanca Range's northern slopes (Weyl, 1957). This depression and its stepping faults are covered by the series of modern volcanic deposits of the Central Volcanic Mountain Range. This mountain range is then the north side of the Central Valley and is composed of several volcanic edifices, separated by a mountain pass, which allows the humid Caribbean flows to penetrate (Gutierrez, 1963a,b). This northern slope is made up of two homogeneous sets: east, the Turrialba–Irazu volcanic complex and west, the Chayote–Poas–Barva volcanic complex. These two sets are separated by the La Palma depression (see the geomorphologic map of the Central Valley, Fig. 6.12).

The south slope of the western Chayote–Poas–Barva volcanic complex is tilted southwest. This is noticeable by observing the hydrographic subparallel network drawn in this piedmont.

The Central Valley is divided into two different areas separated by the small La Carpintera volcanic-sedimentary mountain range of the Upper Tertiary (Malavassi and Suarez, 1953; Meza, 1979). La Carpintera Range divides this valley, giving it a NE–SW orientation (see the map of morphological units of the Central

Valley, Fig. 5.2). The eastern Central Valley is inscribed by the Reventazon River, which drains its waters to the Caribbean Sea, and the western Central Valley is crossed by the Virilla-Tarcoles River, whose waters go to the Pacific Ocean.

La Carpintera (6165 ft) then constitutes the continental watershed divide line. The aborted volcanism of this mountain range belongs to the Upper Pliocene–Lower Quaternary (monzogabbro dated to 2.25 MY by the K/Ar method). It is likely that, during this period, a small volcanic focus allowed magma to rise, associated with the rising Irazu Volcano, deforming and faulting the Miocene sandstone, shale, and limestone of the Coris and San Miguel formations that make up the southern part of this small mountain range (Franco, 1978; Carballo and Fischer, 1978). It was then, at the beginning of the Quaternary period, at the same time that the Central Volcanic Mountain Range started building, that the Carpintera volcanism took place. The construction of this small mountain range determined the individualization of the two different watersheds, draining in opposite directions, toward the Pacific Ocean and toward the Caribbean Sea (see Fig. 5.2).

1. EASTERN CENTRAL VALLEY

From the Ochomogo Pass, northeast of La Carpintera, following the San José–Cartago National Road, we enter into the eastern area of the Central Valley. It is much larger and vaster than the western sector. The Ochomogo Pass previously housed three small lakes, which disappeared after the new road was finished. The southwest limit of the Irazu Volcano slope also constitutes for La Carpintera a division water line.

In contrast to the western slope, the main drainage system is located in the southeastern sector. The main river in this part of the Central Valley is the Reventazon River, which is born on the Cuerici summit, in the foothills of the northeast Talamanca Range. The main river is displaced to the south, because the most important material constituting the eastern Central Valley

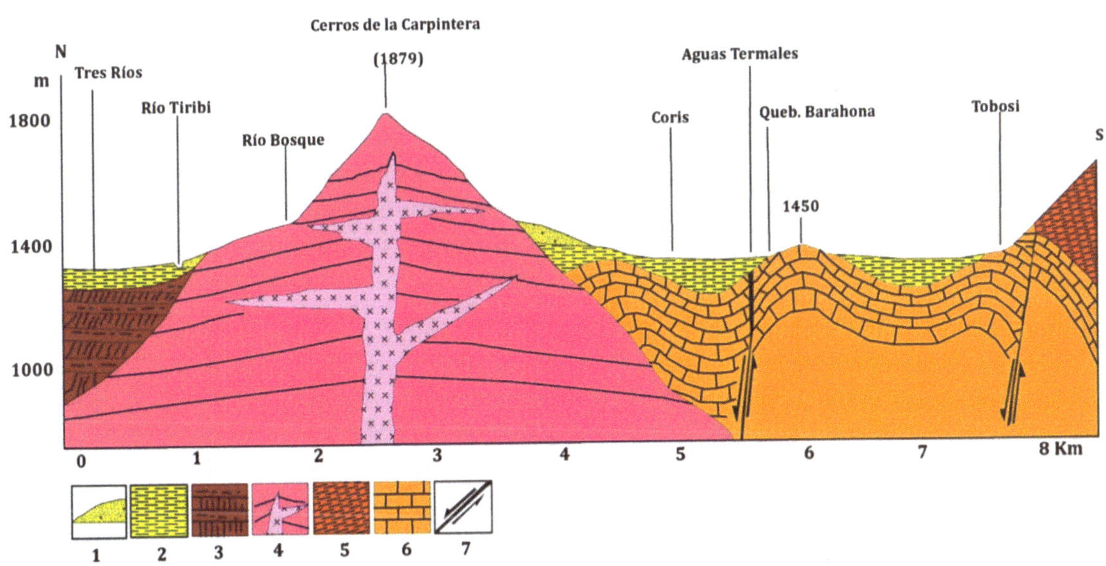

FIGURE 5.2 Geological profile of La Carpintera Mountain Range in Alto de Ochomogo sector (Bergoeing, 2012a). (1) Alluvial fans. (2) Fluvial–lacustrine deposits. (3) Irazu lava flows. (4) Magmatic ascents. (5) Talamanca's volcano-sedimentary mountain range. (6) Miocene shale and sandstones of the Coris Formation. (7) Tectonic faults.

comes from the northern Irazu and Turrialba volcanoes. The Reventazon River, after its inception, changes its name in several sections, like many Costa Rican rivers. In its upper course, it is called the Cuerici River, and then the Orosí River until the Cachí hydroelectric dam. After this dam the river takes its real name, Reventazon, and even arriving at the plain, before its Caribbean Sea mouth, its last tributary, the Parismima River, gives its name. Here the main drainage system is located to the south, where important tributaries are drawn: the Naranjo, Pejibaye, Atirro, and Tuis rivers.

The northern volcanic slope is better exploited from the agricultural point of view, but this is true only for the immediately north sector of Cartago City. This is on the Irazu Volcano southern flank, where horticultural crops are very developed. This volcanic slope is also exploited largely as pastures for dairy cattle, because the climate lends itself to that.

Between the Irazu Volcano summit and Cartago City, we recognize two morphologically different sectors:

- **Between the volcano summit and hamlet of Cot**: This sector has a wavy, undulating relief, consisting of varviable thicknesses of ash covering ancient lava flows. It is a very porous soil but extremely fertile, exploited for pastures and crops (potatoes, onions).
- **Between Cot and Cartago City**: This sector is a land of more rugged terrain, formed by lahars, of which two generations can be easily distinguished. These lahars are coated on the surface by a thickness of 0.30 ft of volcanic ash.
  - **Old yellow-ochre lahars**, with fragments, very altered, of pumice and lapilli, are exposed on the slopes of the Reventado River Bridge road, with 16 ft of apparent thickness.
  - **Young grayish pumice lahars**, in good condition, are mixed with a chaotic rolled material. They are upstream of the older lahars. Their thickness is also 13–16 ft.

The Reventado River of fatal fame drains the northern volcanic slopes in this sector. Cartago City is located on the banks of this river, built on a large alluvial fan formed in part by lahars, mud, and torrential lava (see Fig. 5.3).

In 1963, Irazu's volcanic eruption rains melted projections and ashes from the Irazu crater, causing a catastrophe. A torrential mud (cold lahar) was displaced through the river thalweg, funnel-shaped with a steep slope, overnight, burying houses and residents who were to the west of the city. The Taras sector was then separated at that moment from Cartago City. Today, this mud wash is the natural tomb of the unfortunates who could not be recovered after the disaster. Today a colony of Honduran pines grows in the clay wash deposit of over 32 ft. Cartago City, because of its location, will always be under the threat of a new similar catastrophe.

In the southern sector of Cartago City, between the small La Carpintera Mountain Range and the southern foothills of the Talamanca Range, lies the Coris basin. Coris is an ancient tectonic depression and former lake, today forming a silted valley. Here was seated the first colonial capital city, or "Mud City," but it was early abandoned and replaced in a more strategic location, where Cartago is currently seated.

The Coris basin is a west–east faulted synclinal, covered by the small Quebrada Barahona River, which gives rise to the Aguascalientes River, another tributary of the Reventazon River, joining downstream with the Purires River. Quebrada Barahona follows the route of an east–west tectonic fault alignment. Hot springs arise in the Quebrada Barahona and Aguascalientes rivers.

On the left bank, there are numerous outcrops of hot springs, some covered by reeds, others without vegetation because of the calcareous crust. This is a regional tectonic fault alignment. Following this path, we find other sectors with the same thermal water: near Orosí town, on

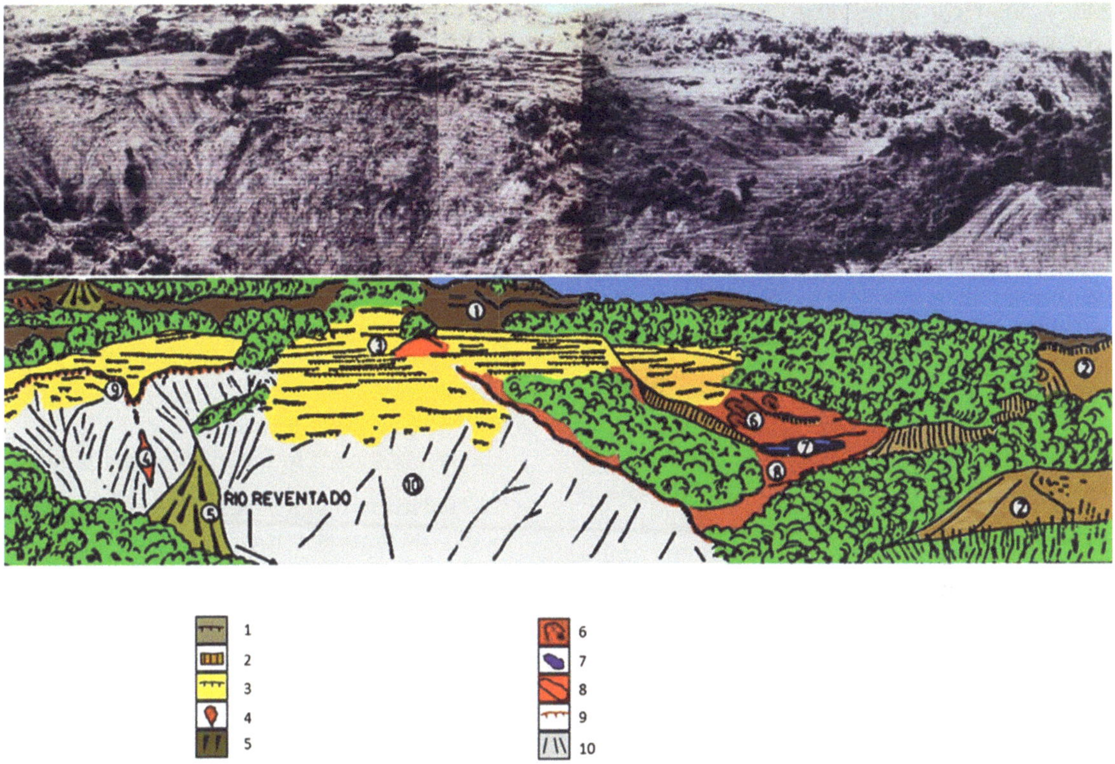

FIGURE 5.3 Cartago's Reventado River western slope showing different landslide types. (1) Erosion talus. (2) Erosion talus and alluvial fan. (3) Sliding terraces. (4) Bedrock outcrops. (5) Alluvial fan. (6) Mass landslide beginning. (7) Lake. (8) Mass landslide. (9) Scar release. (10) Main erosion talus. *From the author's own photography of 1998, Bergoeing (2012a).*

the Orosí River left bank; on the Naranjo riverbanks; in the western sector of the Central Valley in the Azul River (former public sanitary landfill of San José); and in the south of Santa Ana City. It is then an important tectonic fault alignment where hot springs arise, probably infiltrating from the northern volcanic piedmont, but this is no more than a presumption. The thermal water that we have analyzed has a temperature not exceeding 142°F. The hot waters are rich in calcium carbonate (40%) and sulfurous elements.

2. REVENTAZON RIVER

The Reventazon River is born on the east side of the Talamanca Mountain Range, on Cerro Cuerici summit (10,974 ft), with the name of Cuerici River. Here the landscape is an altitudinal "moor," whose soil gives glimpses of the remains of an old snow ice age (Würm–Wisconsinian), similar to and contemporary with the Cerro de la Muerte high summit parking. The pattern of this sector of mountain drainages is clearly dendritic, as it is often guided by important NW–SE tectonic fault alignments. The upper course of the Cuerici River, owing to steep slopes (more than 40 degrees in some parts), is characterized by torrents of a high-flow regime. It is the most wooded of the eastern sector of the Central Valley; because it constitutes the upper part of the Cachí hydroelectric dam basin, it has been declared a forest reserve by the government. Right before entering Orosí

Valley, the river adopts the name Orosí River. The valley extends, and the slopes of the thalweg decrease considerably, which results in a braking force on the flow (load-break), thus creating a braided fluvial system of the river bed, formed by islets and large pebbles (1–2 ft in diameter) and blocks, all well worn. The material is generally volcanic or intrusive. On the margins of the Orosí River, mostly on the left bank, you can see two fluvial terrace levels (see Fig. 5.4).

The top fluvial level, where stands the village of Orosí, is 66 ft above the largest riverbed. The lower level is about 33 ft down from the first one. There are probably other higher levels, but if they exist they are buried under colluvial coalescent fans, very numerous on the left bank of this valley. For example, we give a small stream, but very strong, that during the Holocene built a major alluvial fan arriving at the Orosí Valley. This river has a strong capacity during exceptional floods. It is capable of transporting large blocks (more than 3 ft in diameter). This material almost completely buried an isolated hill, which lies precisely in this sector of confluence. Orosí Valley is purely of tectonic origin (Berrange, 1979), because it is an extension of one major tectonic accident; the Navarro River fault.

The Orosí River describes a curve to the east, where it constitutes an important extension of lake water; here in 1954 was built the hydroelectric Cachí Dam. This dam was constructed in a naturally favorable place, because on two occasions, volcanic flows of the Irazu obstructed the site, thus creating the first conditions for the current hydroelectric dam. The Orosí River, in its middle course, built four staggered fluvial–lacustrine levels during the Upper Pleistocene, located before the dam, on the left bank, close to the tectonic fault escarpment of the Paraiso volcanic plateau, which today overlooks the artificial lake. There are only three visible levels, because the waters of the dam submerge the lower one (see Figs. 5.5 and 5.6).

The levels shown in Fig. 5.6 are as follows:

T1 lacustrine level: This level is visible only when the dam is empty, to avoid the proliferation of water hyacinth *Eichhornia azurea* (gamalote).

T2 lacustrine level: This level is composed of little boulders with altered edges on the surface, medium sized, 0.65–1 ft in diameter. They are mixed with a rather grotesque yellow clay mass.

T3 lacustrine level: This level contains petrographic material similar to that of the T4 level, though it is a little less altered and is thinner. It is 13–16 ft up from the T2 level.

FIGURE 5.4 Orosí River upper course. First fluvial terrace level with pebbles and boulders. *Photo Bergoeing (2012a).*

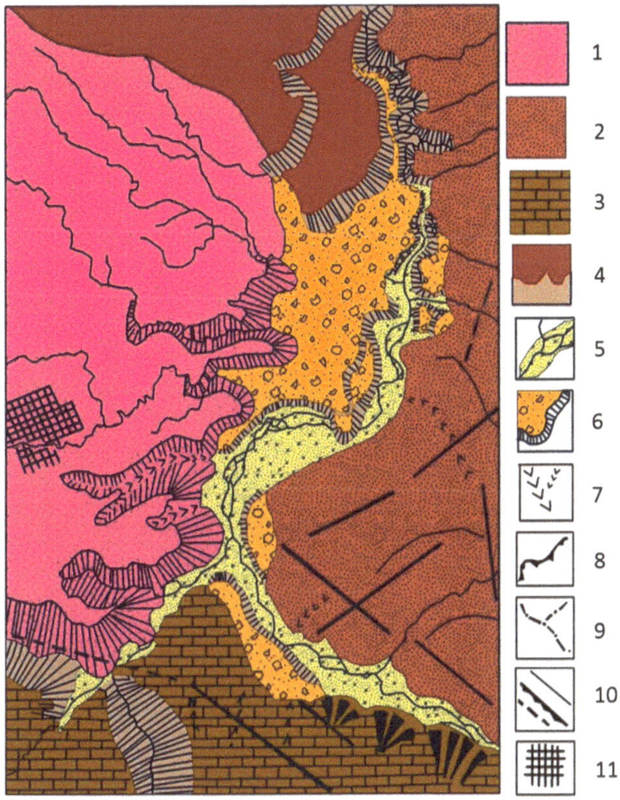

FIGURE 5.5 Cachí Dam morphology sector, eastern Central Valley (Bergoeing, 2007). (1) Low Pleistocene, Paraiso Formation lava flows. (2) Pliocene, Aguacate Formation lava flows. (3) Tertiary sedimentary outcrops. (4) Cervantes lava flows. (5) Reventazon riverbed. (6) Fluvial–lacustrine terraces. (7) V-shaped valleys. (8) Scarps. (9) Water divide lines. (10) Tectonic faults: a, (upper) normal fault; b, (center) fault with scarp; c, (down) probably fault. (11) Urban areas.

T4 lacustrine level: This is the vast superior level occupied by crops of Cucurbitaceae (chayotes) and coffee. The top of this level is made up of ochre sandy clay, gravel, and a strong density of rolled boulders of good size (1.3 ft diameter), altered in surface. The T4 level overlooks the lake and is an important level that joins the volcanic surface before the dam.

The T1 level is not visible on the right bank of the dam but must be submerged. The other levels, if they exist in this margin, are fossilized under colluvial fans formed by the current rivers (Escalante, 1965a,b).

The Reventazon River from Cachí Dam (from which it takes its name) sinks into a canyon that is 330ft or more deep, fitted in the basaltic–andesitic lavas of the powerful Paraiso flows (Lower Quaternary). These lava flows appear here at the base of Irazu Volcano. Over the first lava flow rests a second, very recent and very fluid, lava flow known as Cervantes, dating from 20,000 years BP (dated by ^{14}C with a piece of wood imprisoned in the lava; Madrigal, 1976). This andesitic lava flow draws on the walls of the northern Reventazon Canyon, forming lava curtains. The Cachí Dam has an average water flow of 256,775,235 gal, which allows the production of 100,000 kWh.

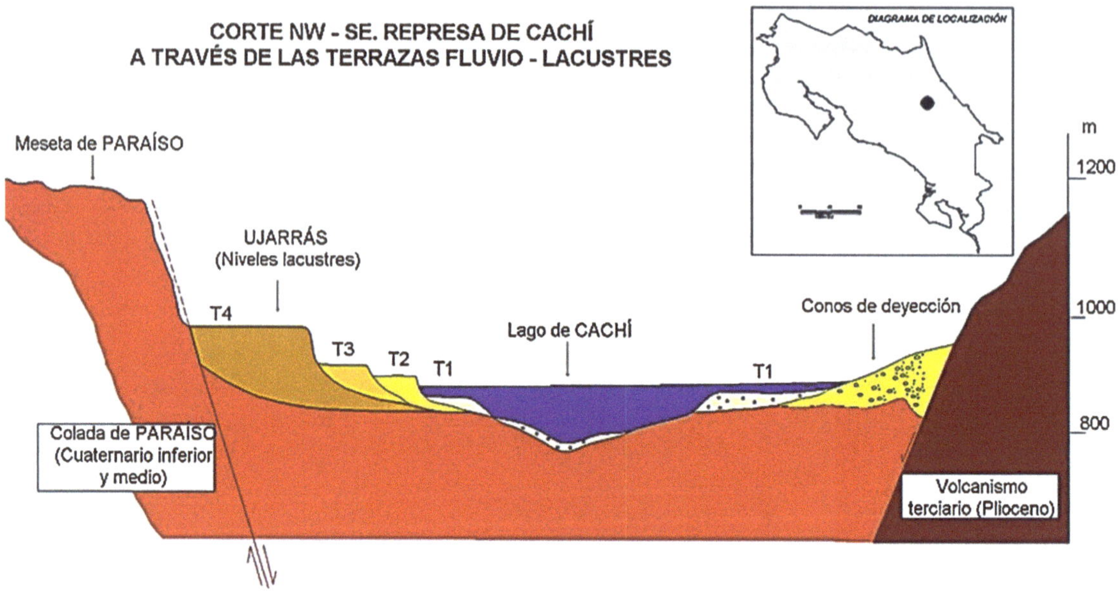

FIGURE 5.6 Fluvial–lacustrine terraces upstream of Cachí Dam (Bergoeing, 2009c).

The Cervantes lava flow (Kruschensky, 1972) is the last great effusive episode of Irazu. The chaotic lava flow morphology suggests its recent age. It extends over 17 square miles and it blocks the Reventazon River course in Cachí, forming a natural dam. It was dated at 23,000 years by the $^{238}Th/^{232}Th-^{238}U/^{232}Th$ method (Allegre and Condomine, 1976), and at over 20,000 years by the ^{14}C method in a fossil wood (Murata et al., 1966). The Cervantes lava flow is in fact constituted by several basaltic–andesitic washes that issued during the same effusive phase from several Strombolian cones (Tournon, 1984).

The Paraiso volcanic plateau, which dominates the artificial Lake Cachí, is formed by a series of basaltic–andesitic lava flows and covered at the top by an important clay alteration thickness of 66–155 ft, as well as by colluvia and volcanic ash (Dondolli, 1970). The plateau is dissected by Reventazon tributary rivers, which are born in Irazu Volcano's southern slope. The main collector of the Coris–Cartago–Paraiso sector is the Aguascalientes River, which cuts the south contact on the bottom of the valley, thus cutting the Miocene sedimentary formations alternating with the most important volcanic deposits of the Talamanca Mountain Range (Bergoeing, 1982a).

It is therefore from Cachí where the river first adopts a sinuous path through the encased meanders of the sector until the Quebrada Grande River sector. Here the river plunges into a 1000-ft-deep canyon. Between the Quebrada Grande River and the isolated Atirro Hill, a sector of many confluences, the Reventazon River follows a straight path, probably of tectonic origin. On the walls of the canyon, we were able to study three well-defined levels of fluvial terraces. These three levels are encased in the lower part of the topography of the volcanic Juan Viñas and Alto de las Mesas plateaus. Marshall, in 2000 (in Alvarado and Gans, 2012), analyzing a sample of lahar matrix dated by the $^{40}Ar/^{39}Ar$ method, obtained an age of 1.42 ± 0.13 MY for the Alto de las Mesas sector, which places it in the Lower Pleistocene (see Figs. 5.7 and 5.8).

FIGURE 5.7 Juan Viñas and Las Mesas volcanic structural plateaus. Probably Middle Pleistocene. *Photo Bergoeing et al. (2010c).*

The Reventazon River, after having crossed the embedded meandering sector, exploits a tectonic hinge between the Central Volcanic Mountain range and the northeast Talamanca Range's apophysis. It is here in the Birrís sector where we have seen and described these three levels, as follows, from the bottom up:

T1 level: This is a small terrace of 13 ft, encased in the Cervantes lava flow and consisting of fresh andesite boulders. The thalweg is here represented by various large rock blocks in the riverbed, where andesite prevails above all. They are well worn and are scattered even in the Atirro sector of many confluences.

T2 level: This is a small narrow terrace developed primarily between the towns of Paraiso and Turrialba. It is encased in the large cones of the lower course. This level is made up of red clay and healthy large boulders, rolled and covered by the Cervantes lava flow (20,000 BP; Madrigal, 1976) fossilizing the vertical walls of the canyon.

T3 level: This is a high deformed terrace formed by andesite boulders altered in parts. The alterations reach several feet deep. The boulders lie in an array of red clays. In the lower Reventazon River, at the mouth with the plain, in the Siquirres village sector, this level becomes a set of suspended coalescing alluvial fans. In addition, they are at the level of the new Siquirres–Guapiles road. Volcanic ashes cover the red clay soil. The Infiernillo terrace (656 ft), partly fossilized by the Juan Viñas lava flow (Tournon, 1984), occupies the T3 level.

The upper volcanic surface, which is the logical continuation of the Cartago tabular level, continues with a series of plateaus (Alto de las Mesas and Juan Viñas), which are, however, younger. However, if we consider that the main surface of the Cartago area belongs to the Middle Quaternary, we can then place the three terrace levels of the Reventazon River between the Middle and the Upper Quaternary.

The Cervantes lava flow, dating from 23,000 years BP, is a limit that serves as a

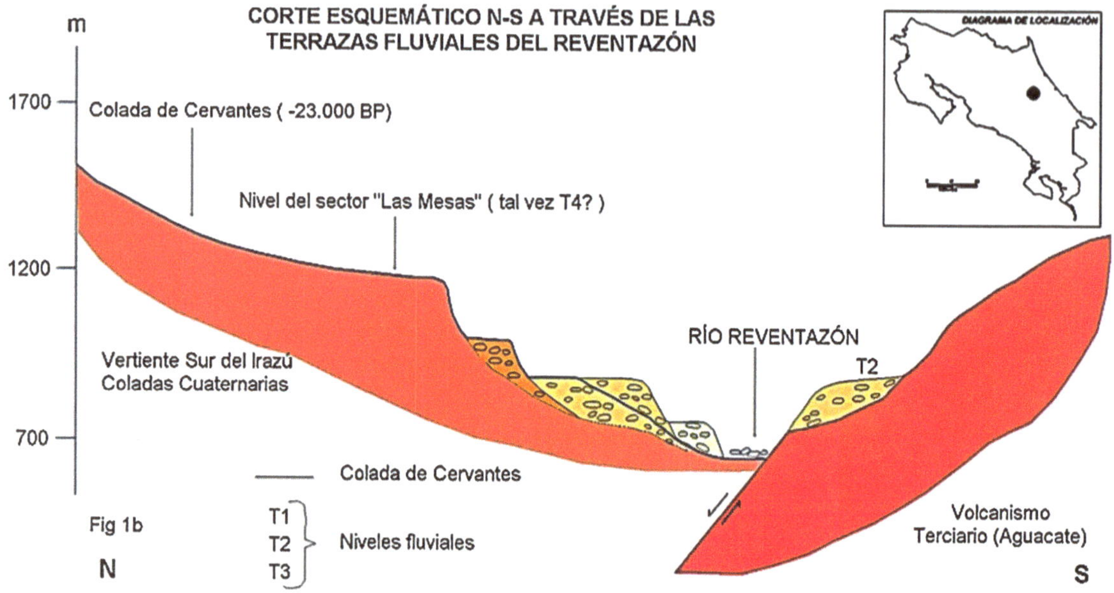

FIGURE 5.8 Reventazon River embedded fluvial terraces covered by the Cervantes lava flows (Bergoeing, 1998).

division between the T2 and T3 levels and the T1 level, which is not covered by the volcanic wash but is embedded in it.

In conclusion, we can say that the Reventazon T1 level is later than 23,000 years BP and can then be terminal Würm–Wisconsinian or Holocene. The T2 and T3 levels are located in a range between the Middle Quaternary and 23,000 years BP, which perfectly agrees with other observations made by us in Costa Rica in terms of the fluvial terraces, and above all in the Pacific slopes of Tilaran and Talamanca.

From the Quebrada Grande River, the Reventazon River continues its east–west tectonic path, deeply sunk into the river canyon. The volcanic Juan Viñas and Alto de las Mesas are exploited by humans (sugarcane in flat areas, coffee crops on steep slopes). On the right bank of the Reventazon, the T2 level is well represented in the Tucurrique terrace, where probably, at a given time, should produce a parking lake, with a possible drain into the Pejibaye River (Bergoeing, 1981).

The Atirro sector of many confluences (Bergoeing, 1981) is a vast tectonic depression. The large number of faults that intersect this sector, and in particular the important Turrialba tectonic fault alignment, confirms this. Here is where they come together, on one hand the Turrialba River, which descends from the flanks of Turrialba Volcano, and on the other hand the southern slope rivers represented by the Pejibaye, Atirro, and Tuis rivers. The Reventazon arrives in this vast, filled depression, which may correspond to a small graben. Indeed, the Pejibaye and Reventazon rivers are aligned following a NE–SW line on the same tectonic axis, whereas the Tuis and Turrialba rivers share a second fault axis whose orientation is southeast, parallel to the Atirro River tectonic alignment. Atirro, the isolated hill, is a small horst fitted on all sides by river erosion. This block is largely composed of volcanic rocks of the Pliocene (Aguacate and Doan formations). Further, a block of vast proportions, crossed by the Reventazon and Pejibaye rivers, emerges also in a horst position with the same volcanic series. The summit of this complex is located at Mount Doan 4921 ft above sea level. Here are drawn the remains of an old caldera, showing a Pliocene volcanic past (Kruschensky et al., 1976).

The Pejibaye River, as well as the Vueltas River, is a typical tropical mountain river with frequent slope ruptures. The lower course of the Pejibaye River, strongly driven by tectonics, ends in a narrow gorge (Canta Rana Sector). The river, by superimposition, was helped in its work by the presence of the faulting crush zone that determines its general layout.

Atirro Hill, as we mentioned earlier, is an isolated hill, in the midst of a multiple-confluence sector. It is the result of the double game of tectonics and river work. It is here where the Costa Rican Institute of Electricity built El Llano Dam (see Fig. 5.9). From this point, the Reventazon River begins its lower course before leaving the Central Valley. This river follows a SW–NE route, parallel to the Pacuare River (which is located to the southeast outside the Central Valley), which is an important fact that well illustrates the importance of plate tectonics in the current route of the rivers in Costa Rica, particularly in the mountain range sector.

The watershed dividing line, separating the Pacuare and the Reventazon rivers, is an extension of the Talamanca volcanic deposits of the end of the Tertiary, which are in high position resting over an Oligocene–Miocene sedimentary series. The summit retains volcanic traces, illustrated by necks, calderas, and andesite lava flows, witnesses to this activity (which may be associated by their shape and their material with the Matama row of Talamanca's Caribbean watershed). The best example is in Pacayitas, where the remains of a caldera forming a semiamphitheater exists, whereas necks and pyroclastic remnants are exposed in the Tres Equis area, following the Turrialba–Siquirres route.

The Reventazon River cuts the sedimentary series in this part of its middle course, before reaching the plain. Sandstone, shale, and graywacke, strongly folded, form riverbanks that correspond to the Uscari and Suretka formations. Shale belongs to the Middle Miocene, and conglomerates to the Pliocene, which particularly emerge in the largest riverbed of the flat and rocky bottom thalweg, where there still rest some severely worn volcanic blocks coming from the top, as well as sandy beaches. On the banks the described sedimentary strata are partly hidden by fluvial deposits, represented here by four different levels, which are situated on the left Reventazon riverbank (see Figs. 5.10 and 5.11).

FIGURE 5.9 El Llano Dam, Turrialba, in the sector of multiple confluences. *Photo Bergoeing (2011a).*

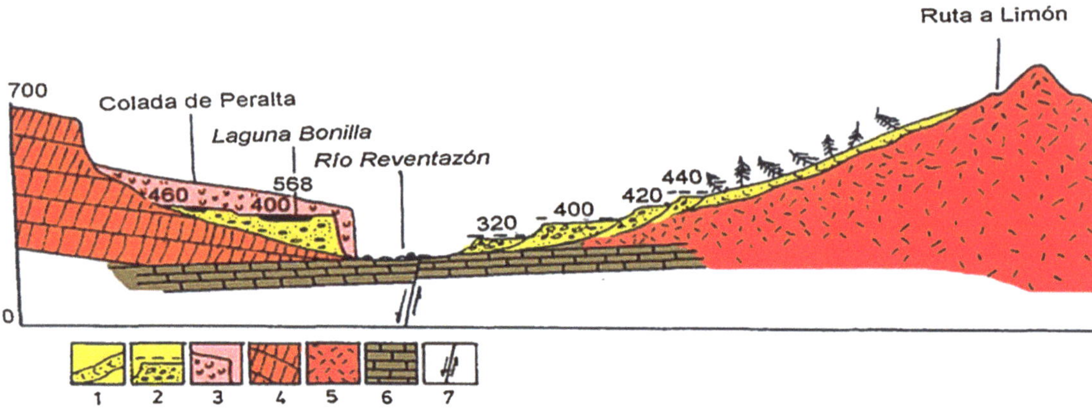

FIGURE 5.10 North–south profile of the Reventazon River (middle course) in Peralta's lava flow sector (Bergoeing, 2007). (1) Solifluction. (2) River levels. (3) Peralta's andesitic lava flow. (4) Turrialba's Quaternary volcanic remnants. (5) Aguacate's Pliocene volcanism. (6) Tertiary sedimentary rocks. (7) Tectonic faults.

Here is a description of the terraces from the lower to the upper level:

T1 level: This is a small terrace of 16 ft from the riverbed formed by well-worn boulders 0.65–1 ft in diameter, mixed with clays and sands. This terrace must be very recent because of its well-preserved forms.

T2 level: This is a well-formed large terrace at 200 ft from the riverbed, or 1312 ft of absolute altitude, with well-worn pebbles and rolled blocks, all strongly compressed but not forming a puddingstone. This level corresponds to the level of Lake Bonilla on the Reventazon River left bank.

T3 level: Located at 260 ft from the T2 level, this is a fluvial terrace formed by large boulders wrapped in dark brown clay. The boulders are marked by a strong deterioration on the surface.

T4 level: This is the top level, at 65 ft from T3, poorly defined, buried in part by colluvia, but identifiable by the presence of good-sized rolled edges (1.3–1.6 ft in diameter), and an altered ochre surface.

This Reventazon River middle–lower course sector is very interesting for several reasons. On one hand, the right bank contains other, poorly defined, river levels, which intersect in some sectors with the four described river levels. This exposes the problem of its origins. We think they are partly the results of ancient natural dams in the Reventazon River middle course due to natural volcanic eruptions damming, with lava flows, enough fluids to reach the thalweg. (This has already occurred with the Cervantes lava flow upstream.) Downstream, the Peralta lava flow seems to indicate the same effect for Lake Bonilla. Damming due to overload of water ruptures has been the origin of deposits of some terraces, especially in places of enclosure constituted by heterometric material (example of the T2 level). Therefore, we cannot say that all river levels correspond necessarily to climate crises, especially in what refers to the layout of the rivers close to the large Quaternary-active volcanoes, as is the case of the Reventazon River.

The Reventazon River's right bank slope presents, in the top sector of the river, abandoned and suspended meanders. They are old thalwegs, either of tributaries of the Reventazon or traces of the river itself from a distant past (Lower Quaternary?). On the right bank of the

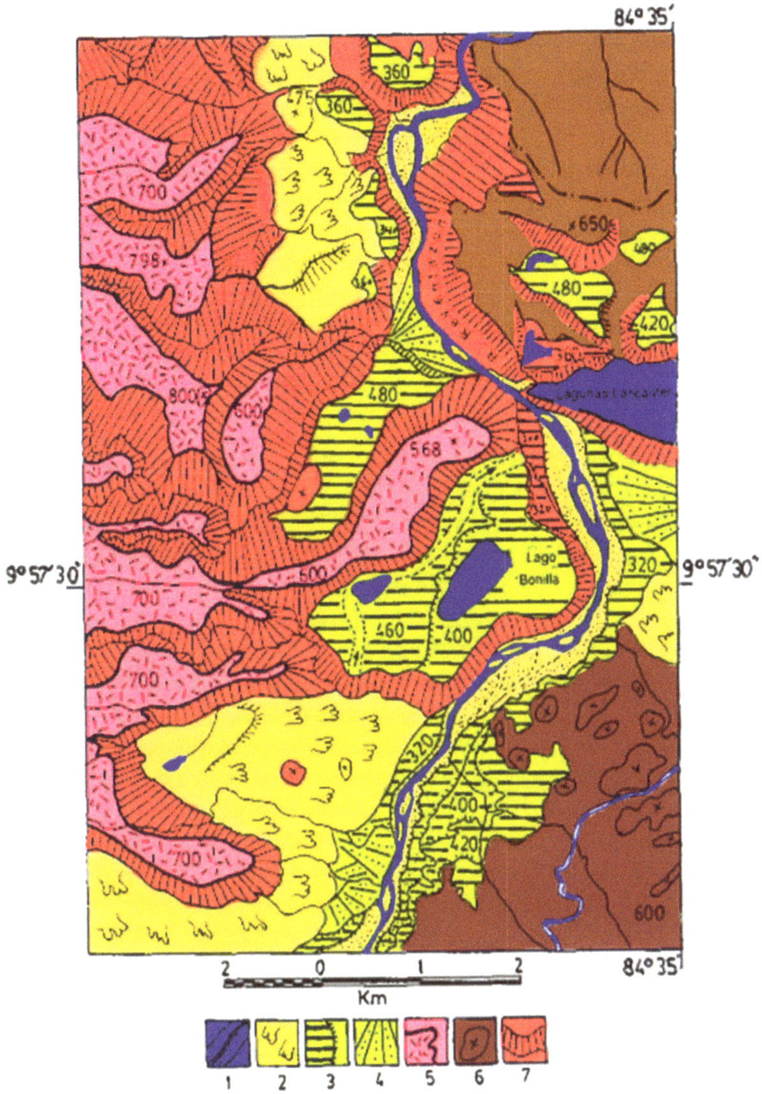

FIGURE 5.11 Peralta's sector, Reventazon River middle course morphology, according to Bergoeing (2007). (1) Reventazon River middle course. (2) Holocene, mass solifluction. (3) Upper Pleistocene lake level fluvial terrace. (4) Cones. (5) Pleistocene, volcanic plateaus. (6) Tertiary isolated reliefs. (7) Slope erosion.

Reventazon River, heading toward the sector of Peralta, the railroad line that connected Cartago City to Limon Port occupies the T2 terrace level (1312 ft). Peralta sector has suffered the same effects of natural lake damming as Cachí, by the presence of Peralta's lava flow that allowed the construction of the river lake levels upstream, where there are still small lakes, particularly Lakes Bonilla and Lancaster.

Since the Turrialba River, volcanic lava flows and other volcanic washes have left triangular basaltic plateaus, deeply dissected by the rivers, which create a different landscape between the north and the south slopes. The triangular basaltic plateaus are exploited as pastures for cattle, but they are not very numerous in this sector. The southern slopes of the eastern Central Valley rivers are characterize by the presence of colluvial cones at the bottom, whereas in the upper part, in the most deforested parts, exist solifluction phenomena and general soil crawling. Deforestation exactly demonstrates the role of a drier climatic crisis, but with a pluvial maximum (4000 mm per year for this sector). Where the forest has disappeared, the phenomenon of clayey soil wash from solifluction phases and mass crawling of land is felt immediately. The slopes of more than 35 degrees help this. After 2 or 3 years, the slopes are deep scars. The ditches are quickly amplifies into ravines. The material torn to shreds by the combined effects of gravity and percolation is transported by runoff to the rivers, where the waters become turbid. Finally, this material is transported in suspension to the sea.

The Tuis River is an example that should be take into consideration. During the rainy season, it often grows up to flood its bed. The hamlet of La Suiza de Turrialba has unfortunately been the victim of this fact on more than one occasion, as it is located on the lower terrace of the river leaving the mountain sector. The Tuis River behaves like a real flood during the rainy season. The slopes of the upper course are deforested and consequently the vegetation cannot control the flow. In the lower course of the Reventazon River reaching the flood plain, in Siquirres, the terrace levels are not different and merge with the volcanic coalescent piedmont alluvial fans.

3. WESTERN CENTRAL VALLEY

The western Central Valley is a basin drained by the waters of the Virilla River and its numerous tributaries that are born, in contrast to the eastern sector, in the north volcanic piedmont. The most important are the Ciruelas, Poas, and Grande rivers, all tributaries of the Virilla River. They are deeply carved in the Quaternary volcanic slopes, forming a subparallel canyon river network that runs in a NE–SW direction.

The northern slopes of the western sector of the Central Valley are uniform thanks to a coalescence of the volcanic cones of Barva, Poas, and Chayote. These are stratovolcanoes consisting of series of andesitic lava flows intercalated with layers of ash and pyroclastic flows. On the surface, thick ash deposits have evolved into layers of red clay, often hiding the lava flows and giving an undulating landscape, where a tropical mountain forest progresses, very conditioned by humans.

The piedmont downstream encounters a lava plateau through a series of alluvial fans, very carved by erosion and formed by lahar deposits and torrential lava. An east–west tectonic fault escarpment, notching the base of the volcanic basalt lava flows of the modern volcanic cones, also characterizes this contact sector.

The Virilla River has deeply notched the volcanic plateau, leaving to appear at the top of the canyon Middle Quaternary ignimbrite deposits (Tournon, 1983). They stretch from Heredia City to the village of Naranjo, along a line that is at the foot of the east–west tectonic fault scarp. They are also present in the towns of Palmares and Atenas, and in the confluence area of the Virilla and Grande rivers, their union gives rise to the Tarcoles River (see geomorphologic map of the Central Valley). West of San José, the Virilla River canyon sinks more than 350 ft into the volcanic plateau, and serves as a boundary between the Quaternary deposits of the plateau and the southern slopes (see Fig. 5.12).

3.1 Northern Central Valley Watershed

Here, volcanic cone summits are preserved by the high-altitude tropical forest, where, in this sector, there are epiphytes and parasitic plants that live off the frequent fogs that prevail in the sector, which are also protected by legislation (Boza, 1978).

In the Barva Volcano sector, the presence of a well-preserved lava flow, most likely of the Upper Pleistocene or even Holocene, more or less (thus a contemporary of the Cervantes lava

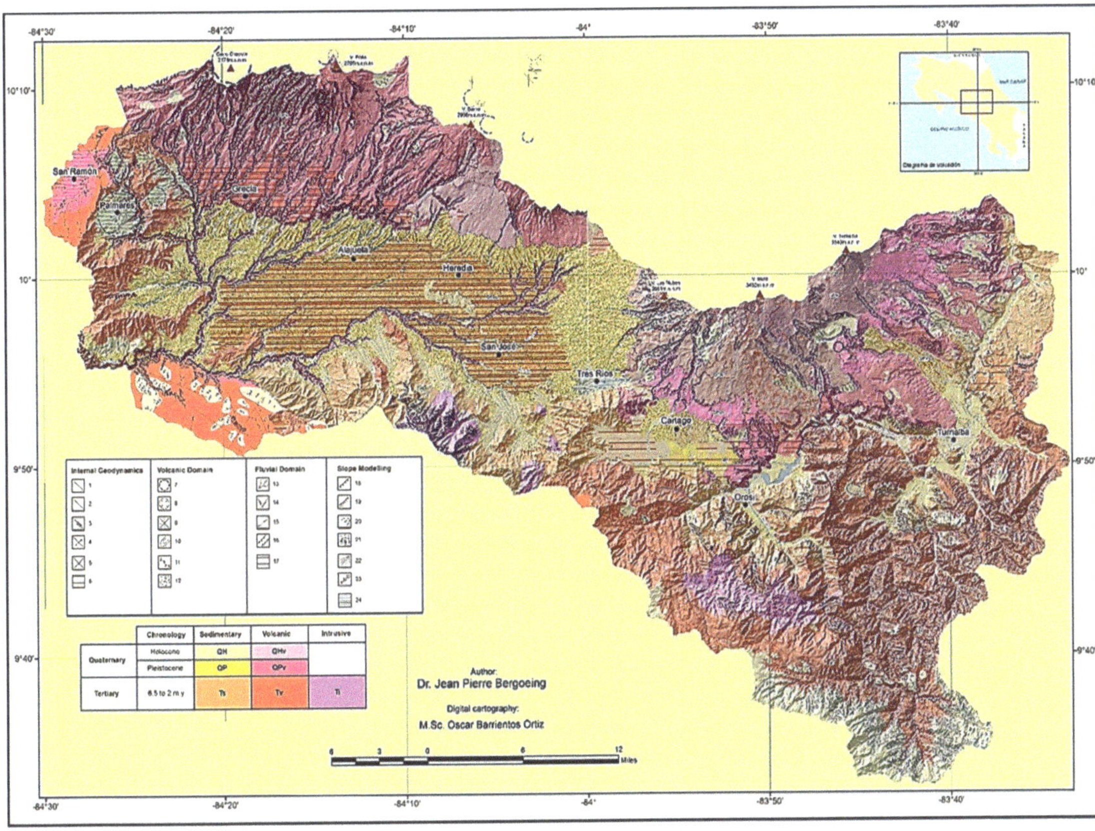

FIGURE 5.12 Central Valley geomorphology (Bergoeing, 2012a). Internal dynamics: (1) Normal tectonic fault. (2) Possible fault. (3) Sliding fault. (4) Anticline. (5) Syncline. (6) Structural plateaus. Volcanic domain: (7) Calderas. (8) Craters. (9) Adventitious cones. (10) Lava flows. (11) Lava flow fronts. (12) Lahars. Fluvial domain: (13) Fluvial network. (14) V valleys. (15) Water divide line. (16) Fluvial canyons. (17) Lacustrine areas. Slope modeling: (18) Rocky escarpments. (19) Erosion talus. (20) Reptation gullies. (21) Piedmont deposits. (22) Alluvial fans. (23) Landslides. (24) Topographic plateaus.

flow), should be noted. This is the Los Angeles lava flow, a result of spills from the Cerro La Cruz parasite cone peaks (Tournon, 1983).

In addition to this lava flow, recent outliers of lava flows in the north volcanic piedmont are rather scarce. To find them, you have to look in the river thalwegs, such as in the Ciruelas River Los Chorros waterfall. In this place, the river notches a powerful series of columnar basaltic outcrops with a 200-ft elevation and forms a beautiful waterfall. Rivers that descend from the volcanic piedmont often run in rapids and small waterfalls when the thalwegs are lava flows, as in this case. This is because the river erosion acts less rapidly than the chemical erosion of the rocks in a humid tropical environment.

Humans have deforested the volcanic piedmont in a systematic manner since the previous century. This has contributed to increasing the effects of soil creeping, mass landslides, and detachment scars. These phenomena are favored by the presence of cattle, concentrated preferably on the northern slopes because of the quality of the pastures. Continuous tamping of the clay soils when wet often contributes to the erosive effects, with the appearance of small erosion terraces, the early stages of wider phenomena reaching to considerable proportions. Once again, is humans who, in search of new lands, pastures, or crops, accelerate the erosion conditions.

The Zurqui volcanic cone, dated at 0.50 MY (Tournon in Saenz, 1981), was broken by a road tunnel to allow the union of the new San José–Guapiles highway. This vast project began in 1977 and ended in 1986, at a cost of $64 million. There was the need to dig a tunnel and build five bridges in the northern volcanic piedmont of the Central Range. Even today, landslides on this road are frequent.

As already noted, the northern slope of the western Central Valley sector encounters the volcanic plateau, in a series of much dissected erosion alluvial fans. They are composed largely of lahar material, lava flows, volcanic ash, mud, and rolled blocks. Alluvial fans often adopt a multiconvex shape, especially in the Naranjo sector. These alluvial fans are eroded by deep thalwegs in the lower sector.

Upstream, the presence of ancient andesitic lava fronts also highlights an important east–west fault (Madrigal and Malavassi, 1967). This escarpment is the only element that cuts the landscape and which allows a free contact to be established between the plateau and the piedmont. North of Alajuela City, this east–west fault alignment is well in evidence. For former geologist Enrique Malavassi, this volcanic lava front would be the raised side of an east–west normal fault. This fault alignment is also evident in the south of Palmares City, where it would be extended in the Grande River canyon (see geomorphologic map of the Central Valley, Fig. 5.12). It is probably in this sector, from a fissure zone (parallel to this fault?), where current ignimbrite deposits of the Lower Quaternary were laid, covering the north and west sectors of the volcanic plateau (Kussmaul et al., 1982).

3.2 The Volcanic Plateau

Between Heredia and Alajuela, the plateau contact is characterized by the presence of ignimbrite that emerges on the top of the Virilla River canyon walls. These outcrops are visible on the Heredia bridge to San José. Such ignimbrite deposits are also visible in the Rosales River (Dengo and Chaverri, 1951), following the national road between Alajuela and Naranjo cities. Here the ignimbrite flourishes on the same slope of the road, porous shaped but compact. Its color, unaltered, is a milky white, but on the surface the alteration gives a gray appearance. These deposits are also present in the Palmares and Atenas urban centers and along the Grande riverbanks before joining the Virilla River. This ignimbrite corresponds to an intermediate stage of the Quaternary volcanism of the Central Volcanic Mountain Range. After building a powerful basaltic base, the basaltic–andesitic magmatic activity subsequently became more acidic, with tuff emissions and pyroclasts (nuées ardentes) that were deposited from a fissure at the foot of the Barva and Poas volcanoes on the volcanic plateau, spreading westward.

On the banks of the Poas and Grande rivers, in their middle courses, in contact with the bottom of the valley, road and construction companies currently exploit powerful pyroclastic outcrops. The quarry leaves exposed a vertical slope of more than 350 ft on the road to Atenas in the Grande River sector. The ignimbrite is also exposed in the eastern sector of the valley, under the volcanic plateau of Cartago, but does not emerge because it is covered by a significant thickness of lahars (Tournon, 1983) (see Fig. 5.13).

The volcanic plateau is a vast extension that rises gradually from west to east, going from 1650 to 3300 ft above sea level (between the confluence of the Grande and Virilla rivers to the San José area, which is a distance of 25 miles). The upper part of the plateau is formed by ignimbrite, some andesite lava flows (Naranjo sector), lahar deposits, and torrential lava. Only the

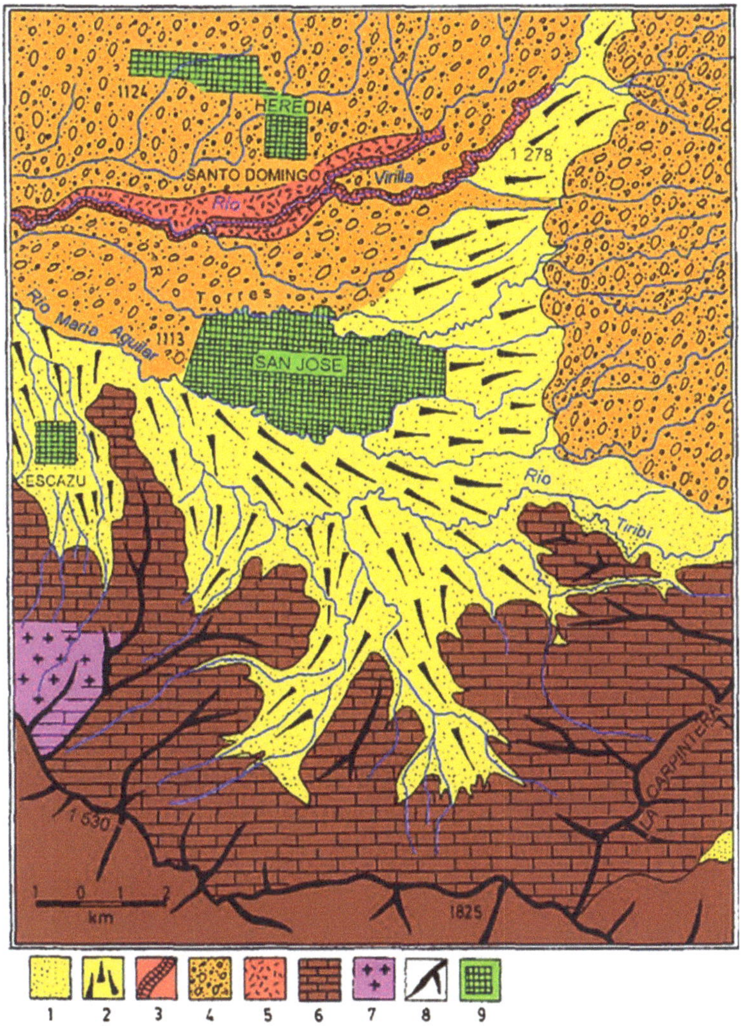

FIGURE 5.13 Southern San José–Escazu geomorphology (Bergoeing and Brenes, 2013). (1) Holocene modern alluvial fans. (2) Upper Pleistocene alluvial fans. (3) Virilla River canyon. (4) Upper Pleistocene lahars and torrential lavas. (5) Middle Pleistocene ignimbrite of the Virilla River canyon. (6) Tertiary Coris and San Miguel sedimentary series (sandstone and limestone). (7) Escazu's intrusive (fractured from the Miocene). (8) Watershed. (9) Urban centers.

most rugged Turrucares sector disturbs the plateau with Tertiary volcano-sedimentary hills (see geomorphologic map of the Central Valley, Fig. 5.12).

The middle part of the plateau is formed by basaltic–andesitic lava flows alternating with deposits of ignimbrite (William, 1952; Echandi, 1981). Geologically this area is known as an "intracanyon lava flow." The base of the plateau is composed of an Oligocene–Miocene volcano-sedimentary series. To the south, the Virilla River canyon cuts the plateau, separating it from southern contact with the Central Valley. The Virilla River follows an east–west path, almost entirely bordering the plateau.

This sector of the volcanic plateau contains the best lands, where coffee crops have been grown since the colonial period, but the proximity of San José City has made sure that these lands have given way to speculation for the sale of urban properties.

3.3 The Southern Contact

The southern watershed contrasts with the northern side, both by morphology and by the diversity of its lithology. It contains low mountains (6000 ft in altitude, which is considerably reduced as seen from the Central Valley). It is very affected by erosion and almost completely deforested, which makes it lose its colors during the dry season. The higher summit is Pico Blanco (6600 ft), formed in the granodioritic massif of Escazu (Rivier, 1983). This southern part of the western valley can be divided into three sectors as a function of its lithology and modeling:

- La Carpintera–Escazu eastern sector
- Escazu–Santa Ana sector
- Jaris River western sector

3.3.1 *La Carpintera–Escazu Eastern Sector*

This sector is a Tertiary sedimentary hill system extension of La Carpintera Mountain Range. Here we find siliceous and calcareous sandstone of the Coris and San Miguel formations, faulty and folded, forming anticlines and synclines (Castillo, 1978). Limestone with pectens of the San Miguel Formation is also affected by cryptokarst, with the formation of calcite in the diaclases or joints (Franco, 1978) (see Fig. 5.13). This folded system is located south of San José in the canyons of tributaries of the Virilla River, established through the expansion of powerful coalescent alluvial fans transforming into glacis (Rojas, 1978). These cones are similar to the Santa Ana cone, which we will describe in detail, and are probably the results of repeated climatic crises of which several generations can be recognized. In the southern suburbs of San José, they become glacis before being carved primarily by the canyons of the Maria Aguilar River.

During the rainy season, the alluvial fans are very active, flooding the vast precarious urbanized sectors with strong population densities of unlucky people. These less privileged classes of Costa Rican society suffer each season from floods and misfortunes that happen especially during the months of September and October, when the stronger rains arrive.

3.3.2 *Escazu–Santa Ana Sector*

This is the central sector of the southern contact. The Escazu massif, consisting largely of granodiorite intrusive rocks, represents the highest part. This intrusive is the youngest in Costa Rica, its granodiorite dated by the K/Ar method at 5.3 MY (Bergoeing, 1982b). The slopes of this sector are mainly composed of fractured grandiorites, monzogabbros, and volcanic and sedimentary rocks of the Middle Tertiary. Intrusive, much-altered rocks on the surface favor a strong erosive action, that is, a series of mass landslides aided by tectonic and seismic activity. The best example is situated in Alto Tapezco on Escazu's western side (see Fig. 5.14). Contact between the fractured granodiorites and sandstones of the Pacacua Formation has given birth to cornubianite rocks, by contact metamorphism, which have been destroyed by the effects of a strong chemical activity and gravity. The summit consists of cornices of 60 to 130 ft. At the base of the same, alluvial fans and colluvia are staggered. Rivers that descend these steep slopes expand in the sector of Santa Ana, creating a wide alluvial fan, which progresses downward, to be cut by the Virilla River canyon.

The Santa Ana Valley alluvial fan is interesting because it teaches us a little more about the past of this sector. During exceptional rains, currently, Santa Ana City suffers from floods. Rivers become very large and can then transport stone blocks and even thick tree logs (see Figs. 5.15–5.17). Two series of alluvial fans exposed on the new highway linking San José to Santa Ana and Ciudad Colon represent the upper part of the same, on a 23-ft slope.

- The top of the alluvial fan is formed by badly rolled boulders with a thickness of 0.5 ft. The material is compacted in a dark red clay mass.

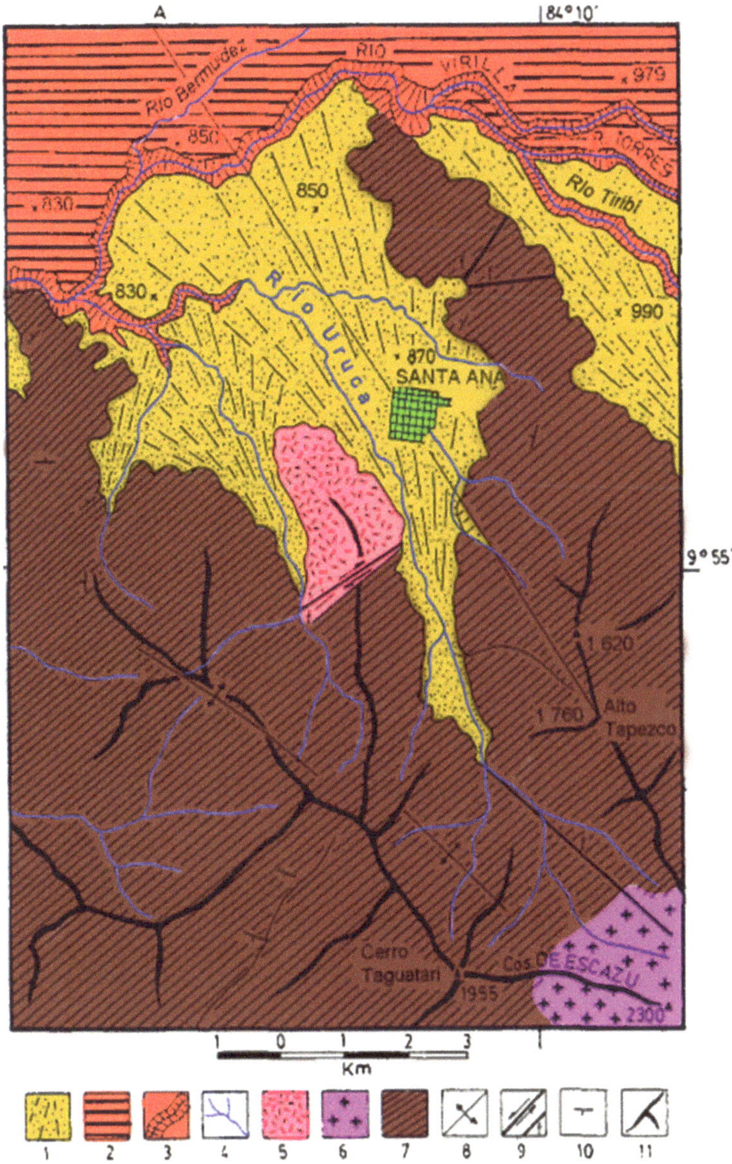

FIGURE 5.14 Geomorphological scheme of Santa Ana's alluvial fans (Bergoeing, 2007). (1) Upper Pleistocene–Holocene alluvial fans of Santa Ana and Escazu. (2) Lower Pleistocene central volcanic plateau. (3) Virilla River canyon and tributaries. (4) Hydrographic network. (5) Tertiary volcanism (Miocene basalts). (6) Escazu intrusive (Miocene monzogabbros and granodiorites). (7) Oligocene–Miocene volcano-sedimentary rocks of the Terraba and Pacacua formations. (8) Anticlines. (9) Tectonic faults. (10) Dips. (11) Dividing watershed line.

It rests in a small heterometric riverbed, dominated by thin elements.

- The lower part is strongly compacted in dark red clays containing blocks of more than 3 ft in diameter with worn, rolled edges, as well as poorly rolled material. Everything is exposed in a cut of 16 ft, which also rests on a thin riverbed. Given the weak alteration of the lithic material, and its compaction forming puddingstone, it is possible to associate this material with the Upper Pleistocene (post-Riss–Illinoian and post-Würm–Wisconsinian?). In Escazu, the new road reveals a cut where material coming from the alluvial fan of Escazu flourishes, and which in some places is mixed with boulders

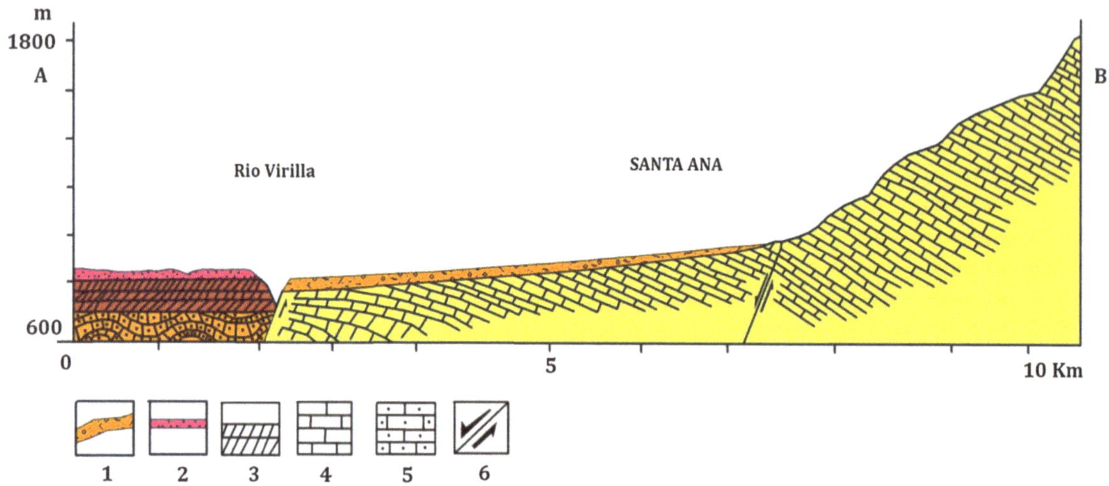

FIGURE 5.15 San Rafael Ojo de Agua—Alto Tapezco geological profile (Bergoeing, 2007). (1) Upper Quaternary; Santa Ana's alluvial fan. (2) Central volcanic plateau, ignimbrite coverage. (3) Lower and Middle Pleistocene; central volcanic plateau basal lavas ("intracanyon lavas"). (4) Shale, sandstone, tuff, and conglomerates of the Pacacua Formation (Miocene). (5) Shale, sandstone, and volcanic intercalations of the Terraba Formation (Oligocene—Miocene). (6) Tectonic faults.

and fluvial material probably from the Maria Aguilar River.

- The Escazu and Santa Ana valleys are separated by a small volcanic hill system (much altered Miocene basalts, but still compact enough to be exploited by a quarry). Santa Ana's alluvial fan presents three encased levels, separated by small lake passages, which testify to a change in deposits and therefore to climatic phases in which the river transport was less important between them.
- The youngest, top, level presents a heterometric material, less important than the lower similar-sized level, which shows less action between the two rexistasic phases that allowed it to settle. We can compare the top level of the Santa Ana River alluvial fan with the T1 level of the Reventazon River, which is younger (23,000 years BP). The lower Santa Ana alluvial fan level should necessarily approach a very dry period, prior to 23,000 years. This makes us go back to a colder period, which might be the Würm—Wisconsinian, that would have affected the high mountains (which is to be demonstrated) and would have resulted in the plain from the dryness. The third alluvial fan level is presumed, because it does not appear in the geological profile, but is limited by the intermediate level.
- The Santa Ana alluvial fan, as well as those of Escazu and southern San José, is the result of a climatic crisis. The surrounding relief does not reach a very important relative altitude. However, the effects of current anthropogenic deforestation allow us to understand the mechanism of a drier phase, because currently these alluvial fans are very active during the rainy season and carry an important amount of material.

3.3.3 *Jaris River West Sector*

The last southern contact sector of the western Central Valley is characterized first by the Jaris River, a tributary of the Virilla River on its left bank of little importance but to notch a sliding tectonic fault that separates sedimentary units (Terraba Formation) from the volcanic units (Aguacate Formation) (see Fig. 5.12, geomorphology map of the Central Valley)

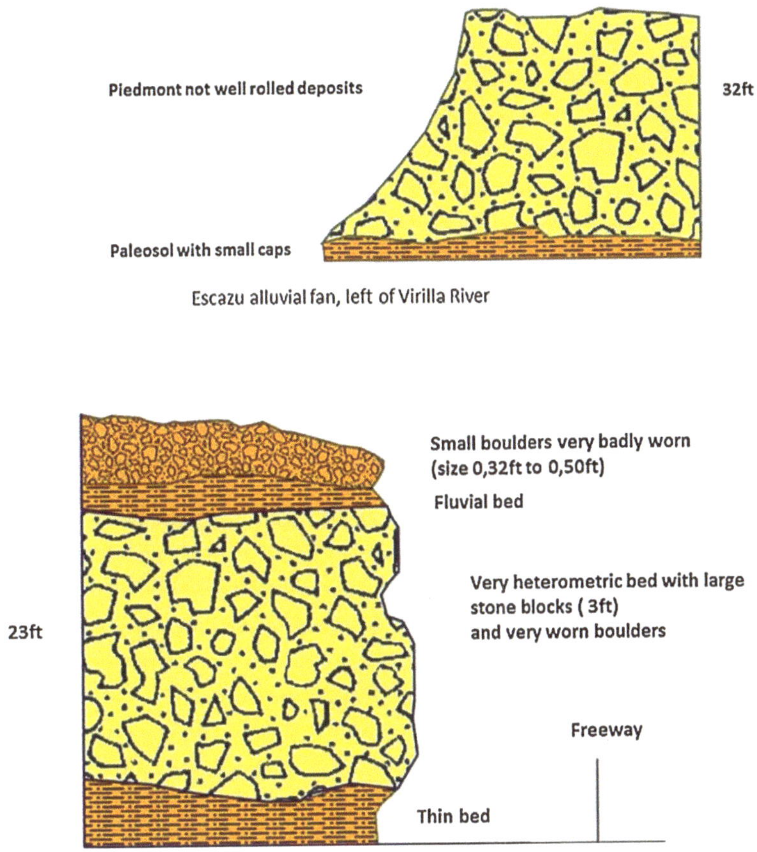

FIGURE 5.16 Alluvial fan cut of Escazu south of the Virilla River (Bergoeing, 2012a).

The southern slope is characterized in this sector by an advanced evolution of clayey soils, with a real red to red-brown color. These soils are 90% devoid of forest cover and this since the beginning of the 20th century. As a result, all the characteristics of advanced erosion are presented. The dry season has effects on the soil moisture buffer (December to April), but the practice of burning vegetation in the steep slopes does not help to slow the progress of erosion. During the rainy season, and with a prolonged dry season, the first downpours in the cracked soil produce erosion disasters.

3.4 West Contact

The western Central Valley is separated to the west from the volcanic plateau by the imposing presence of the Río Grande. This sector is composed of Pliocene volcanic material from the Aguacate Formation. The northwestern part is represented by a circular depression of 2.5–4 miles diameter, the Palmares collapsed caldera (see Fig. 5.18). This volcanic depression is the product of the magma chamber collapse of an ancient Pliocene volcano (Dondoli, 1951; Bergoeing and Malavassi, 1981a).

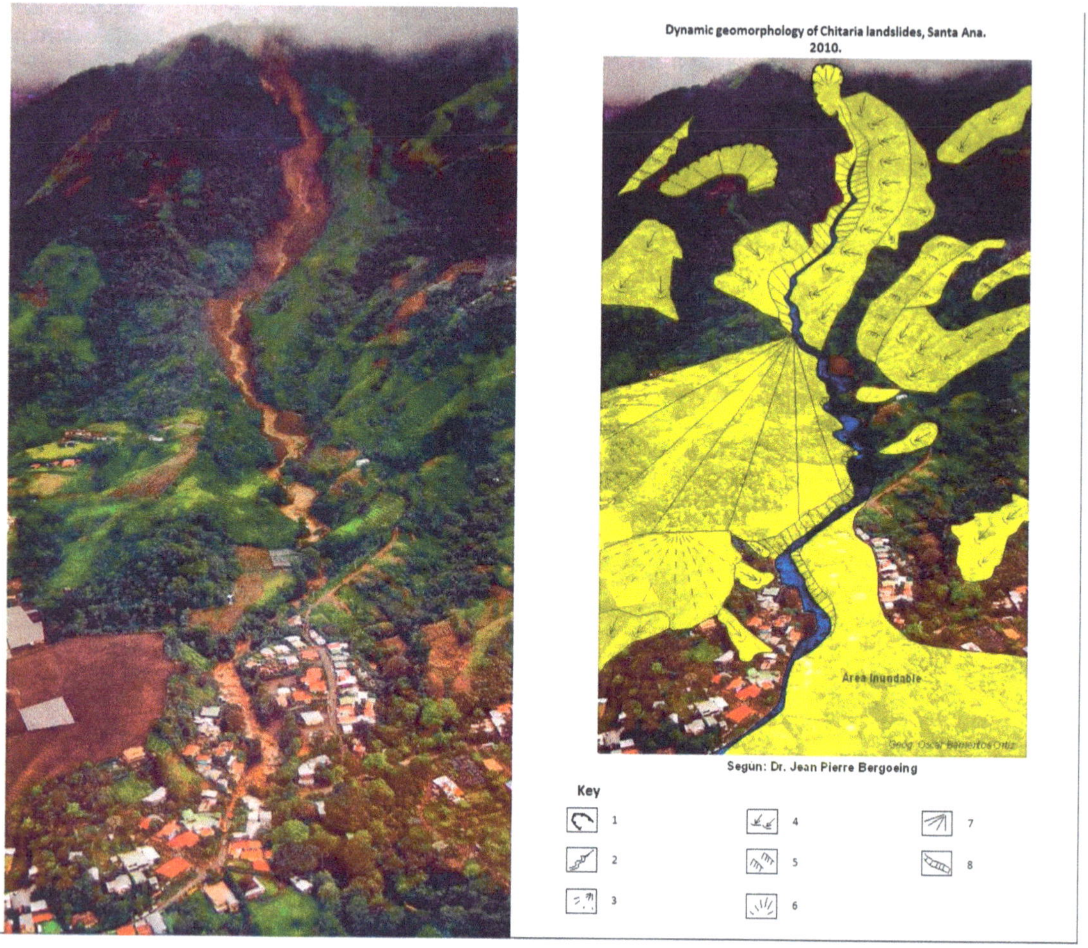

FIGURE 5.17 Chitaria (Santa Ana) landslides: dynamic geomorphology. In 2010 it took victims and left many people without their homes due to a lack of foresight by authorities. (1) Landslide amphitheater. (2) Mud from torrential flood. (3) Glacis. (4) Mass landslides. (5) Creeping soil. (6) Rupture area. (7) Alluvial fans. (8) Erosion talus. *Photo courtesy of La Nacion Newspaper (2010).*

Palmares City is located in a circular depression of 4 km in diameter, surrounded by all sides of volcanic reliefs. Dondoli showed that it's a collapsed caldera partly silted by lacustrine sediments providing Mastodon remains. The conservation of this type of morphology suggests a recent age (Upper Pliocene–Low Pleistocene). ***Tournon (1984).***

After the volcano collapsed, some magmatic activity continued, creating peripherally since then a postcaldera cone that emerged to the north. It is the Espiritu Santo volcanic cone. During the Lower Quaternary rain and river waters on the northern slope formed a lake that later came to be San Ramon City, gradually filling the caldera or volcanic depression. Later, the Grande River took the opportunity to dig the current river canyon, following an important east–west tectonic fault alignment that extends to Alajuela, benefiting no doubt from very active neotectonic effects.

FIGURE 5.18 Pliocene Palmares collapsed caldera. It later housed a lake that was emptied by the Grande River tectonic fault. *Photo Bergoeing (2013).*

The lake lasted at least until the Middle Pleistocene, because important deposits of ignimbrite, probably from fissures, spill at the foot of the Poas and Barva volcanoes, deposited following aqueous forms. Later, the lake was filled with lacustrine sediments, forming the summit of this stratigraphic column. They are intercalated with silts, clays, and more or less compact diatomite layers.

It is interesting to note, in the quarry north of San Ramon City, where these lake series are exposed on a slope of 150 ft, and contained inside the series, small pellets the size of marbles, formed of clay. These brown balls rest in a whitish lacustrine matrix. They are formed, apparently, from the aerial arrival of small fragments of pumice and tuff during the Lower and Middle Quaternary volcanic crisis. These small fragments, once deposited in the water by whirlwinds, were wrapped with clay from the bottom of the lake. On the walls of the quarry, these clay beads with nuclei of pumice are well exposed, and their density covers 150 square feet. The upper part of this profile is crowned by an ignimbrite deposit (Bergoeing, 1981b).

This proves two things: the first is that the Middle Quaternary volcanic crisis affected the San Ramon–Palmares region. The second is the presence of a lake at the time of this crisis, from lake deposits and ignimbrite that crowned them. South of Palmares also extends the depression of Atenas (Fernandez and Rodriguez, 1952), stuck to the caldera depression of Palmares. It is probably a subsidence area due to neotectonic effects and mostly to east–west tectonic faults.

The floor of this basin is built of deep layers of red decomposition clays, from Pliocene volcanic rocks of the Aguacate Formation (Dondoli and Torres, 1954). The collapsed Palmares caldera's northern contact is marked by the presence of small banks, 3 ft in thickness, of white ignimbrite. Downstream, small colluvial fans serve as a union with the bottom of the ravine. The southern Atenas basin edge consists of a small string of volcanic Pliocene–Pleistocene domes (Alvarado et al., 1980), which dominate 300 ft above the bottom of the basin. It is shaped by rounded and stricken laminar erosion and by the crawling soils, a little like everywhere on this side of the Aguacate Mountain Range. To the south, this small rosary of domes dominates by 1300 ft the Tarcoles River canyon. Among them is the Macho Chingo volcano, for which Ar/Ar radiometric dating in an ignimbrite gave an age of 6.05 ± 0.04 MY (Alvarado et al., 2006). All this dating confirms that the volcanic activity of this sector goes back from the Upper Miocene to the beginning of the Pliocene.

3.5 Virilla and Tarcoles Rivers

The Virilla River notches the Western volcanic plateau at its southern limit. It is an antecedent river that dug its narrow and deep canyon after the Lower Quaternary when the latest lava flows spilled on the plateau. The Virilla River is born inside the destroyed Las Nubes crater, on the southwest Irazu conical flank. Down this river is the southern boundary of the central volcanic plateau, notching in about 300 ft. The current thalweg cuts volcano-sedimentary rocks of the Terraba Formation resting in the interbedded basaltic–andesitic lava flows of the plateau.

Other than the alluvial fan terraces of the southern canyon river rim (Escazu–Santa Ana–Ciudad Colon sector), forming a high level, the riverbank in its upper course has not left visible traces of river lots. This high level is equally present in the sector of confluence with the Grande River. Here, this level is covered by ignimbrite, probably hiding a fluvial material (see Fig. 5.19).

The Grande and Virilla rivers adopt the name of the Tarcoles River after their confluence (Bergoeing, 1983). It is in this middle course where the high terrace level is better developed. By both its composition (volcanic lava flows) and its altitude (1300 ft), it joins perfectly with the central plateau level. This level holds the remains of a park lake in the San Pablo de Turrubares sector, probably from the Lower Quaternary, because that is the age of the ignimbrite that covers part of the fluvial level (1.39 MY; Battistini and Bergoeing, 1980). This T4 upper level is on both banks of the Tarcoles River downstream. Also in this sector are the lower levels T3 (1200–1180 ft), T2 (920 ft), and T1 (722–650 ft). This final level is poorly represented. It is necessary to reach the lower course of the Tarcoles River to find well-exposed T1 and T2 levels They are well defined in Lake Lagarto, downstream from the Agres bridge, stuck to the right riverbank (Orotina piedmont) (see Fig. 5.20).

T1 level: This is a very vast level, flooded several times by overflow from the increased bed during exceptional Tarcoles River floods. Its relative height is 10–16 ft. The river below mingles with further bedding. Level T1 is formed by large healthy boulders, gravel, sand, and silt.

T2 level: This level is attached to the Orotina plateau escarpment from 23 to 33 ft of height and supports the former small Lake Lagarto. This level is covered with an escarpment colluvium. Some profiles allow one to sense a material consisting of heterogeneous boulders, well worn, of medium size (0.65–1 ft diameter wide), and little altered. The interstitial clay is dark red and very hydrated.

T2 and T1 levels are common on the banks of the lower course of the Tarcoles River. However, they are often covered by tropical vegetation that hides them, sometimes completely.

4. CENTRAL VALLEY MORPHOGENESIS

- Costa Rica's Central Valley is a vast tectonic and volcanic depression, oriented east–west, on the southern edge the Nicaragua graben, and coated with Quaternary volcanic deposits.
- During the Lower Quaternary (1.8–0.3 MY) the Central Mountain Range volcanoes first issued powerful basaltic lava flows and then basaltic–andesitic lava flows. These formed the basis of the Central Volcanic Mountain Range as well as that of the central plateau. This base has been uplifted and tilted by orogenic and tectonic effects (Kussmaul et al., 1982).
- The Middle Quaternary environment (0.3–0.15 MY) is marked in part by the issuance of spills that overlapped the basal lavas (Tournon, 1980). They emerge in the western sector from Heredia to San Ramon and underlie the eastern sector in Cartago.
- The Upper Quaternary (0.15 MY to present) is represented by the construction of the current stratum volcanic cones. They are slightly tilted to the southwest, which explains the parallel hydrographic network of the southwest Poas and Barva volcano façade. This period is also characterized by the deep

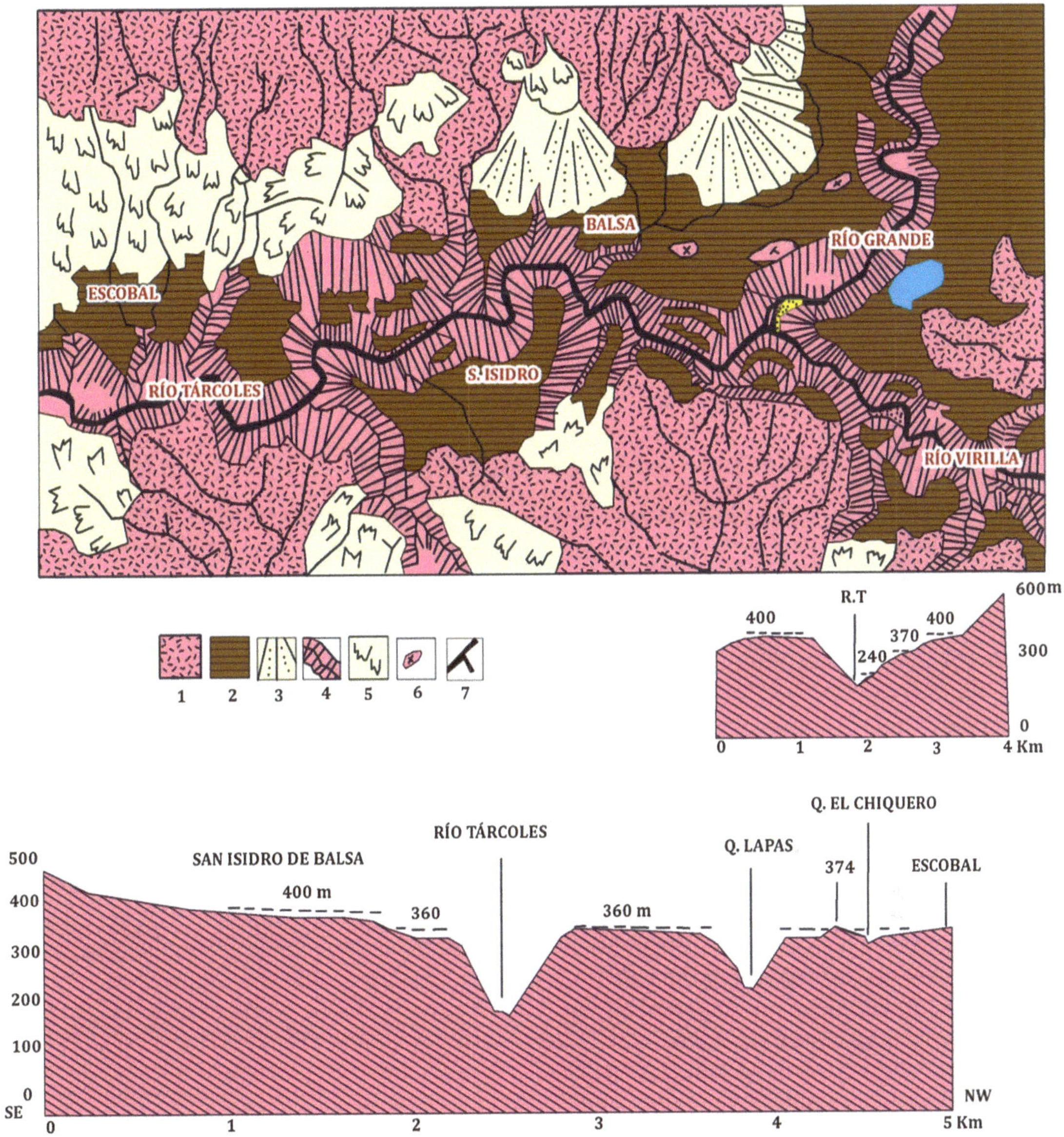

FIGURE 5.19 Confluence area of Virilla and Grande rivers: geomorphology. Their union forms the Terraba River. Suspended fluvial levels are on both sides of the Tarcoles River canyon (Bergoeing, 2012a). (1) Pliocene volcanism area (Aguacate Formation). (2) River levels covered by Middle Quaternary ignimbrite. (3) Alluvial fans. (4) Erosion slopes. (5) Mass and laminar solifluction. (6) Isolated reliefs. (7) Watershed dividing line.

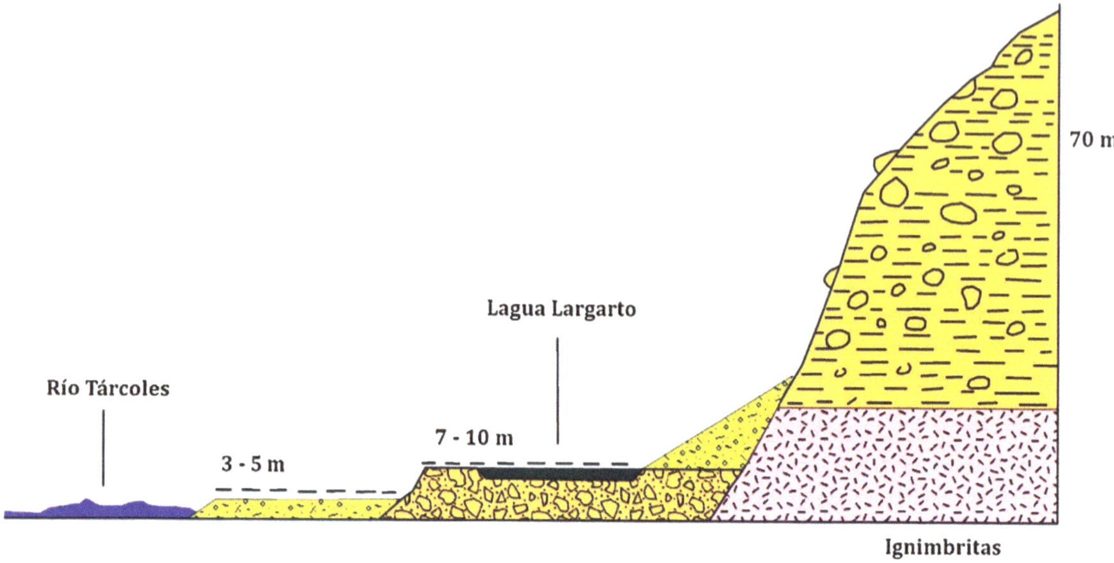

FIGURE 5.20 Fluvial terraces of the lower course of the Tarcoles River on its northern bank (Bergoeing, 2007).

river carving of the central plateau, which began during the Lower Quaternary and gave the narrow and deep canyons of the Virilla and Reventazon rivers.

- The Upper Tertiary volcanic deposits are displaced to the south and west of the Central Valley. They are close to the Miocene sedimentary deposits. These older areas show a multifaceted transition to a multiconvex relief pattern.
- The sedimentary formations of the Central Valley's southern slope are preferentially in the western sector and are situated chronologically between the Oligocene and the Pliocene. Almost all are of marine or coastal origin with continental contributions in the Pliocene. These formations are more resistant to chemicals and hydrothermal volcanic rock effects, which, however, are younger, but crumble faster.
- Some intrusives of the Upper Miocene are evident. They are part of the plutonic rock complex of the Talamanca Mountain Range. One is located in Escazu (Rivier, 1983). The other belongs to the southwestern Talamanca flank (eastern section of the Central Valley) (Berrange, 1979). These rocks are very friable and provide a fractured granodioritic material that accumulates in small colluvial cones of fans before being washed by pluvial erosion and transported by the rivers.
- The northern side of the Central Valley, to leeward, is largely covered by a thick layer of volcanic ash that alternates with other friable volcanic deposits. This allowed the hydrographic network to create deep carvings.
- Upper Quaternary and Holocene climatic crises are evident in the Central Valley. They are represented by the T2 and T1 levels of the Reventazon and Tarcoles river terraces as well as the cliffs of several alluvial fans in Santa Ana and Escazu.
- The Reventazon River levels T3 and T2 are located in a range that goes from the Middle to the Upper Quaternary (level of the plateau). They are covered by Cervantes lava flows (Madrigal, 1976) dating back to 23,000 years BP that fit the T1 level.

- The T4 level in the Virilla and Tarcoles rivers is represented by the San Pablo de Turrubares fluvial terrace surface, dated at 1.38 MY, a level that matches the central plateau surface. The lower terraces are thus placed in the same range of time as the Reventazon River valley. They in principle must be contemporaries, if the climatic effects that created them suffered no substantial local changes due to volcanic eruptions.
- The activity of the Central Volcanic Mountain Range is always present, as evidenced by the Los Angeles lava flow (Barva volcano) and Cervantes lava flow (Irazu) and by the activity of the recent Irazu vulcanian-type eruption (1963) or the Poas volcano activity (1952). The latest lava flows are almost contemporary with one another (Kussmaul et al., 1982).
- Currently, the erosion can be see in all Central Valley slopes. At the northern end, piedmont agricultural effects activate erosion processes, whereas those in the southern piedmont (west sector) are rather spontaneous and disorderly effects on urbanization.

CHAPTER

6

Tilaran Mountain Range

OUTLINE

Tilaran is a mountainous unit, less high and covering a much smaller area than the Talamanca Mountain Range or the Central Volcanic Mountain Range. It is located northwest of the Central Valley. Its natural boundaries are Lake Arenal to the north; the San Lorenzo River to the northeast, which serves as a border with the Central Volcanic Mountain Range; and the Aguacate Mountain Range to the south, in contact with the Talamanca Mountain Range (see Fig. 6.1).

Tilaran extends south until the Tarcoles River, if we consider the Aguacate Mountain Range as forming an integral part of this unit. Indeed, by its geological history and the nature of the rocks and their ages, we can also include the mountain range that lies to the south of the Tarcoles River and that part of Santiago de Puriscal. This part ends with the volcanic peak of Turrubares Volcano (5761 ft) overlooking the Pacific coast. The Tilaran Mountain Range is the almost exclusive product of a Pliocene volcanic activity whose deposits have been regrouped under the name of the Aguacate Group or Aguacate Formation (Madrigal, 1972), which includes a large series of volcanic focuses of this mountain range in which most effusive rocks have disappeared or are unrecognizable because of erosion effects. However, some elements of domes, craters, necks, and calderas remain, such as the Palmares collapsed caldera or the triple volcanic unit Pelon–Mondongo–Tinajita (Begoeing et al., 1978a), as well as ancient volcanic spots: Cerro Tilaran, Cerro Pelado-Delicias, Los Perdidos, Poco Sol, Pelados-Herrera, Espíritu Santo (Mora, 1977), and Cerro Macho Chingo–Pelon (Alvarado et al., 1980). The aforementioned authors place these ancient volcanic emission centers on the "Pliocene–Pleistocene volcanism map."

Geomorphology and Volcanology of Costa Rica
http://dx.doi.org/10.1016/B978-0-12-812067-5.00006-X

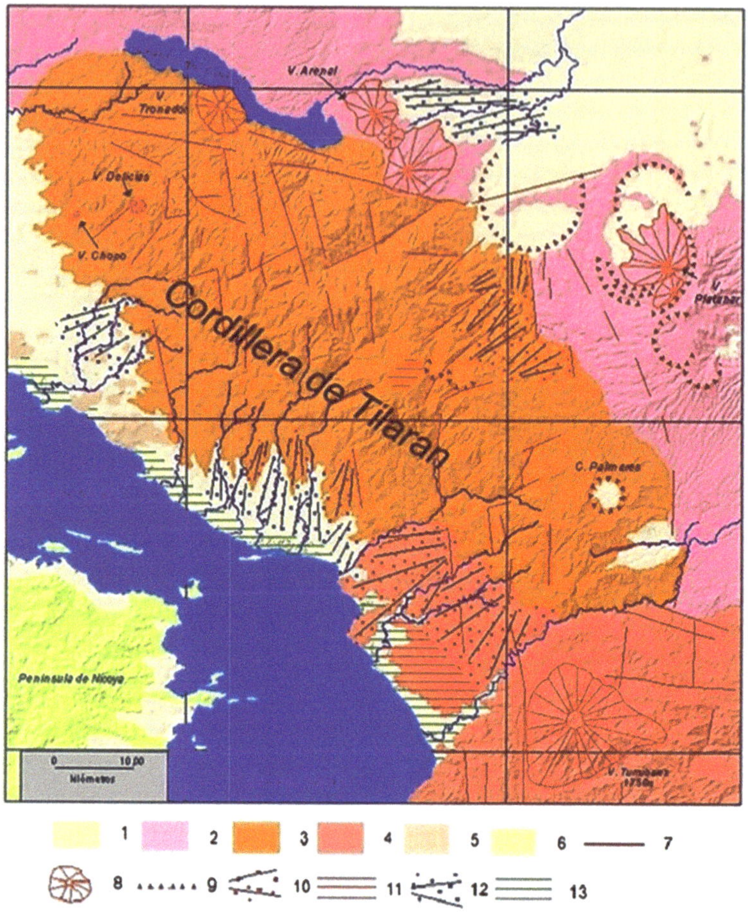

FIGURE 6.1 Tilaran's Tertiary mountain range geomorphology. Limits are to the east with the Quaternary Central Volcanic Mountain Range (Platanar–Porvenir volcanic complex) and with the Holocene volcanism of El Arenal–Chato and to the north with the last foothills of the Guanacaste Mountain Range (Bergoeing, 2011a). (1) Quaternary sedimentary area. (2) Quaternary volcanism area. (3) Pliocene volcanic area of Tilaran. (4) Cretaceous–Tertiary volcanism area. (5) Tertiary sedimentary area. (6) Cretaceous sedimentary area. (7) Undifferentiated tectonic faults. (8) Volcanic cones. (9) Calderas. (10) Lahar cone flows. (11) Structural plateaus. (12) Alluvial fans. (13) Flood plains.

This volcanic range follows the general direction of the country, whose axis is NW–SE. It has two well-define watersheds:

- **Northeast watershed**, or **San Carlos slope**, which pours its waters into the northern floodplain, where rivers, following a sinuous path, are tributaries of the San Juan River
- **Southwest watershed** or Pacific slope, which is subdivided into three sectors:
  - **Northern sector**, between the Cañas and Barranca rivers

- **Southern sector**, between the Barranca and Tarcoles rivers
- **Turrubares volcanic cone sector**

1. NORTHEAST WATERSHED OR SAN CARLOS SLOPE

This is much smaller than the Pacific slope sector, because of the imposition of the huge Platanar–Porvenir–Viejo volcanic complex. As a result, this slope rises gently to the southwest. It is a slope formed by volcanic, altered rocks at the surface (mainly andesite and basalt); however, at the summit it is possible to observe the outcrops of an intrusive mass of monzonite type, composed of monzodiorite, diorite, and adamelite rocks. This intrusive is known by the name of the Guacimal Formation (Chavez and Sáenz, 1974a). Near the top, it is not uncommon to find forms of volcanic dissected plateaus that give this region a quite rugged aspect, with cornices formed by volcanic or intrusive rocks and ravines that are the products of chemical alteration of the rocks and gravity, which together with the action of the rains often produce soil detachments of considerable size.

This slope base consists of a formidable filling, formed by lahars and piedmont deposits that built coalescent alluvial fans. At the end they are gradually transformed into glacis and then disappear, confused with the floodplain. The confluence of the Arenal, Peñas Blancas, La Esperanza, Jamaical, and San Lorenzo rivers forms the San Lorenzo River's lower course, which, once it reaches the floodplain, takes on the name San Carlos River. These rivers have a large capacity and are responsible for an important filling of the depressed sector. The climatic crises in Costa Rica during the Quaternary period are reflected once again in this part of the territory (see Fig. 6.2).

The northeast watershed is wetter than the Pacific side, because of its geographical position. It is exposed throughout the year to the humid northeast flows, coming from the Caribbean Sea, that cross the entire Northern Plain to finally penetrate the flanks of the mountain and precipitate in the form of orographic rains. Thus, this slope is not subjected to contrasting seasonal weather, as is the case for the Pacific slope of this mountain range.

Between the Jamaical River and the San Lorenzo River's lower course we find alluvial cones, probably deformed by neotectonics, whose outstanding values are greater than 20 degrees. They are dissected by the aforementioned rivers (Vargas, 1978). Neotectonics can explain the slopes of this single unit. On the other hand, the rivers run parallel between them, following a northeasterly direction. As all are tributaries of the San Lorenzo River downstream, it clips them squarely, because its thalweg fits in a lineup of NW–SE tectonic faults. The dissected cones were formed by torrential lava, whose matrix is composed of compact clays, enclosing rolled and more or less worn blocks of varying sizes, ranging from 0.65 to 3 ft in diameter. The chaotic rock material is essentially volcanic and much eroded at the surface. This reminds one slightly of the coalescent cones of the General Valley, which authorizes the hypothesis of approaching identification with the Lower Quaternary.

Piedmont deposits are encased in two levels. Downstream, the San Lorenzo River leaves exposed two fluvial terrace levels, more modern than the fluvial cones, with a homometric material. The San Lorenzo River profile is as follows: the thalweg is flat, with a width of 65 ft, and covered with a background of boulders and well-worn blocks of all sizes. The blocks can reach up to 6 ft in diameter, which gives to the San Lorenzo River the appearance of a torrential alpine river (see Fig. 6.3).

T1 level: This is situated between 16 and 23 ft in height over the riverbed. It consists of healthy pebbles and worn boulders of igneous origin, as in the level above it. The pebbles are mixed with a very sandy

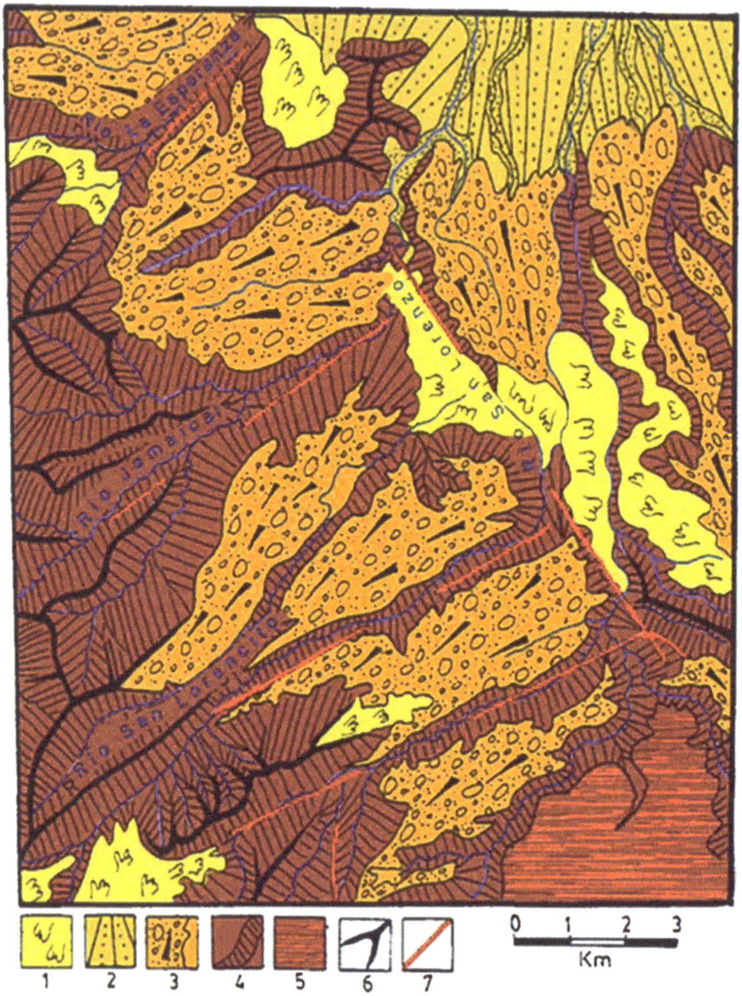

FIGURE 6.2 Eastern Tilaran slope geomorphology limited by the San Lorenzo River (Bergoeing, 2007). (1) Holocene; generalized solifluction. (2) Holocene; alluvial fan/glacis of the contact with the plain. (3) Upper Pleistocene; dissected alluvial cones. (4) Erosion talus. (5) Miocene volcanic plateaus. (6) Water divide lines. (7) Tectonic faults.

matrix. However, there is the presence of a dark red piedmont, probably of modified clays.

T2 level: This is 66 ft in height. In certain places, it could be confused with the slope and is often covered by tropical vegetation. It is formed by rolled edges 0.6–1 ft in diameter. It is altered on the surface, including a red sandy–clay compact mass.

Finally, in this last sector is part of a former collapsed caldera. About 6 miles in diameter, it is probably from the end of the Tertiary and has been named the San Lorenzo caldera. It is visible only in satellite images.

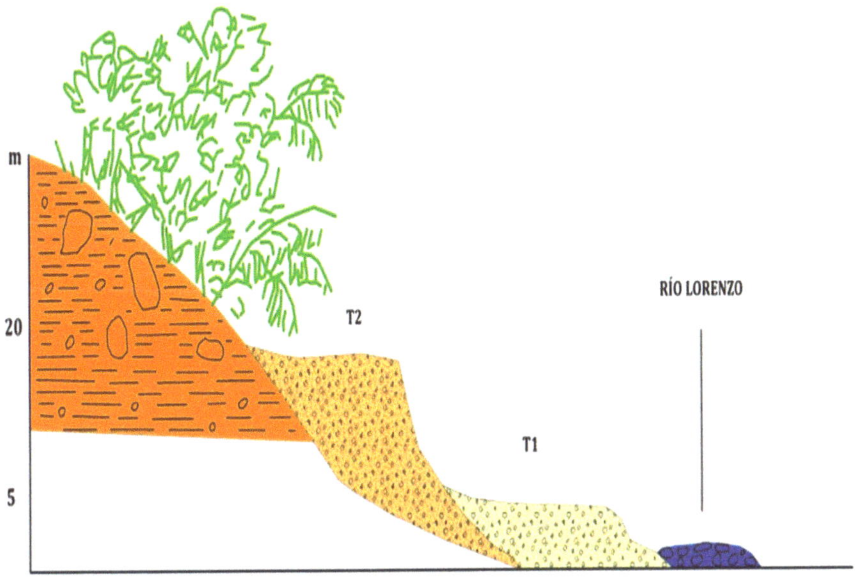

FIGURE 6.3 San Lorenzo River fluvial terrace shed of the eastern Tilaran Mountain Range (Bergoeing, 2007).

2. TILARAN SOUTHWEST WATERSHED

2.1 Northern Sector, Between the Cañas and Barranca Rivers

This is a much more evolved slope. The rivers fit it deeply and have a longer path around sometimes isolated hills, particularly in their middle courses before reaching the plain. This aspect can be divided into three different parts:

- **High mountain sector**, consisting of small volcanic plateaus and major erosion slopes (more than 650 ft of uneven ground)
- **Piedmont sector**, where is an important coalescent alluvial fan with rolled igneous origin elements, large, altered on the surface, but whose core is still recognizable
- **The floodplain**, near the coastline, recently filled by the piedmont deposits, which often form deltaic cones at the end, protected by the current mangroves

Volcanism in this sector (Chaves and Sáenz, 1974a) should have extended everywhere during the Lower Pleistocene. Near the Cañas River, following the road that leads to Tilaran Village, emerge the remains of a beautiful pyroclastic cone. It corresponds to the Villafranchian volcanism of Cerro Chopo or Mount Asuncion (Mora, 1977). A second focus that is also observable 3 miles from Cañas is Pelon Volcano (see Figs. 6.4 and 6.5), a ruined relict of Pliocene–Pleistocene volcanic activity (Alvarado et al., 1980). Thermal water springs are present 1.2 miles northwest of the ruined volcanic mass. The lower sector of the Tilaran Pacific slope, from Colorado de Abangares, is covered by isolated sedimentary hills. The outliers are composed of folded anticline and syncline stratifications formed by rocks of the Upper Cretaceous and Lower Tertiary. These rocks are greywacke, sandstone, and shale associated with calcareous material.

Here the weather is of contrasting seasons, which makes these isolated hills largely devoid of arboreal vegetation at the top. From the Lagarto River, a series of rivers, such as the Guacimal, Sardinal, Aranjuez,

FIGURE 6.4 Pelon volcano relict, where it is possible to see an ancient crater and two volcanic domes. *Photo courtesy of Gilbert Vargas Ulate (2010).*

FIGURE 6.5 Cerro Chopo or Asuncion volcanic relict. Tertiary. *Photo courtesy of Guillermo Alvarado (2010).*

Quebrada del Palo, Seco, Ciruelas, and Naranjo, trimmed the important piedmont deposits downstream, to give rise to flattened alluvial fans with deltaic episodes in contact with the sea. The village of Miramar de Puntarenas is seated on a powerful piedmont alluvial fan deposit, which extends more than 3 miles from its point of origin, which is Cerro Negro (2625 ft), located upstream. This cone is spanned on its sides by two important rivers: the Ciruelas and the Naranjo. These two rivers left the chaotic material of the slope exposed on the banks not occupied by the embedded fluvial terraces.

The terraces have three well-defined levels, as in the Ciruelas riverbank (see Fig. 6.6). More bedding consists of a heterogeneous mass of large and healthy, well-polished boulders. This river has a strong current in times of flood during the rainy season. Then the largest bed is completely flooded, and the river erodes the T1 level.

T1 level: This is formed by a small terrace constituted by healthy and good-sized boulders (1.3 ft diameter) of volcanic origin, on a brown sandy—clayey matrix.
T2 level: This is a medium-sized terrace, embedded in the preceding level. It is not present everywhere. It consists of good-sized boulders, well worn but altered at the surface as in the T3 level.
T3 level: This level is related to the piedmont, which is located at 460 ft above sea level in

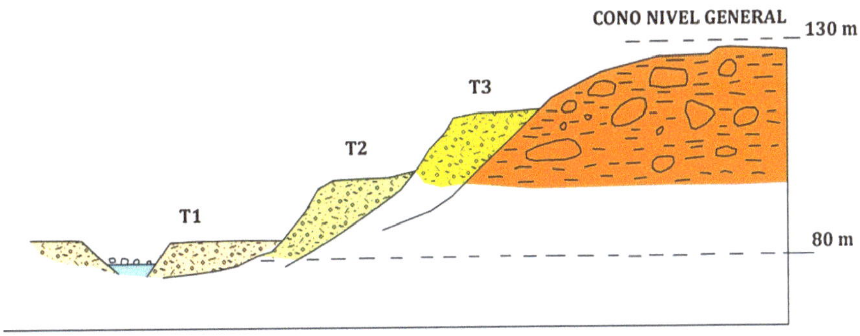

FIGURE 6.6 Three cyclic boxed fluvial terrace levels of the Aranjuez River of the Pacific watershed of the Tilaran Mountain Range (Bergoeing, 2007).

this sector. Rolled boulders of great size, of volcanic origin (andesites), that lie in a dark red clay mass form it. This level is located 130 ft above the major riverbed and is present in both riverbanks.

The Aranjuez River is also interesting, because it confirms the aforementioned levels, which are otherwise on the banks of the rivers nearby.

Following the road to the small Chomes fishing cove, the description of the profile is as follows:

The largest fluvial bed is wide and covered by very well worn boulders of important size, ranging from 0.33 to 1.6 ft in diameter, with some parts comprising rolled blocks. In the summer the river describes twisted meanders.

T1 level: This is 33 ft from the T2 level, made up of large boulders, very similar to the older levels, but much more compact and of higher density.

T2 level: This is a high and well-developed terrace consisting of rolled edges 0.6–1 ft in diameter, altered on the surface, and which rests on a dark red clay matrix. The material is well rolled and well compacted.

T3 level: This upper level is combined with alluvial cones and is 66 ft high with respect to the T2 level.

Small forest patches upstream cover alluvial cones, where humans considered that it was not necessary to continue deforesting to build grasslands. As a result, this piedmont is strongly exposed to erosion. A large volume of boulders covers all the rivers here, with a generally flat bottom. They are torrential regime rivers. The previously described three levels of the Ciruelas and Aranjuez rivers are irrefutable proof of the successive climatic crises of Costa Rica during the Upper Pleistocene. They do not present tectonic deformations despite their location in a region with important neotectonic pushes, and are far from the current volcanic emission centers.

The T1 lower level, by its size and good conservation of boulders, allows us to relate it to the latest Würm–Wisconsinian terminal climate crisis, known in Costa Rica by the Chirripó II glacier shapes.

T2 is less well preserved; however, it has enough fresh forms and reveals a certain youth that might belong to the Upper Quaternary. It is clearly older than the T1 level, though. It could be associated with the Riss–Illinoian climatic crisis.

Finally we see level T3, where the alteration of boulders is advanced and clays that make up the matrix are also advanced, which may be related to a drier period, which would have occurred in the Middle Quaternary.

Farther south, the Barranca River is one of the major rivers of this watershed. The Barranca River is born near the town of San Ramon, on an ancient dissected volcanic surface. Downstream rivers form important fluvial terraces, including that of the San Jeronimo, which is covered by piedmont deposits formed by large blocks.

The upper river level downstream is crossed by an old railway line, which linked Puntarenas Port to San José. Barranca is a shallow but torrential river. The lower bed presents braided forms, compounded by sand and boulder banks. The river has three fluvial terrace levels that are largely on the left bank. An important escarpment fitted in Tertiary sedimentary rocks dominates the right bank (see Fig. 6.7).

The Barranca river mouth has resulted in the construction of the Puntarenas coastal arrow (Battistini and Bergoeing, 1982a,b). Important sediments, swept away by the river and by the littoral drift, served to build this arrow, which owes its survival to the construction at the end of the 19th century of the railway line to Puntarenas.

The photograph of the lower course of the Barranca River displays the sandy bar formed by river sediments hauled in contact with an important coastal drift in this sector (see Fig. 6.8).

2.2 Southern Sector, Between the Barranca and the Tarcoles Rivers

From the Barranca River and up to the Tarcoles River, the piedmont changes radically. Indeed, the part downstream from the Aguacate Mountains suffered major tectonic deformations at the same time that it was filled with detritus and volcanic sediments (ignimbrite). The Aguacate mountain piedmont from there takes a topographic plateau form, dissected by the rivers that fit it in deep gorges and canyons. Neotectonics has conferred it a form divided into compartments, like "piano keys" (Battistini and Bergoeing, 1984). The Costa Rican Great Reverse Fault crosses the edge of the slopes' rupture area between the plateau and the piedmont, following a straight line, which is drawn between the towns of Esparza and Orotina.

This faulted plateau was formed by a double fracture system following a main NW–SE tectonic alignment, parallel to the Costa Rican great natural accident. It is cropped at right angles by a second SW–NE system that has allowed the emergence of small horsts and grabens, as for example, the Mata de Limón, Tivives, and Guacalillos tectonic depressions, which contrast with the Alto de Carballo, Alto de Las Mesas, or Guachipelín horsts. The main rivers of the sector have taken advantage of the crushing tectonic zones, forming deep river canyons with vertical walls up the thalwegs.

The structure of this plateau is broken down into three fundamental levels (see Fig. 6.9).

- **Top level**: This is the most modern and consists of decomposed red clays transported from the piedmont, which contain large volcanic heterometric blocks, more or less worn and with much eroded surfaces. The thickness of this level is variable and it is located between 32 and 164 ft. This level is based on fluvial deposits.
- **Middle level**: This level consists of 130–200 ft of ignimbrite deposits, sometimes crossed by basaltic lava flows (3 miles south of Esparza village). This material comes from the Pliocene terminal activity of the Aguacate Mountains. An ignimbrite sample taken at the Agres Bridge on the Tarcoles River was dated, giving an age of 1.28 MY. At the piedmont base ignimbrites rest on a gray welded tuff level enclosing rolled volcanic elements of large size (3 ft in diameter).
- **Lower level**: This is the current base on which rests the Quaternary piedmont. Consisting of 330- to 660-ft-high Miocene sedimentary rocks, known as the Punta Carballo Formation (sandstone in the summit and

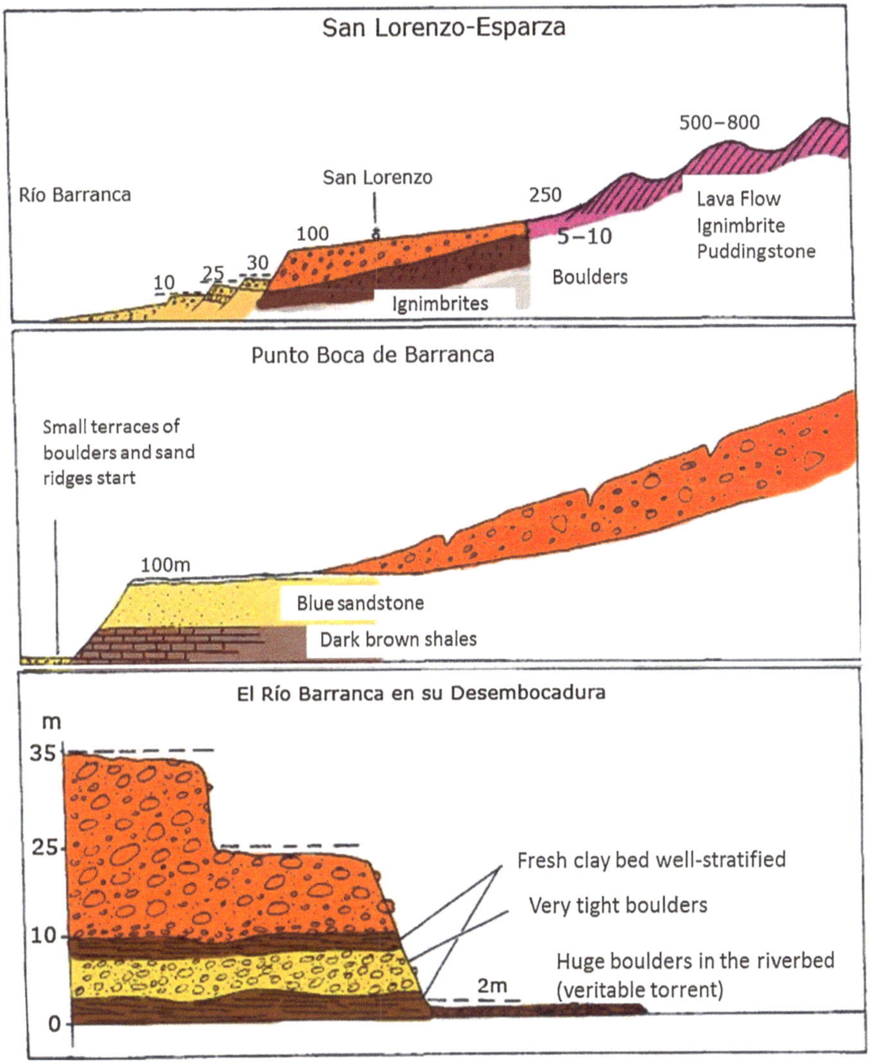

FIGURE 6.7 Barranca River fluvial terraces near Port Puntarenas (Bergoeing, 2007).

brown shale at the base), this unit is in a monocline position tilted to the northeast.

It is evident that the most important river in the sector, and which serves as a reference, is the Tarcoles River. It marks the southern limit of the plateau and is the natural drain of the western sector of the Central Valley.

At Agres Bridge, just 6 miles from Orotina village, the Tarcoles River presents a very interesting cut showing in detail the main elements of the piedmont (see Fig. 6.10).

This allows one to observe that, after the ignimbrite volcanic event, the river deposited a riverbed of thin and thick layers represented by puddingstone and well-rolled and cohesive

FIGURE 6.8 Barranca River mouth in the Pacific Ocean forming a sedimentary coastline that feeds the arrow of Puntarenas with the littoral drift current. *Photo Jean Pierre Bergoeing (2009c).*

boulders. The upper part of the profile is represented by an 82-ft-thick lacustrine horizon on which piedmont torrential lavas were finally deposited. Downstream, the Tarcoles River's lower course shows three small river levels of 16, 33, and 130 ft height. The last level is the most interesting, because this upper level consists of two well-worn medium-sized boulder strata, separated by an ochre sandy bank. This terrace is faulted and tilts slightly toward the north (see Fig. 6.11).

The Tarcoles River today maintains a large capacity manifested in its lower course by the presence of rocks of 0.65–1 ft diameter that emerge during periods of low water and are present in the sector of the bridge that joins Orotina village with Quepos Port (Costanera Sur road). During the floods in the rainy season, the Tarcoles River floods its greater bed and sometimes overflows onto the lower terrace. At its mouth, the river has recently built a delta with a second arm. The suspended material transported by the river, in contact with the salty seawater, peptizes and is deposited on the bottom by decantation (flocculation phenomenon).

Orotina's plateau is fitted by a second important river that lies between the Tarcoles and the Barranca rivers. It is the Jesús María River, which has carved a deep canyon. This river ends its journey in the Tivives tectonic depression, forming meanders in a vast mangrove swamp (the most beautiful of the central sector).

The Jesús María River slips under the Damas Bridge. This bridge serves as a union for the Esparza–Orotina highway. Here the river divides the ignimbrite plateau. In this place, the profile shows the same series described earlier for Agres Bridge (see Fig. 6.12); however, here, the ignimbrite series is interbedded by basaltic lava flows (which perhaps should be related to the end of the Pliocene volcanism of the Pelon–Mondongo–Tinajita volcanic sector; Bergoeing et al., 1978a, op. cit.).

Other stratigraphic profiles are visible on the Palo Seco riverbanks, near Damas Bridge and after Vueltas sector. This small river has built a terrace of 33 ft that dominates a vast flat clay sector filled with palm trees. The 33-ft level is composed of rocks of medium size (1–1.3 ft in diameter) resting on a red clay mass. The 33-ft level is dissected and eroded into a ravine, the terrace is attached to a rocky outcrop formed with basal piedmont sandstone, very altered and evolving into clay in certain places. The basal rock presents similar altitudes, forming a 115- to 130-ft level where the profiles are visible in the recess of the Palo Seco River, near Damas Bridge and after Vueltas sector (see Fig. 6.13).

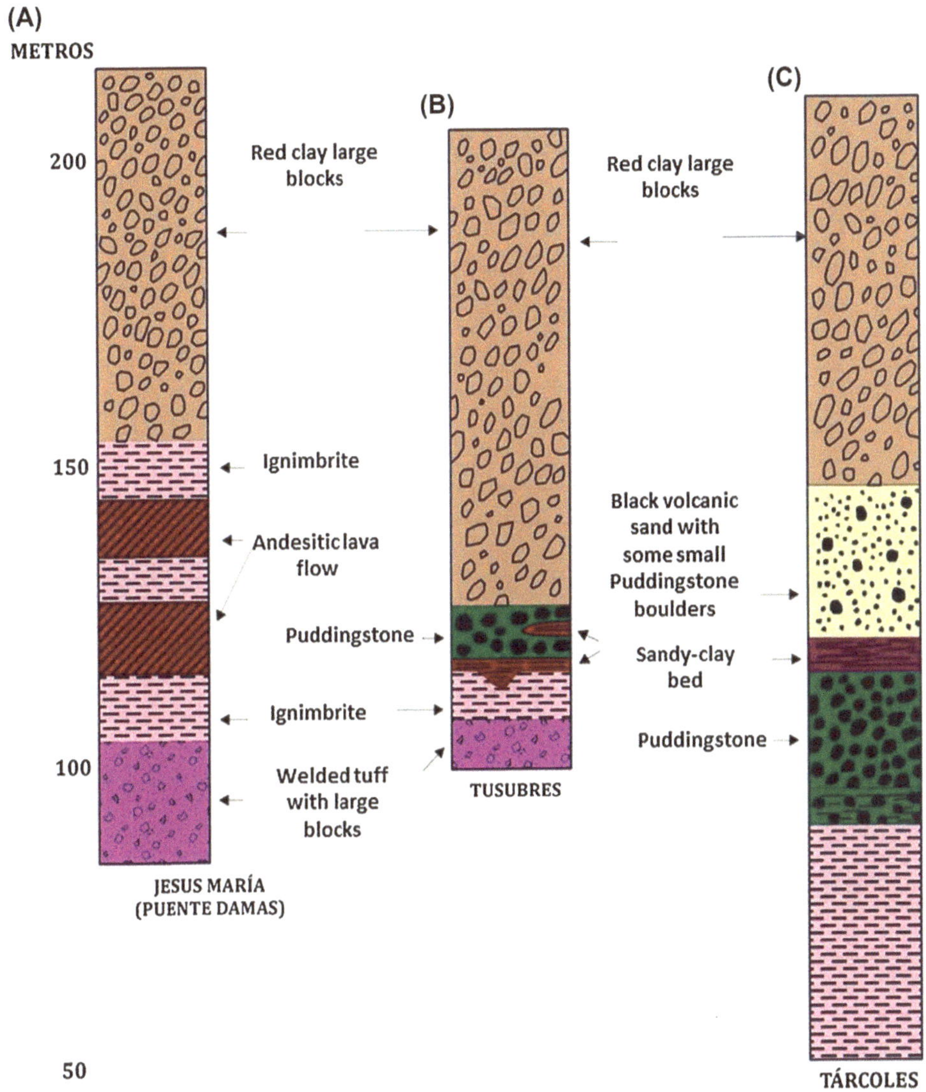

FIGURE 6.9 Three stratigraphic profiles of the Orotina–Esparaza plateau and fluvial sectors of the Jesús María, Tusubres, and Tarcoles Rivers (Bergoeing, 2007).

3. MOUNT TURRUBARES VOLCANO SECTOR

This last sector is characterized by a piedmont, which gradually descends toward the coast from Santiago de Puriscal village (3300 ft above sea level). It is strongly devoid of vegetation, a consequence of deforestation at the beginning of the 20th century, which took place in search of land for the planting of red kidney beans (puriscos), to the detriment of the cloud forest. As a result, today the vocation of this

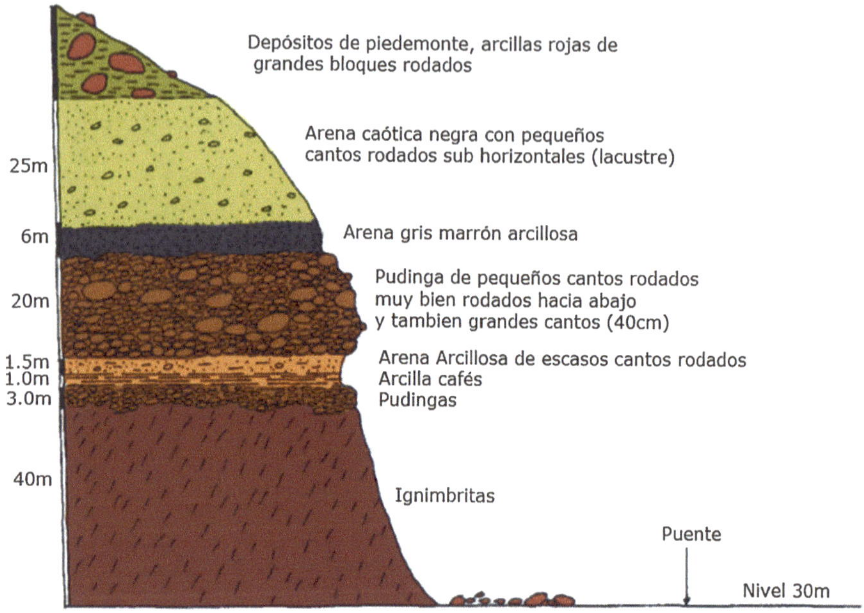

FIGURE 6.10 Agres Bridge translation: (From the top to the base) (A) Piedmont deposits. Red clays in big boulders. (B) Lacustrine black sands with sub-horizontals boulders. (C) Sandy brown clay. (D) Puddingstone formed with little boulders of 13 ft diameter. (E) Sandy clay. (F) Brown clay. (G) Puddingstone. (H) Ignimbrites.

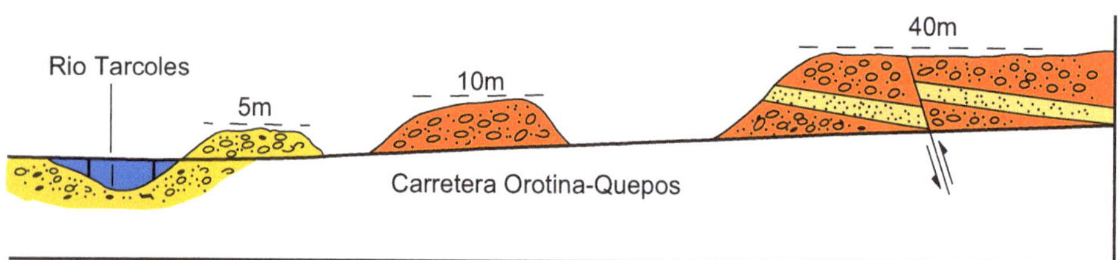

FIGURE 6.11 Fluvial terraces of the lower course of the Tarcoles River on the Orotina Quepos Bridge (Bergoeing, 2007).

land is grassland, but in the steep slopes (more than 45 degrees in some sectors) appear deep traces of solifluction. The same occurs in all stages: laminar, scars, mass landslides, and packages of landslides that testify to a more advanced and inexorable erosion of soil phases (see Fig. 6.14).

Mount Turrubares dominates a vast depressed sector where the highest peak rises to only 2300 ft. The basal lavas of Turrubares are much altered basalts, which are associated with the Upper Cretaceous–Lower Tertiary volcanism phase (Dengo, 1962a). However, we can say that this volcanic edifice is the result of a constant activity that lasted at least until the beginning of the Quaternary. The freshness of its forms, its dimensions, and the little advanced alteration of the lavas of the summit support this hypothesis. Turrubares can be compared with El Hacha Volcano (Aguacate volcanism at the northern end of the Guanacaste Mountain Range): the first has intact forms and may be therefore Quaternary (Middle or Upper); the second, Pliocene, is reduced to a heap of volcanic ruins.

FIGURE 6.12 Tarcoles River lower course, during low-water period: first fluvial terrace. In the background are the Turrubares Volcano foothills (Cretaceous–Tertiary). *Photo Bergoeing et al. (2010d).*

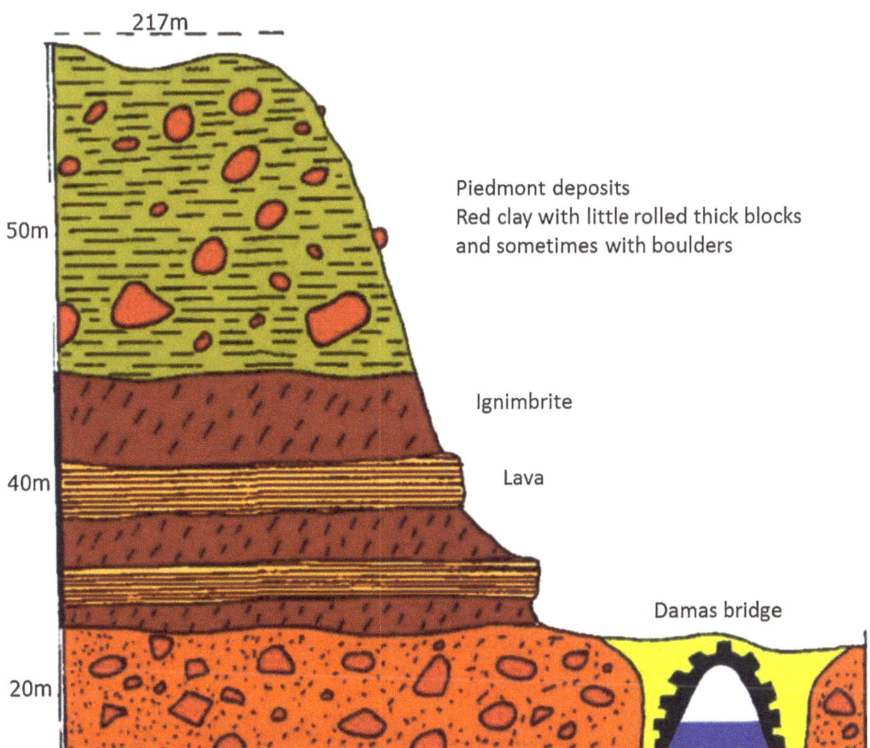

FIGURE 6.13 Stratigraphic profile of Jesús María River at Damas Bridge (Bergoeing, 2007).

FIGURE 6.14 Turrubares Volcano, 5761 ft altitude. This is a Tertiary volcano that emerged in a midoceanic area and then was lifted up with the Quaternary orogeny. *Photo Bergoeing (2012).*

The Orotina piedmont is of obvious interest both from the point of view of the river and because of the presence of ignimbrite deposits. Indeed, and as we describe it, the ignimbrites are present everywhere, provided as an intermediate layer between the summit piedmont deposits and the sedimentary rocks at the base (sandstone monocline of the Punta Carballo Formation) (see Fig. 6.15).

For a long time it was believed that Orotina ignimbrite was associated with western sector Central Valley volcanic flows, from a fissure at the base of the Poas and Barva volcanoes. Now, these overflows are widely spread in the Orotina plateau, present in the north of the plateau, between the Naranjo and the Barranca rivers, forming small groups on both sides of the Inter-American Highway. What is even

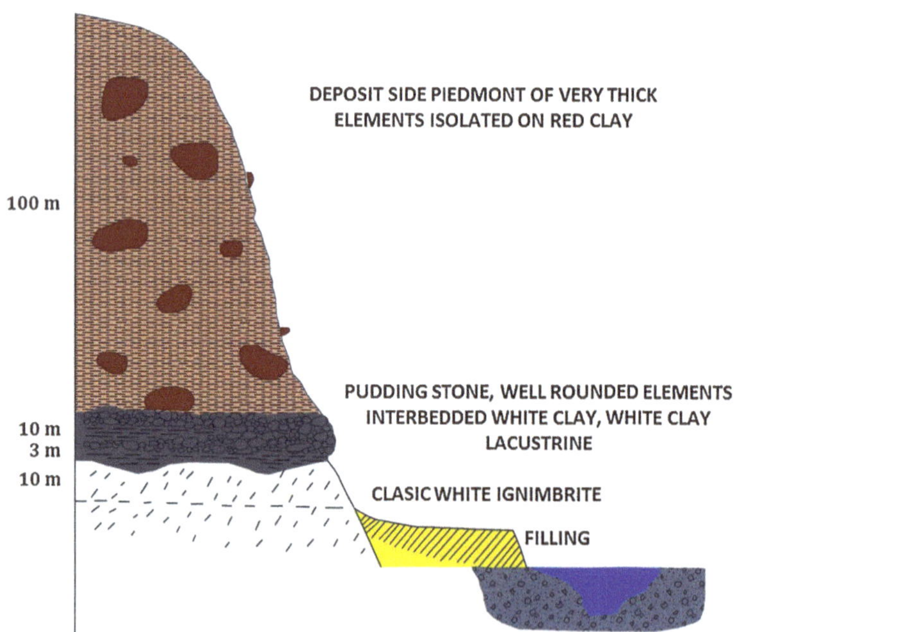

FIGURE 6.15 Stratigraphic profile of Tusubres riverbank (Bergoeing, 2007).

FIGURE 6.16 Pelon–Mondongo–Tinajita Volcanic Complex in the Aguacate Mountain Range. *Photo Bergoeing (1980).*

more telling is to find ignimbrite south of the Tarcoles River, in the lower course of the Tusubres River, in its classical form: white ignimbrite, crowned with piedmont deposits (Battistini and Bergoeing, 1982a,b).

Now, this new fact requires us to assume that these ignimbrites (dated to 1.38 MY from San Pablo de Turrubares and 1.23 MY from Agres Bridge, and consequently from the Lower Quaternary) cannot come from only one focus located either in the Aguacate Mountain or close to the Turrubares volcano. Given their thickness in the Orotina piedmont and its position at Alto de las Mesas Summit, overlooking Caldera Port on the Pacific Ocean coast, these ignimbrite flows could have come from active volcanic cones of Pelon–Mondongo–Tinajita at that time (Bergoeing et al. op. cit.) or else an acidic volcanic activity that can be located at the foot of Turrubares (see Fig. 6.16).

On the other hand, in the Central Valley an ignimbrite activity existed more or less contemporarily with the volcanic Orotina activity. However, despite the fact that the volcanic activity of Aguacate Mountain was at the end of the Tertiary (Pliocene), the Palmares and Atenas tuff deposits in the western sector of the Central Valley are more recent and date back to the Middle Quaternary. They would have originated in the Barva and probably Poas volcanoes when they built their current stratovolcano cones. The ignimbrites of the Turrubares, Tarcoles sector, until almost reaching the Pacific coast, have an age of 1.4 MY according to Ar/Ar dating (Pérez, 2000b). A similar age was found by Battistini and Bergoeing (1983a,b) for the fluvial–ignimbrite Turrubares terrace deposits.

The Tilaran–Aguacate Mountains' Pacific piedmont fluvial terraces are also of great interest because of their position and their material (three river levels with alternating rolled and worn material down from the upper levels). They are the result of two or three alternate climatic crises that would have occurred between the Middle and the Upper Quaternary (given their position with respect to the ignimbrite that rests at the top level of San Pablo de Turrubares).

In the sector of the Tilaran Mountain Range, the volcanic cones are not well preserved enough to attribute recent activity to them. The grounds are very disturbed by erosion that has contributed to the formation of fluvial terraces of the Ciruelas and Aranjuez rivers. Three river levels on the Pacific slope of Tilaran are present in the lower banks of the Barranca and Tarcoles rivers. The Tarcoles River has, however, upstream, in the sector of San Pablo de Turrubares, three fluvial levels, which fit narrow and deep slopes. Is it possible that these levels are concordant with the lower level, but uplifted by neotectonics?

The top level of San Pablo de Turrubares is covered by an ignimbrite tuff that was dated at 1.38 MY. Topographically this level serves as a junction between the Central Valley ignimbrites upstream and those of Orotina, downstream. This higher level of San Pablo de Turrubares then gives us the chronological basis of the Virilla–Tarcoles river thalweg's beginning incision, and consequently, the lower river levels are necessarily much younger than the Lower Quaternary.

The volcanic base of Turrubares is in contact with the Cretaceous Nicoya Peninsula Complex rocks. They are staggered at the base of the old volcano, forming a plateau much dissected by rivers that often adopt directions following the alignments of ancient faults, because in general they do not correspond to the NW–SE orientation of the country. The Tusubritos River, the major tributary of the Tusubres River, is born in the west Turrubares piedmont and digs its own course in an important NW–SE tectonic fault alignment, which extends into the Pacific to Puntarenas Port. This alignment is part of the Pacific slope of Turrubares Volcano and is parallel to the Costa Rican Great Reverse Fault that trims the Turrubares in the northeast.

To the south, this sector ends in a Paleocene–Eocene sedimentary series outcrop (Brito Formation), which is juxtaposed to the Nicoya Complex rocks. These series, less cohesive, have allowed a more advanced evolution of the coastal plains; the basaltic ravines give way to small sandy bays (Punta Leona, Herradura). After the fluvial terrace of the lower course of the Tusubres River, some river levels are still visible, from 17 to 50 ft high, and correspond to parts of the great alluvial fan built by the Tusubres River. Boulders of small or medium size, among which some are badly rolled, form the lower level. The rivers here have short journeys, because they come from the Turrubares volcanic piedmont; nonetheless they are less active than the others.

Farther south, in Parrita, other examples of old alluvial fan remnants are present. We can therefore say that the Tilaran mountain piedmont is the result of the following events:

- **A volcanic activity** beginning in the Miocene and continuing until the end of the Pliocene installed the imposing mass of Tilaran, Aguacate Mountain, and also isolated volcanoes such as the de Turrubares cone.
- **Once the volcanic activity** that gave rise to basalt lava flows was interrupted, more acidic activity evolved toward andesite flows. Quaternary erosion attacked the slopes, leaving some pieces of volcanic plateaus on the summits and downstream gave rise to valleys in a multifaceted general modeling.
- **Mountain landslides** created a more important piedmont on the western watershed than on the Caribbean slopes of Tilaran Mountain Range, creating alluvial fans. The filling of these slopes should be related to repeated Quaternary climatic crises, which uprooted considerable amounts of material from the mountains. Alluvial fan accumulations between the Lagarto and the Barranca rivers cover an area of approximately 39 square miles. On the other hand, the distance in a straight line between the peaks of Tilarán (2600–3300 ft) and the mangrove coast is only 14 miles. This may explain, in part, the easy slope washing by pouring rains during climatic crises, taking into account the igneous material, previously put under these conditions, because plutonic

rocks have an accelerated decay under the influence of the humid tropical weather.

- **The sector between the Barranca and Tarcoles rivers** (Orotina's piedmont) was completed during the Quaternary period, by means of powerful ignimbrite flows and deposits alternating in some places with basaltic lava flows. The whole was crowned by a torrential lava. This piedmont, taking neotectonics into account, was deformed, creating faulty marquetry. The plateau, lifted up by the tectonics, was fitted by the rivers that gave rise to deep canyons.
- **In the Tilaran–Aguacate sector, Pacific slope rivers** have built on their banks levels that are encased in the alluvial fans and of which we find very evident traces of three levels. Currently, if we take into account the steep slope, they behave as torrents, during large floods, depositing in their beds boulders and cobble blocks. These three levels are staggered chronologically in a range that goes from Lower to Middle Quaternary (deposits of ignimbrite dating the Orotina piedmont to 1.28 MY) to Upper Quaternary. The T1 and T2 levels are the youngest; they are witnesses to subsequent climatic crises, drier and close together, if we stick to the relative positions of the levels as well as the degree of deterioration of the worn material (for example, the Tarcoles River terraces). The T3 level, in some sectors, is in the highest position (Tarcoles River) and bears witness to a tectonic deformation (uprising). Given its degree of the alteration, we can attribute to it a Middle Quaternary age.
- **After the Flandrian transgression**, these rivers built their coastal plains (deltaic cones), where riparian and mangrove species thrive today.

CHAPTER

7

Guanacaste Volcanic Mountain Range

OUTLINE

The Guanacaste Volcanic Mountain Range is the northernmost Costa Rican volcanic range, where the volcanoes follow a NW–SE direction. Almost every volcanic edifice is well individualized, permitting almost the time joining ways between the Northern Plain and the Guanacaste plateau and plain. In the Pacific watershed, it is possible to observe little Pliocene volcanic focus ruins, showing a volcanic activity prior to the Quaternary (Lloyd, 1963) (see Fig. 7.1). It is likely that the current volcanic edifices stand on older volcanic structures (Neogene) (Malavassi and Madrigal, 1970). The mountain range is composed of the following volcanic units from the northwest to the southeast: Orosí, Cacao, Rincon de la Vieja, Miravalles, and Tenorio. All these volcanoes are in a horst position, because on one hand, to the northeast, they are on the edge of the Nicaragua graben's southern rim, and on the other hand, to the southwest, they are bordered by the Costa Rica tectonic inversed fault partly hidden under younger ignimbrite deposits. Almost all volcanoes of this mountain range have their craters open southwesterly (because of the frequent northeast trade winds, especially at the time of eruptions), and cones usually sharp and very regular. They are stratovolcanoes (Weyl, 1960) built up from a magmatic fissure activity. The majority of these volcanoes have provided large amounts of lava (basalt, andesite), but also pyroclastic material, ashes, and over all ignimbrite. From the Pliocene and especially during the Lower Pleistocene, and certainly from a fissure-type activity, the Pacific slope between Bagaces and La Cruz villages was buried by large ignimbrite inflows (Malavassi and Chaves, 1970). Middle Quaternary

Geomorphology and Volcanology of Costa Rica
http://dx.doi.org/10.1016/B978-0-12-812067-5.00007-1

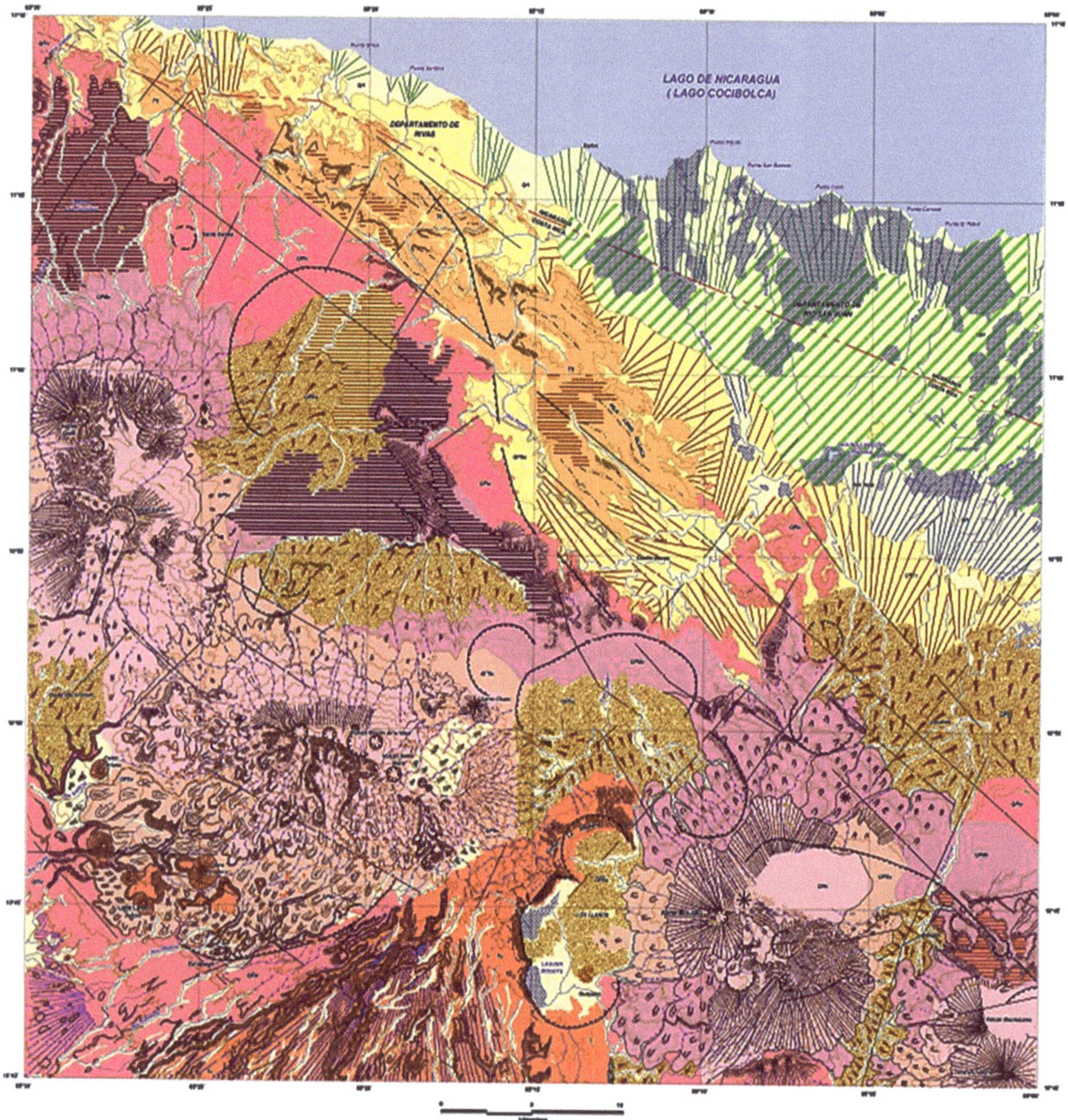

FIGURE 7.1 Geomorphology of the Quaternary Guanacaste Volcanic Mountain Range Orosí–Rincon de la Vieja–Miravalles sector. There are craters and especially the Guayabo collapsed caldera. *Taken from the Caribbean Geomorphological Atlas, Bergoeing et al. (2010c).*

ignimbrite deposits concern only the region of Liberia. The plateau formed thus corresponds to the Bagaces Formation, whereas the modern ignimbrite cones correspond to the Liberia Formation.

According to Tournon (1972): "Is to the Middle and Upper Pleistocene which should attribute the upper rhyolitic tuff robes. (Ech. Mu10 = 0.6 M.y. and Ech. TM 14 = 0.5 my.) Most of the andesitic strata-volcanoes must been built during the Upper Pleistocene. Historical volcanic activity was only explosive (Rincon de la Vieja, 1967; Tournon, 1972)."

At the base of these volcanoes, and especially among the Orosí and Rincon de la Vieja volcanoes, the old Pliocene-Quaternary volcanic structure remains. They may be associated with the Pliocene volcanism (Aguacate Formation). At least this is what we can conclude from the studies of Dengo (1962b) and the Costa Rican Institute of Electricity (ICE).

1. EL HACHA VOLCANIC RUINS

One of these old volcanic ruins at the foot of Orosí Volcano is El Hacha Volcano, with an altitude of 1969 ft above sea level. It emerges through the ignimbrite plateau of the Bagaces Formation at some 300 ft and, toward the north, leaves a unique important andesite lava flow, which rests on the ignimbrite plateau. By its form, one could suppose that it is a small shield volcano (see Fig. 7.2). However, it is not more than a small cone, whose activity should not have extended beyond the Lower Pleistocene. On the other hand, at the foot of the Rincon de la Vieja, on the Pacific slope, old dacitic domes are also testaments of peripheral volcanic activity to the axis of the current large volcanoes. For a better understanding, we detail each volcanic unit, according to its particularities and geomorphologic modeling.

2. OROSÍ VOLCANO

Orosí Volcano, 4724 ft above sea level, is the northernmost of the volcanoes of the Guanacaste Mountain Range and has a perfect cone. From its crater, it is possible to see Lake Nicaragua and the Pacific Ocean. Its open-mouthed crater is exposed to the southwest, probably because of the combination of eruptions of ash and lapilli associated with frequent northeast trade winds, forcing the eruption column to tilt toward the southwest. It has numerous lava flows, visible to the naked eye; some of them reach the

FIGURE 7.2 El Hacha Pliocene volcano emerging from the ignimbrite plateau, north of La Cruz village. Observe the cone head and some parasitic dacitic domes. *Photo courtesy of Alvarado et al. (2011).*

northern hills to the edges of Lake Nicaragua. To be precise, the lava flows go up to a place called the Sitio Santa Manuela, marking the eastern boundary of the ignimbrite plateau. In these places, following the route that leads to Hacienda Inocentes, it is possible to observe the ignimbrite resting on a lava flow, which proves that this lava came prior to the issuance of ignimbrite. The humid tropical vegetation covers part of the summit of the cone. The bottom was removed and dedicated to cattle pastures. In this sector, and due to deforestation, it is possible to perceive nude volcanic rocks, products of the most recent lava flows.

The ignimbrite plateau's surface is dissected by the rivers from the eastern sector that fit it easily. The plateau runs along the Orosí Volcano foot, in its northern and western part. The eastern part, although it remains a plateau, differs from the first, because here there is an absence of ignimbrite (at least on the surface, and if it exists it must be quite deep; no field cuts let us perceive such a phenomenon). This second plateau is covered by lahars and above all by torrential lava. It is formed by blocks, of volcanic origin of course, badly rolled and resting in a dark clay structure (see Fig. 7.3).

The east plateau façade starts at the Haciendas River and stretches to the foot of the Rincon de la Vieja volcano. This plateau is probably what is left of the old pre-Quaternary surface. To the north and east of Orosí Volcano, the plateau ends with a series of hills whose summits do not exceed 1640 ft and are fitted in the folded sedimentary series of the Cretaceous and Paleocene (Rivas and Sabana Grande formations). It is precisely in these hills, northeast and east of Orosí Volcano, that the satellite SEASAT radar image allowed us to observe alignments that correspond to the south rim of the Nicaragua graben. This edge draws a tectonic northwest alignment.

The southern Orosí Volcano sector is a watershed dividing line, linking this crater with a second one, whose mouth is unleashed to the southwest and is called Cerro Orosilito Volcano (3675 ft). It is the twin crater of Orosí Volcano. Following the watershed, we finally reach a third crater, the Cacao Volcano (see Fig. 7.4).

FIGURE 7.3 Santa Rosa ignimbrite plateau covered by xerophytic vegetation of the dry Pacific area. In the background are the cones of Orosí and Cacao volcanoes. *Photo courtesy of Gilbert Vargas Ulate (2010).*

FIGURE 7.4 Orosí and Orosilito volcano cones seen from the southeast. *Photo courtesy of Gloria Muñoz (2009).*

3. CACAO VOLCANO

Cacao Volcano, 5443 ft, marks the southern edge of the first volcanic edifice complex of the Guanacaste Mountain Range (Healy, 1969). Like the others, its runaway crater is exposed southward, and in part has been destroyed. It has a small adventitious cone whose summit reaches 3504 ft above sea level. It is carved by the Gongora River, which is born in the same Cacao crater. From the crater, a huge lahar flow formed the actual cone during the Middle Pleistocene. It covers the entire southwest flank and reaches the hamlet of Quebrada Grande. This lahar flow allows us to observe that the primordial material was composed of pumice, andesitic boulder fragments, and lapilli, all mixed with ashes ground to a yellowish colored clay, very little compacted. The top of the cone is covered by piedmont deposits or by torrential lava, which covers the lahar flow, showing that it is the most recent. These deposits give the top of the cone the partial aspect of a wavy piedmont with rounded spots of green pastures. It is among Cacao Volcano and the Orosilito crater that the Tempisque River, the main river of the Guanacaste Pacific watershed, is born and receives the waters of almost all the rivers that originate in the western flank of the volcanic Guanacaste Mountain Range. After a long journey, this river empties into the Nicoya Gulf.

The contact of Cacao Volcano with the base of the Pacific slope carried out to 660 ft above sea level corresponds to the Santa Rosa ignimbrite plateau (Bagaces Formation). This ignimbrite plateau is very broad in this sector and partly covers the Santa Elena Peninsula peridotite (see geological profiles a–a′ and b–b′). In general, the plateau has variable thicknesses ranging from 130 to 330 ft and butts up against the Eocene–Paleocene sedimentary series of the Rivas and Sabana Grande formations. The rivers that fit its surface form either small canyons or riverbeds of flat bottoms with stepped erosion terraces. The rivers carry little rolling pieces of material and the particle size of this material is rather thin (sand and gravel for the coarse fraction). The waters are generally clear, except at the time of floods during the rainy season, when the material in suspension, clays and silt, comes directly from the volcanic slopes. The SEASAT (radar image) satellite has made it possible to discover an important NE–SW tectonic fault among the Cacao and Rincon de la Vieja volcanoes, which has notched the plateau and extends into the Pacific Ocean. We had already observed this fault in what is referred to as the Santa Rosa plateau, because a large fault scarp separates it from the Liberia sector (sunken block). This great fault line gives an imposing escarpment showing the ignimbrite stand on the Nicoya Complex basalts, at the edges of Culebra Bay (see Fig. 7.5). This is also the place where the Tempisque River leaves the plateau by a narrow gorge, starting its tour in the tectonic depression that bears its name and is known commonly as "bajura," inshore. Erosion effects are increased by the strong dry season (January to June) on these first two volcanoes that share the same structure. However, vegetation remains dense on their flanks (altitude tropical dry forest), which contrasts markedly with the scattered vegetation of the ignimbrite plateau. The plateau plays the role of sponge

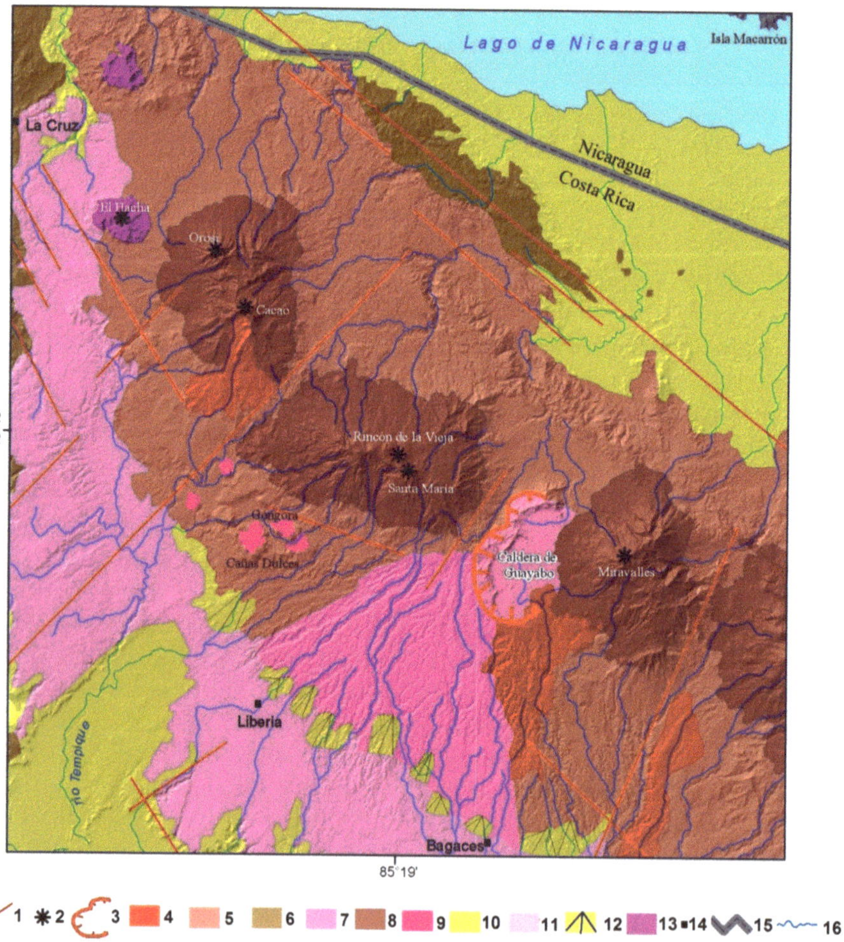

FIGURE 7.5 Morphology of the Guanacaste Volcanic Mountain Range from Orosí to Miravalles volcanoes by Bergoeing (2012a). (1) Tectonic faults. (2) Craters and volcanic cones. (3) Collapsed calderas. (4) Lahar fields. (5) Quaternary altered volcanic flows. (6) Tertiary sedimentary hills (Paleocene–Eocene). (7) Ignimbritic cone/glacis of the Guayabo collapsed caldera. (8) Middle Pleistocene cones and volcanic structures. (9) Cañas Dulces dacitic domes. (10) Holocene sediments. (11) Ignimbritic plateau. (12) Alluvial fans. (13) Pliocene volcanoes. (14) Urban areas. (15) International border. (16) Hydric network.

and soon absorbs the water from rainfall, allowing only a few inclined trees to grow in the ignimbrite piedmont irrigated by the humid northeast trade winds. These winds often take the aspect of hot, very strong winds, once they have gone through the volcanoes' pass and peaks, and therefore dry up even more the Pacific piedmont and ignimbrite plateau (foehn effect). We found some tall grass that serves as pasture for cattle. More dry vegetation of the plateau is due to the humid tropical climate even with the dry season, but it is also the result of the nature of the soil (porous ignimbrite) (Fig. 7.5). The Orosí and Cacao volcanoes have had no known activity (emission of lava flows, eruptions) in historical times.

4. RINCON DE LA VIEJA VOLCANO

Rincon de la Vieja is the name of the volcanic set 6217 ft above sea level, as well as of one of the craters of this structure. It is perhaps the most active and important volcanic edifice of the sector. It houses nine craters aligned and well preserved; among the most important are Rincon de la Vieja (5925 ft), Santa Maria (5599 ft), and Von Seebach (6217 ft). It is a young stratovolcano, whose structure was built in the past 500 years. The active crater, about 984 ft in diameter, has a pluvial water pond with intense fumarole activity. It is a very active volcano with 12 large eruptions occurring since 1750 and characterized by gas emissions, large clouds of ash and pyroclastic rocks, and hot laharic flows. The crater lake waters are evacuated through the Azul and Penjamo rivers (Boudon et al., 1995).

The Rincon de la Vieja crater presents activity manifested by small hot water and gas vapor eruptions of magmatic origin, as well as, during the recent eruptions, ash and slag. Santa Maria is an inactive crater, occupied by a small pluvial lake. However, in the last century, the inhabitants of the sector agree on seeing vapors and smoke out of this crater, unless they have confused it with that of Rincon de la Vieja. The other craters have not manifested activity and are partly colonized by vegetation. Other craters have collapsed because of rain erosion (see Figs. 7.6 and 7.7).

On the Pacific slopes of this volcanic edifice emerges a series of dacitic domes and among them the most important are the following: Cerro Fortuna (1572 ft), Cerro San Roque (1788 ft). Cañas Dulces Hill (2149 ft). Cerro Gorgona (2520 ft), Las Tablas (2005 ft), and Cerro San Vicente (1988 ft). These dacitic domes are dated and belong to the Lower Quaternary in the range of around 1 MY. At the foot of San Vicente Hill, following a southeast fissure alignment that some consider the rim of an old collapsed caldera (Bohnberger et al., 1966), in Hacienda Guachipelin, there are small mud craters or hot springs, "mofetales." It is an area of small circular depressions from 3 to 33 ft in diameter, where there is a constant thermal bubbling. This thermal activity is directly related to the NE–SW fault alignment at the foot of the volcano. "The ignimbrites have compositions often close of domes lavas and seem to correspond to the establishment of degassed magma in which early quartz crystals process have been completely dissolved at low pressure" (Tournon, 1984) (see Figs. 7.8 and 7.9).

The southwest flank of the Pacific slope, from 2300 ft above sea level, is covered by ignimbrite of the Liberia Formation. The origin of this ignimbritic spill is without doubt volcanoes in the formation of the Rincon de la Vieja sector.

The ignimbrite of the Liberia Formation has been dated, giving an age that varies from 1.8 to 1.2 MY by different authors. Thus Gillot et al. (1994) measured by K/Ar a rhyolitic flow deposit, rich in biotite, from Hacienda Ciruelas Way, which gave an age of 1.83 ± 0.03 MY. Gardner and Turin (in Alvarado and Gans, 2012) measured by $^{39}Ar/^{40}Ar$ a deposit of pumice flow, also rich in biotite, taken up from the Bagaces–Guayabo road, which gave an age of

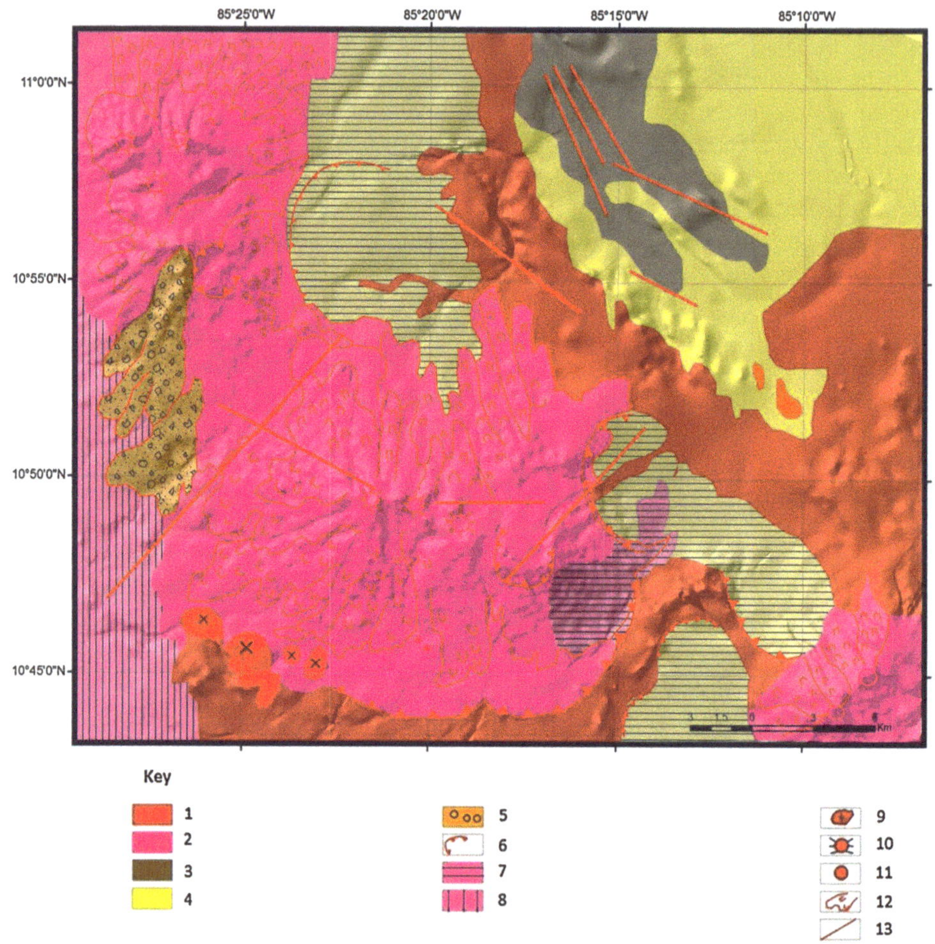

FIGURE 7.6 Geomorphology of Rincon de la Vieja Volcano, southeast sector, and Pliocene dacitic cones (Bergoeing, 2007). (1) Pliocene–Pleistocene volcanism. (2) Pleistocene–Holocene volcanism. (3) Tertiary sedimentary outcrops. (4) Quaternary sedimentary fillings. (5) Lahars. (6) Calderas. (7) Quaternary oriental structural plateau. (8) Pliocene–Pleistocene ignimbrite plateau. (9) Dacitic domes. (10) Craters. (11) Thermal springs. (12) Lava flows. (13) Tectonic faults. *From Bergoeing, J.P., 2007. Geomorfologia de Costa Rica. Libreria Francesa, San José, Costa Rica.*

1.45 MY. The collapsed Guayabo caldera is a flat depression located on the southwestern edge of the Rincon de la Vieja volcanic edifice forming a vast piedmont in the form of an erosion cone/glacis, with numerous cracks or flaws, visible in vertical aerial photographs and satellite images. Mainieri dated this event in 1976 (Alvarado op. cit. 2012) by the K/Ar method in a pyroxene andesite from Cerro Mogote, obtaining an age of 2.3 ± 1 MY. Given that it is a rigid but not very resistant material, it is easy to understand that it is route fault, in a region of active tectonics. The ignimbrite is white, which contrasts with the gray color of the Bagaces Formation that covers the entire plateau.

There are major rivers that are born in the Pacific watershed of Rincon de la Vieja Volcano. We will quote, for example, the Blanco and

FIGURE 7.7 Rincon de la Vieja volcanic complex summit craters. *Photo courtesy of Oscar Barrientos (2012).*

FIGURE 7.8 Las Pailas sector, hot springs of mud ("mofetales") at the foot of Rincon de la Vieja Volcano. *Photo courtesy of Oscar Barrientos (2010).*

Colorado rivers. Most of these rivers cut deeply into the ignimbrite piedmont forming small canyons, which are arranged according to a subparallel network. It is also in this sector where the ignimbrite plateau is more extensive, making contact finally with the coast by a high cliff over 640 ft that dominates the Papagayo Gulf, leaving only two sectors of beaches: north of Playa Naranjo Beach, in contact with the Santa Elena Peninsula (on this beach turtles come to spawn annually), and to the south, the small Cabuyal Beach in contact with Culebra Peninsula.

On this plateau, the Tempisque River has greatly expanded its bed, allowing to issue the hypothesis that at a given moment, this river built a natural water dam, creating lake conditions. This extension goes between Hacienda Guapote and Irigaray hamlet. This depression must be related to an ancient natural dam, which allowed the construction of a special shape with lake erosion levels in its banks. Right after Irigaray, the river sinks into

FIGURE 7.9 San Roque dacitic dome cut in its southern face by pluvial and fluvial erosion. A hornblendic dacite was dated by Bellon and Tournon (1978) by the K/Ar method, giving an age of 1.6 MY. *Photo courtesy of Oscar Barrientos (2012).*

a narrow and deep canyon for more than 320 ft, crossing and cutting a NE–SW direction tectonic alignment before reaching the alluvial plain. This is an important tectonic alignment contact between the plateau and the plain, observed by SEASAT satellite radar image.

5. MIRAVALLES VOLCANO AND GUAYABO COLLAPSED CALDERA

Although considered active, the Miravalles volcanic edifice does not currently present external manifestations. Its crater rises 6654 ft above sea level and is attached to the northwest side, in contrast to the others. Two important andesitic lava flows are on its northwest flank, but they are not dated (see Fig. 7.10). The first one descends to the west, going toward the sector of Hornillas, with which the ICE is developing an ambitious geothermal project. The second flow goes to the north, near the site of Las Tablas, and descends into the Guayabo collapsed caldera.

Miravalles Volcano has parasitic cones that are located near the crater. On the watershed, they emerge to respective altitudes of 3280, 5059, and 5098 ft above sea level. Miravalles has a twin cones known as Cerro la Giganta (4888 ft). In 2012, Alvarado (op. cit.) dated by $^{39}Ar/^{40}Ar$ an andesite of Cerro La Giganta and gave an age of 0.416 ± 0.008 MY. On the south façade of the volcano a lineup of three small cones is visible. They include the Espiritu Santo volcanic cone of 3212 ft altitude and Gota de Agua, 3550 ft above sea level, which should be related to the Guayabo caldera, given its peripheral position to the west. The Espiritu Santo cone has a basaltic lava flow dated by the ICE at 1.07 ± 0.09 MY. The southwest flank and especially the northern slope are covered by lahars and torrential lavas that form a regular piedmont and progressively descend over the plain. The hydrographic network is consequently radial. However, the southeast sector is deeply marked by a tectonic fault whose orientation is NE–SW, similar to Cacao Volcano and which separates Miravalles from the Tenorio Volcano edifices (Elizondo and Bergoeing, 1980). This tectonic fault is fitted by two rivers, which have dug their respective valleys following the crushing area. The current aspects are developing a colluvial coating. To the northeast the Zapote River runs, following a straight path, before joining the other rivers of the Great Northern Plain. Up

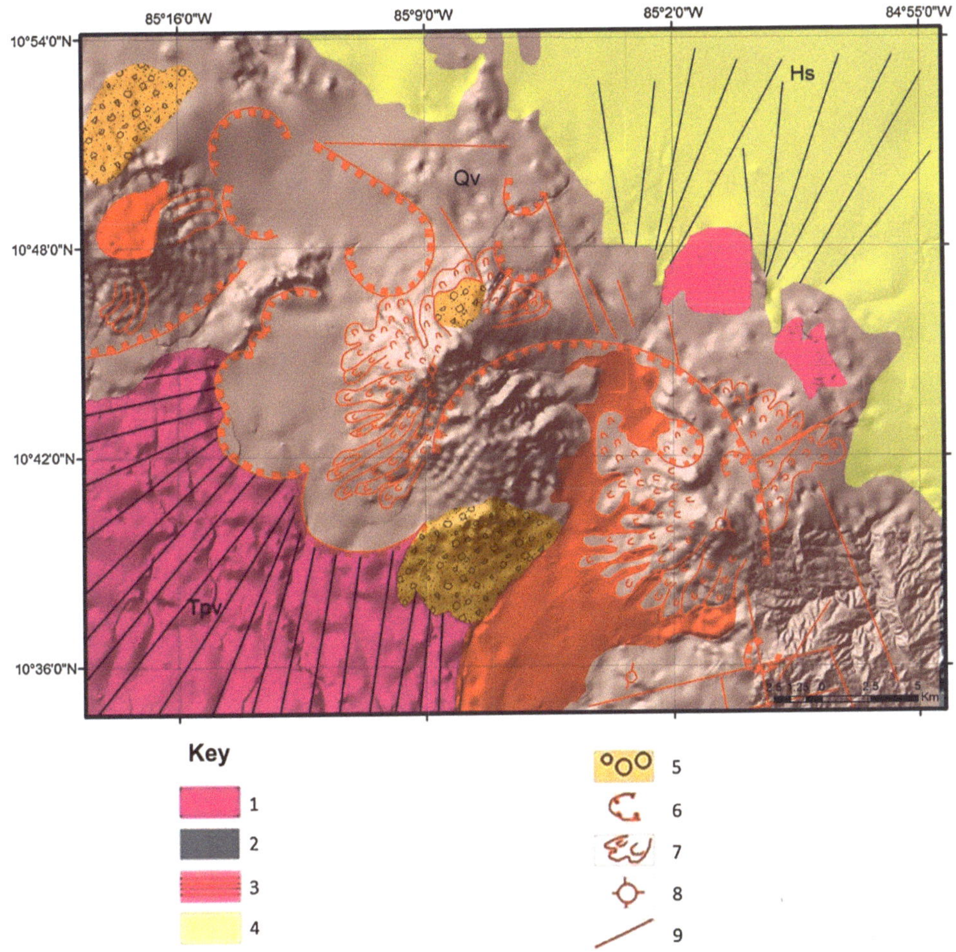

FIGURE 7.10 Geomorphology of Guayabo collapsed caldera and Miravalles Volcano (Bergoeing, 2012). (1) Guayabo's ignimbrite cone. (2) Volcanic shapes. (3) Planezes. (4) Glacis. (5) Lahars. (6) Collapsed caldera rim. (7) Lava flows. (8) Craters. (9) Tectonic faults.

this river, the hamlet of Bijagua is the only inhabited place in the sector (see Fig. 7.11).

The southwestern sector, on the other hand, is the result of the Tenorio River's hydrological work, which finds its origin near the Montezuma Volcano crater. This valley has an important village known as Naranjo River village, which serves as a meeting point for the inhabitants of scattered farms dedicated to the breeding of cattle.

Miravalles Volcano has been the subject of scientific research by the ICE with a view to the exploitation of geothermal energy, which manifests in the sector of Las Hornillas (northeast flank of the volcano). Here Carr et al. (2007) in Alvarado (2000) dated, by $^{40}Ar/^{39}Ar$, a basaltic–andesitic sample giving an age of 0.416 ± 0.008 MY (Alvarado, op. cit.). Using the technical work of Ricardo Granados (1979) to sketch a classification of the volcanic forms of this sector, the Miravalles has a total volume of 30 cubic miles, whereas the Rincón de la Vieja is slightly

FIGURE 7.11 (A) Miravalles Volcano, northwest flank, where there are ancient overlapping lava flows. (B) Crater containing a pluvial lake. (C) Down Las Hornillas sulfur springs. (D) Ignimbrite plateau. *Photos courtesy of Gilbert Vargas (2009).*

larger, with 31.7 cubic miles. These volumes show that these two volcanic masses are the most important in the volcanic Guanacaste Mountain Range. According to Tournon (op. cit.) the age of the Miravalles lava is around 0.9 MY, whereas the Liberia ignimbrite coming from the Guayabo collapsed caldera is between 0.60 and 0.14 MY.

The current Miravalles structure rests on a more ancient volcanic core, probably of Lower Pleistocene age. Ancient lavas contrast with recent lavas, because they are andesites with dark gray augite. The lahars have thicknesses ranging from 32 to 300 ft and are associated, in certain places, with piedmont deposits and with lake temporal events. It is this kind of material that filled Guayabo Volcano after its collapse, becoming a caldera (see Fig. 7.12).

6. GUAYABO CALDERA

This is a vast circular depression, 1.8 × 3 miles in diameter. Its western edge is formed by a rim escarpment of a few hundred feet known as La Montañosa Hill, of which Granados (op. cit.) describes the presence of two different lavas series.

Cerro Mogote lavas: These are less altered, formed by andesite with pyroxenes and tuff. The andesite's absolute age is approximately 2.3 MY (Mainieri and Protti, 1976).

La Montañosa lavas: These are very disturbed and show an advanced evolution to lateritic soils. They are also andesite with pyroxenes, but are probably of older age given the degree of alteration of the rock (see Fig. 7.13). After the collapse of this caldera a postcaldera activity, intensely

FIGURE 7.12 Miravalles Volcano and La Giganta Volcano planezes. In the center is one of the geothermal plants and in the foreground are Las Pailas hot springs. *Photo courtesy of Gilbert Vargas (2009).*

FIGURE 7.13 Guayabo collapsed caldera (Upper Pliocene–Lower Quaternary), Cerros la Montañosa escarpments, and old cone remains. In the background are the slopes of Rincon de la Vieja Volcano. *Photo courtesy of Jeffrey Ruiz Valverde (2012).*

acidic, hydrothermal, and gaseous, developed, giving the cone of vast white deposits of ignimbrite poured toward the west, and known under the name of the Liberia Formation. This is what we can conclude from radiometric dates of the sector. Here is a chronological summary of the main events in this sector:

Guayabo volcano formation and perhaps construction of the old cone of the Miravalles (andesite of Cerro Mogote: 2.3 MY)

Collapse of the Guayabo volcano, with formation of the caldera by vacuum of the magma chamber

Strong acid caldera activity and ignimbrite spill on old ignimbrite deposits of the Bagaces Formation (Liberia ignimbrite: 0.6 to 0.14 MY) and, at the same time, construction of the current Miravalles volcanic cone as a stratovolcano with ash, lava deposits, lapilli, andesitic lava, ignimbrite flows, and later alluvial deposits on the slopes

Filling of Lake Guayabo, leaving as the only witness the current expanse of water known as Laguna Mogote, filled mainly by waters from the northwest flanks of Miravalles; current erosive effects on the inherited volcanic forms, especially soil creeping, and generalized mass solifluction effects in the sectors where the clay layer is more evolved

7. TENORIO VOLCANO

Tenorio Volcano is the last volcanic edifice of the volcanic Guanacaste Mountain Range and is located southeast of Miravalles Volcano, separated by a tectonic fault alignment already described. This enormous structure is composed of two distinct cones: northwest is the Montezuma crater (4987 ft), which detaches from the Tenorio crater (6286 ft) that lies to the southeast. These two craters are open mouthed, the first to the southeast and the second toward the southwest. Tenorio Volcano forms a very regular cone that gradually descends toward the ignimbrite plateau (Gutierrez, 1962). It is a volcano that has had recent Holocene activity. Tenorio lavas were dated by Carr et al. (2007; in Alvarado, op. cit.) and sit atop a margin ranging from 0.740 to 0.090 MY (first lava flow at 14 miles from Tierras Morenas and second in Hacienda Tierras Morenas) (see Figs. 7.14 and 7.15).

The Bagaces ignimbrite covers Tenorio Volcano's Pacific piedmont. This piedmont is deeply notched by many rivers that, after having fitted the ignimbrite plateau, extend on the plain before joining the Tempisque River. The most notable of these rivers is the Tenorio River, given its important volume. The southeast flank of the Tenorio cone is interesting, because it houses a number of small parasitic cones, today inactive and forming rounded shapes. This sector lies between the villages of Palmira and Tierras Morenas.

From this last point and following the watershed dividing line (Fila Dormida row), we reach a vast circular depression, which is Lake Cote. This lake is the site of a former volcanic focus of the Pliocene, which by collapse or explosion left this caldera as a vestige. Farther south, a second elongated tectonic depression appears, which is Lake Arenal.

This last Pliocene volcanic depressed region marks the southern limit of the volcanic Guanacaste Mountain Range. Red clay soils are deep

FIGURE 7.14 Tenorio Volcano cone with a destroyed crater summit. In the foreground is Lake Camelias. *Photo courtesy of Francisco Solano (2010).*

FIGURE 7.15 Celeste River waterfall on the northeastern slopes of Tenorio–Montezuma volcanoes. The waters take on a light blue color due to sulfur springs. *Photo, Bergoeing (2008).*

and occupy most of the Cote–Arenal sector. The Arenal and Chato volcanoes are still farther to the south of the Guanacaste volcanic range alignment and are not included in this volcanic range, because they are more recent (Holocene), peripheral to this and marking a new phase in the volcanic activity, which will be studied separately (see Fig. 7.16).

8. LAKE ARENAL SECTOR

The Lake Arenal sector is a region of low hills whose average altitude is about 2625 ft. It is a transition area between the volcanic Guanacaste Mountain Range and the Tilaran Mountain Range. Here the northeast trade winds easily penetrate. As a result, an exuberant tropical vegetation has spread, and here is a place where humans still have not dismantled the forest for pastoral lands or agricultural purposes, as is the general case in the country (see Fig. 7.17).

High humidity in the sector has caused these rocks to be very disturbed, leaving deep altered red clay soils. They are partly lateritic and sometimes lixiviated. The profound alterations can reach 30–130 ft deep and, because of this, it is sometimes difficult to identify the bedrock.

On these Pliocene volcanic soils, ancient lava flows are still visible. They consist primarily of andesite and are present as an alignment of slabs resting on a clay soil. Indeed, on the surface is a line of altered slabs where the nature of the andesitic rock is recognizable, whereas in depth, the rocks have disappeared completely, evolving into decomposed clays. The surface slabs still have the sense of the lava flow. This phenomenon supports the hypothesis that the surface rocks have been better preserved than the deep rocks, the latter covered by clay supersaturated with water.

This sector is a Tertiary volcanic extension (Aguacate Formation), very affected by plate tectonics, and Lake Arenal is located in a

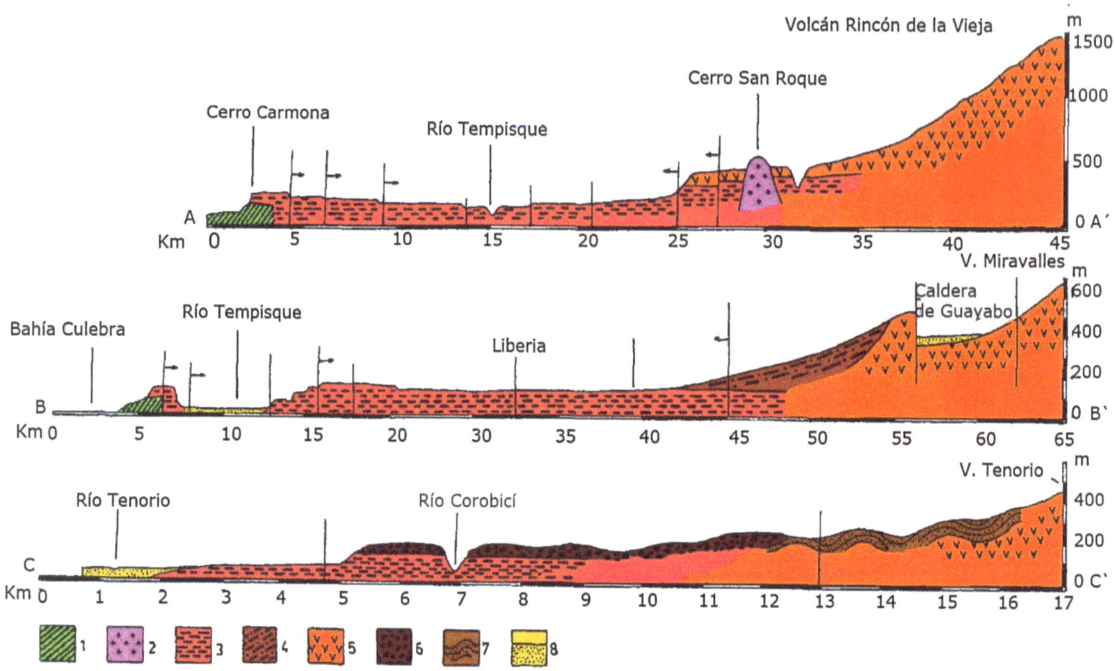

FIGURE 7.16 Three geological profiles through the Liberia ignimbritic plateau and Rincon de la Vieja, Miravalles, and Tenorio volcanic cones, Pacific watershed. (1) Cretaceous–Eocene volcano-sedimentary series. (2) Dacitic domes. (3) La Cruz–Santa Rosa–Liberia ignimbritic plateau. (4) Guayabo caldera cone/glacis ignimbrite. (5) Andesitic Pleistocene lava flows and projections. (6) Lahars. (7) Altered cinders. (8) Recent alluvium. *From Bergoeing (2007).*

graben, whose direction is roughly NW–SE. This area was a vast swamp that had had a previous lake due to natural damming of the Arenal Volcano lavas. The drainage of this depressed sector goes toward the Northern Plain. Today Lake Arenal has been converted into a dam, blocking the waters of the river gorge at the foot of Arenal Volcano (see Fig. 7.18).

The lake created today allows the irrigation of an important sector of the Tempisque River plain, especially during the dry season (Moracia irrigation area), and in a second phase, already completed, it provides a supplementary source of electricity to the country (Corobici project). These two projects were completed in 1982. However, the Arenal Dam, built on a fault alignment and resting on the southern flank of Arenal Volcano, is subject to the whims of nature. The drainage of this depressed sector, going toward the Northern Plain, today has been converted into a dam, blocking the river gorge at the foot of the Arenal cone.

> Under Arenal there are numerous seismic focus [sic] less than 40 km deep, most of the sources are even less than 10 Km. On the other hand, deeper focus are distributed in a subvertical Benioff zone. *Tournon (1984).*

Pliocene volcanic activity was unimportant, compared with that which gave rise to the current volumes of the Tilaran Mountain Range, farther south, with which it is closely linked. The Quaternary was for the sector a period of

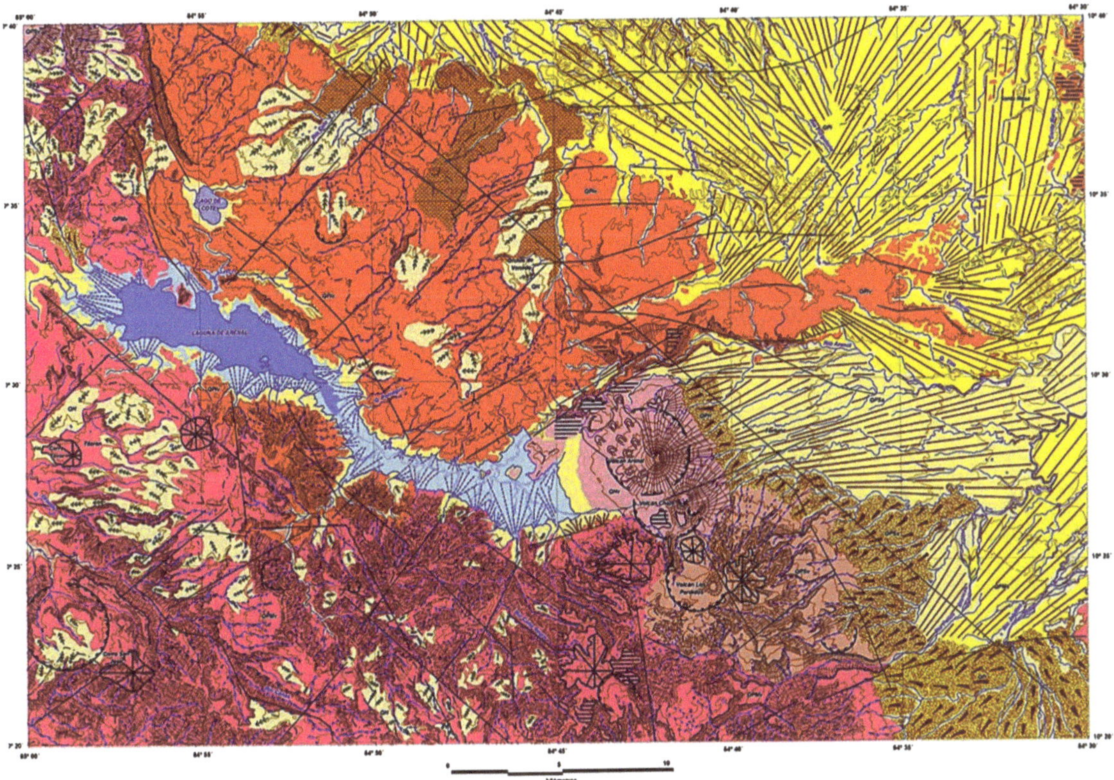

FIGURE 7.17 Arenal tectonic depression: Holocene Arenal Chato and Perdidos volcanoes. In the north is a Pleistocene volcanism resting on the Tertiary sedimentary series of the Venado Formation and lahars evolving into alluvial fans. In the south is the Tilaran Range Tertiary volcanism. *From Costa Rica Geomorphologic Caribbean Atlas, Bergoeing et al. (2010c).*

relative calm, because outside of the Espiritu Santo small cone (Mora, 1977) and Arenal–Chato, other volcanic foci of this period are not known. The Holocene is marked by the appearance and building of the Arenal–Chato volcano. This is a small isolated volcanic massif, which consists of two modern cones. Chato Volcano is 48,000 years old, according to an andesite ($^{40}Ar/^{39}Ar$) dating by Alvarado and Gans (2012); also, the Quebrada Flores River andesite of the Perdidos Volcano massif was dated at 84,000 years, and another of the Cerro Los Perdidos at 99,000 years (Gillot et al., 1994), which places this area in the Pleistocene–Holocene. Northwest, Arenal Volcano is a perfect, active cone, which emits lava. During the last eruption in 1968, Arenal's cone was broken by a side fissure eruption, leaving the andesitic lava to flow from the side crater that opened in the middle of the cone, while ash and water vapor went up through the chimney to come to the surface by the upper crater. Lava flows were directed toward the northwest in the direction of Arenal Dam. According to Borgia et al. (1988), Chato Volcano had its last activity 35,000 years ago, and about 3000 years ago began the construction of the Arenal Volcano cone.

Arenal eruption of 1968, accompanied by the emission of pyroclastic burning clouds (nuées ardentes), claimed 500 victims, despite being a rural area (Fig. 7.19).

FIGURE 7.18 Arenal Volcano in activity. This is a young perfect cone built by volcanic pyroclastic flows. *Photo Bergoeing et al. (2010d).*

From Tournon (op. cit.): "Arenal that small isolated massif which was built during the last thousands of years (MELSON and SAENZ, 1974). It's the only Costa Rica volcano, which has issued flows since the Spanish conquest (since 1968 emission of andesite lava flows)."

Chato Volcano, southeast of Arenal, is a truncated cone. It is an explosion caldera with a pluvial lake origin. Dense tropical vegetation covers its slopes, as well as the Arenal Caribbean slopes off of the top side, unaffected by the latest eruption. The sector of Venado, situated northeast of the Arenal–Chato volcanic complex, is the only area where there are Miocene limestone outcrops, known as the Venado Formation (see Fig. 7.20).

These limestones form a karstic shape, at 1000 ft, of underground galleries at Finca La Luisa. On the limestone surface type, it is possible to perceive in some places sinkholes, or "dolines." Some, in contrast, sink completely, leaving a hole opened on the site (collapsed cavity). These Miocene limestones are also testimony to the general horst position of this entire sector (despite the Arenal graben). They have suffered the effects of the Plio–Quaternary orogeny, which has raised them with the Pliocene volcanic set that is around the sedimentary outcrop.

In conclusion, we can say that the volcanic Guanacaste Mountain Range is primarily the result of Quaternary volcanic activity, which is inscribed on the margins of one major accident that corresponds to the great volcanic–tectonic alignment of southern Central America. It begins in northern El Salvador, continues through the Nicaraguan Pacific coast, and ends in Costa Rica in the Central Volcanic Mountain Range. This great tectonic accident is the product of the clash of the Cocos and Caribbean tectonic plates. It is through the subduction zone that an important magmatic rise has produced, on one hand, through fissures, ignimbrite spills, much younger, that have built the Liberia plateau, and on the other hand, extrusions of basalt and andesite, much older, that have formed the basis of the volcanic Guanacaste Mountain Range and later allowed the construction of the current stratovolcanoes. This testifies to the vigor and youth of volcanism and shows strong neotectonics very active in our days. In conclusion, the volcanic Guanacaste Mountain Range has the following characteristics.

They are well-individualized volcanic edifices, separated by passages wide enough that the northeast trade winds filter through. These passages are tectonic–volcanic faults whose orientation is NE–SW. These modern volcanic edifices are stratovolcanoes built between the Middle Quaternary and the Upper Quaternary, from an older volcanic base (Plio–Quaternary?). The set of these structures is in horst position. The mountain range is bordered by two major

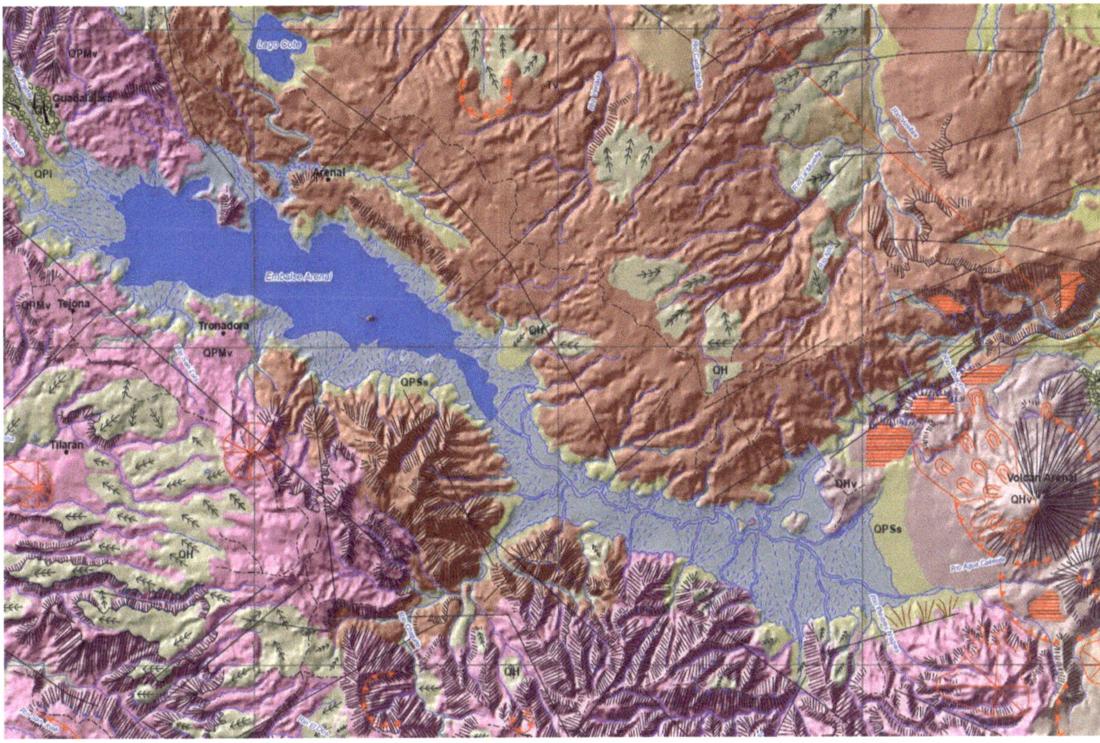

FIGURE 7.19 Lake Arenal's former alluvial fans before being dammed by Bergoeing (2016). Geomorphology Atlas of Costa Rica. UNA.

FIGURE 7.20 Chato Volcano's truncated cone (explosion caldera). The crater hosts a pluvial lake. *Photo Bergoeing, J.P., 2008. Interpretación Geomorfológica del volcán Barú- Panamá. Revista Geográfica IPGH N° 143, México.*

regional tectonic alignments: to the northeast the Nicaragua graben south rim and to the southwest the Reverse Tectonic Fault of Costa Rica. The northeast piedmont of this mountain range is mainly formed by lava, with rare deposits of lahars, whereas the southwest piedmont forms a uniform topographic plateau, consisting of Bagaces Formation ignimbrite. The volcanic activity that manifested during the Quaternary period on an ongoing basis was rather acid (according to spill cast types). Climate shapes that have been formed over these volcanic piedmonts are largely of the erosive type (V valleys, gorges, canyons, and fluvial erosion terraces in the ignimbrite) and partly sedimentary (colluvial cones, filling of lakes and puddles, and formation of alluvial fans and glacis in contact with the plains).

CHAPTER

8

Geomorphology of Guanacaste

OUTLINE

1. NICOYA PENINSULA

The Nicoya Peninsula is a Pacific Ocean rectangular land area, oriented NW–SE, 81 miles long and 25 miles wide, whose limits are Papagayo Gulf to the northwest and Nicoya Gulf to the east. It is characterized by heavy mountainous relief areas, whose average elevation is 2000 ft, and whose main summit does not exceed 3300 ft in elevation. A hilly relief imposes everywhere, which leaves little place for vast coastal plains development. These hills determine the main features of the coastal relief in the filling or flooding of valleys and notching of the relief in the cliffs (see Fig. 8.1).

The Nicoya Gulf coastline in contrast presents vast plains where lie extensive mangroves, sheltered from the swell of the Pacific (Battistini and Bergoeing, 1982a,b). From a lithologic point of view, the Nicoya Complex flourishes everywhere, even on the coast (Henningsen, 1964; Tournon, 1971; Kuyjpers, 1980). It is presented in the form of columnar basalts as well pillow lavas. There are many basalts dating from 139 million (Hoernle et al., 2008) to 72 million years (Barr and Escalante, 1969), which places them as Cretaceous volcanic island outcrops, mainly constituted at their base by sandstone, limestone, and greywacke represented by the Rivas and Sabana Grande formations, partly covering the Cretaceous complex. The Rivas and Sabana Grande formations are found mainly on the Nicoya Peninsula's central west coast in Punta Guiones, Samara, and Garza beaches, as well as in the southeastern coast in Bahia Ballena Bay and in Cabo Blanco

Geomorphology and Volcanology of Costa Rica
http://dx.doi.org/10.1016/B978-0-12-812067-5.00008-3

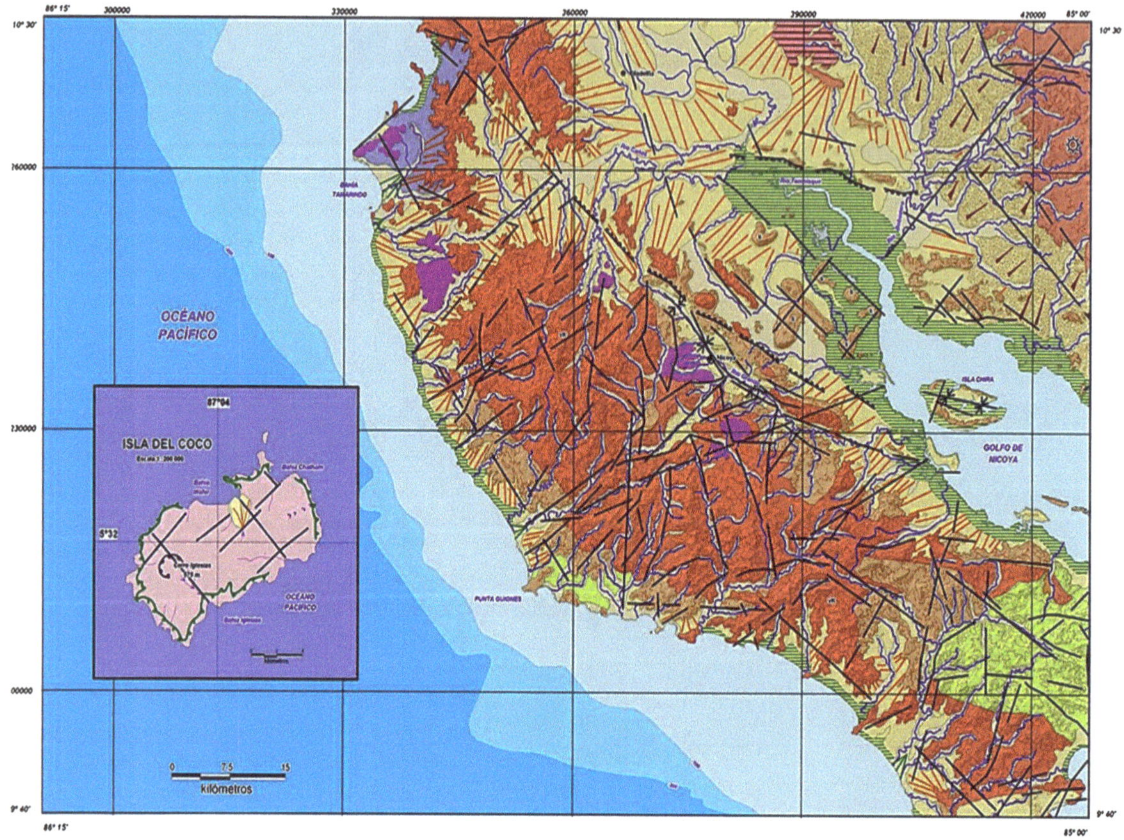

FIGURE 8.1 Nicoya Peninsula geomorphology constituted by Cretaceous and Tertiary volcanic and sedimentary outcrops of the Nicoya Complex. To the north lies the Tempisque River graben, going through the Nicoya Gulf. *From Costa Rica Gemorphology Atlas, 1:350.000. Bergoeing et al., Instituto Costarricense de Electricidad ICE (2010).*

Point in the extreme southwest of Nicoya Gulf. Here cliffs are composed of hard limestone and sandstone that lies directly in the rocks of the Nicoya Complex (Punta Carballo and Montezuma formations). General tectonic movements determine the general configuration of the Nicoya Peninsula. Of the large regional fault alignments, there are many others whose main directions are northwest–southeast, east–west, northeast–southwest, and north–south. They play a morphological role of great importance because the hydrographic network register paths follow these directions (Dengo, 1962b).

Because of this, the erosion has also allowed in some cases the isolation or submersion of some reliefs, as well as the filling of the coastline. The interior of the peninsula is characterized by hills and mountains, consisting of important layers of clay alteration associated with bedrock outliers, which give a wavy look to the landscape in some places. There, where limestone emerges, are escarpments with cornices dominating the landscape, for example, in the Barra Honda sector. The bottom valleys in general are wide and the rivers run in flat thalwegs. Contact between hills or mountains and the plain is formed by powerful alluvial fans evolving in glacis, witnesses to the upper Quaternary climatic crisis (Santa Cruz sector). The large valleys that humans have managed to use as way of penetration and communication are largely related to tectonics (Nicoya synclinal). Rain and Quaternary interpluvial periods have enabled the fit and then the filling of these valleys.

1.1 Nicoya Peninsula Southern Extreme

This is a triangle where Cabo Blanco Cape constitutes the southern point and San Lucas Island the western extremity. This part of the peninsula is characterized by a landscape of impressive relief where Cretaceous rocks of the Nicoya Complex cohabit with the major Tertiary sedimentary outcrops of the Montezuma and Las Palmas formations. This sector is controlled by numerous NW–SE parallel alignments of tectonic origin. It is a very folded region to the northeast from Bahia Ballena Bay sector to San Lucas Island, whereas southeast the Cobano sector presents a tabular structure formed of limestone and sandstone of the Montezuma Formation (Miocene–Pliocene) (see Fig. 8.2).

The Cobano sector is very interesting because we discovered there several erosion levels with no correlative deposits, which are located at the following altitudes: 722, 525 (which is the Cobano level), and 260–330 ft, which corresponds to the top of the Montezuma cliff. This plateau gives testimony of a strong tectonic uplift, uninterrupted since the Pliocene, and which had been active from the Flandrian transgression. Here exists a Flandrian level of 20 ft on

FIGURE 8.2 Nicoya Gulf: Guayabo Island in front of Paquera between Cedros and San Lucas islands. Guayabo Island, photographed on all four sides, shows an anticline structure on a Tertiary sedimentary series (Eocene) with a fault sliding scarp and overhanging block. *Photo Bergoeing (2010).*

Montezuma Beach that is witness to a 3-ft rising value by the millennium (Battistini and Bergoeing, 1983a,b) (see Fig. 8.3).

The landscape of this region is rich and varied in forms. The southeast Cobano sector is characterized by a plateau, which ends in inactive cliffs following tectonic alignments. The Tambor–San Lucas northwest sector presents a series of parallel valleys filled during the Flandrian that lead to the coastal sector bay, internal funds (see Fig. 8.4). Numerous sedimentary outcrops are featured in cornices in the surrounding hills. Because of a prolonged dry season, and a deciduous vegetation and forest cover, they are sometimes very degraded. At the foot of the slopes are observed small debris cones or alluvial fans composed of angular rocks at the mouths of small creeks active only during the rainy season.

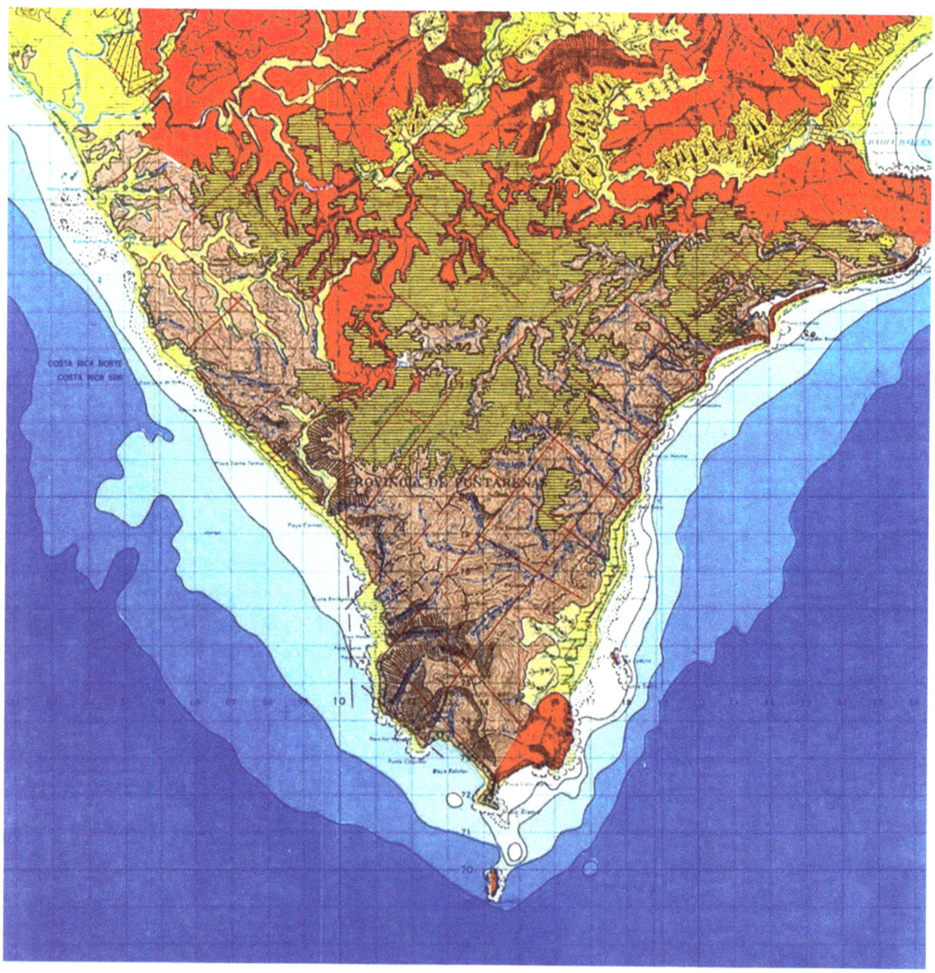

FIGURE 8.3 Cobano structural plateau in the Tertiary sedimentary series. North sector, Cretaceous volcanic outcrops. East sector, Montezuma cliffs. The sector is subject to a widespread faulting and tectonic lifting process. It is in this sector where a 20-ft lumachelle terrace from the current level of the sea was dated and gave a Flandrian age of 7500 years. *Taken from the geomorphologic map of Costa Rica, North Pacific. From Bergoeing et al. (1982b).*

FIGURE 8.4 Montezuma Beach and Cobano cliffs area, where neotectonics lifted up a Flandrian marine lumachelle terrace. *Photo Bergoeing (2010).*

Here we do not see important fluvial terraces. However, there are cyclic erosion levels (Tambor, Paquera, and San Lucas) that we will explain in the chapter on the coastline. These levels of erosion are located at around 26–33, 82, and 130–160 ft.

On the coast, between Cabo Blanco Cape and Bahía Ballena Bay, powerful cliffs (up to 330 ft) were raised, as in Montezuma. They are composed of limestone with fossils and sandstone, everything resting on the rocks of the Nicoya Complex base, which is in contact with the sea. These cliffs are closely related to active tectonics. The network's main faults focus here, SW–NE, and they are arranged in parallel to the cliffs. The first faulting network is clipped by a second nearly perpendicular system, which gives a squared tectonic graph, partly responsible for the rectilinear form of this part of the coast between Blanco Cape and Coyote Point. Between these, the sedimentary formations disappear progressively to give way to the Nicoya Complex rocks. These rocks, less resistant to erosion, have allowed the development of a coastal plain bordered by coastal Flandrian cords, hosting some mangroves (*Avicennia, Rhizophora mangle*), at the mouths of river streams active only during the rainy season.

Between Ballena Bay and San Lucas Island, this southern part of the Nicoya Peninsula, which follows Cobano plateau, is characterized by a vigorous relief (1300 ft) formed from sedimentary, folded and faulted, spurs that are arranged along a WNW–ESE axis. In fact it is a series of sedimentary parallel folds (sandstone and shale), which form synclines and anticlines and are cut by tectonic faults that remain in the folded axis. This relief has enabled the construction of a rugged coastal landscape, formed of semiclosed bays and active cliffs following tectonic paths. Among the reliefs forming the coastal plain points, sometimes narrow and deep, are the Curu, San Rafael, Paquera, and Gigante rivers, these being fed after the Flandrian transgression.

In some islets, like the Tortugas, Negritos, and Cedros islands, Nicoya Complex rocks predominate. The seabed also testifies to the vigor of the folds of the coastal hills. This rugged seabed demonstrates the continuity of inside land folded forms. This is at least what one can conclude from observations of the aerial photography of the Costa Rica Geographic Institute (IGN), Costa Rica.

1.2 Nicoya Peninsula Pacific Sector

The Nicoya Peninsula Pacific sector consists of a series of small compact hills formed in Cretaceous igneous rocks (Kuyjpers, 1980) with much altered clay, fitted by very active erosion mainly for two reasons:

- Uncontrolled deforestation at the beginning of the 20th century and continuing deforestation by the agricultural method of burning during the dry season, which is responsible for the almost total disappearance of the tropical forest.
- A dry season extending from December to May–June, followed by a period of intense rainfall that reaches its peak in September–October and which washes the clay material so that it is prepared for erosion. The soils then crack during the drought season.

This part of the Nicoya Pacific coast has the following characteristics:

Between Samara Beach and Punta Guiones, Tertiary sandstone outcrops form the coastline. They form small horseshoe bays with a sandstone coastal foreshore with well-developed front ends. The valley courses, in the interior of the peninsula, are commanded by tectonics. These valleys empty into the coastal plain, often forming flattened deltaic fans. The tectonic guidelines often focus here following a preferential SW–NE path (see Fig. 8.5). The Quiriman River has eroded more deeply the compact mass of the Nicoya Complex. Very active during the rainy season, the river formed a broad valley framed by two fluvial terrace levels, by colluvial cones and solifluction spills. In general, solifluction packages are scattered in the clay hills where the slopes are very steep, leaving scars of landslides upstream.

The Quiriman River is one of the major rivers of this sector of the Pacific Nicoya. It runs through important regional tectonic fault alignments before going into the Pacific Ocean. The base level of the Samara sector, despite being a tectonically subsiding local area, suffered a 1.6-ft rising due to the September 5, 2012, earthquake, which reached 7.6 magnitude on the Richter scale, due to tectonic plate subduction movements, which continue to cause a general uprising of the peninsula.

1.3 Nicoya Gulf East Coast

This coast is less monotonous than the previous one, because of a more varied lithology. There are isolated hills, forming mesas or "mogotes," whose base presents powerful debris alluvial fans transforming into vast glacis. It is a deeply tectonic area, drowned by NW–SE faults and clipped by others in the SW–NE orientation (Mora, 1978b). The sedimentary relief presents folds and faults. They were formed from a Miocene limestone and old shale reef uplifted in the Quaternary and today contain vertical karstic caverns. At the Cerro Barra Honda base, on the southern slope, Quebrada San Diego is a karstic water source that draws staggered limestone travertine by saturation of calcium magnesium carbonate (see Fig. 8.6).

It is in the Morote River synclinal area (Dengo, 1962b), 4 miles south of the town of Nicoya, where on the road there is an alluvial cone cut where two periods can be recognized:

- **Upper sector**: 16 ft thick, constituted by red clays formed by numerous and very compacted limestone debris.
- **Bottom sector** (observable part at 10 ft): formed by the same material but much more compact. The red clays are much darker.

FIGURE 8.5 Samara Beach, a sector of negative tectonics. One can see the important beach erosion where a human-made terrace has been dismantled by sea erosion. *Photo Bergoeing (2010).*

It looks like to be two alluvial fans successively overlapping, with short periods of two-stage rexistasis alternating with diluvial rains, which have enabled the washing of the sedimentary slopes at the top. North there were drier paleoclimates and more contrasted samples than at present, founded in the Santa Rosa sector and especially in Nicaragua. A gap of the climate system to the south is not impossible, for one or several times in the past.

We relate these two encased alluvial fans to the Middle Quaternary environment given the good conservation of material that constitutes them and their relatively low altitude. On the other hand, in the sector of Colorado de Abangares, 6 miles from the Tempisque River, it is also possible to see a beautiful erosion glacis, which is at 33 ft of altitude of the current Flandrian plain, giving its position, apparently, as situated between the Eemian and the Flandrian. These drier and wetter alternating phases in the past were probably responsible for the filling of current flat-bottom valleys in the whole Nicoya sector. In some places, the filling is not over, and the valleys have a concave shape. The Morote River empties into the Nicoya Gulf. Before reaching its inhabited mangrove delta, the river, after crossing a last obstacle of narrow relief, describes some meanders on the plain, framed by a large fluvial terrace of 13 ft. This terrace flooded some isolated reliefs such as Loma Chumico (Cretaceous sedimentary residual relief), where Schmidt-Effing (1975) found the first known ammonites from Costa Rica.

FIGURE 8.6 Stalactites formed by dissolution of calcium carbonate content in the limestone of Barra Honda (Damas Cavern), by rainwater that infiltrated into the labyrinth of caverns. The carbonate saturation reconstitutes the material by continuous drip. In the sector where the water emerges staggered travertine forms. *Photos courtesy of Gustavo Quesada, Anthros Costa Rica grotto, 2010.*

1.4 Nicoya Peninsula Northern Contact

Nicoya Peninsula reliefs, at its northern end, gradually decrease, leaving some isolated hills exclusively fitted in the Nicoya Complex rocks (Barr and Escalante, 1969). However, there is continuity between the compact mass of Nicoya and the Liberia plateau, made by a small set of hills bordering the Pacific coast.

Northern contact is also characterized by vast alluvial fans/glacis extending from the mountains' feet, which can reach several miles' radius. They constitute the contact area between the Nicoya Peninsula and the Tempisque River plain, and are additional proof of climate changes that occurred in Costa Rica during the Middle and Upper Quaternary. The most depressed part of this area is occupied by Cañas River, which is the main drain of the northern hydrographic system of Nicoya Peninsula. This river and its tributaries fit the old Quaternary alluvial fans/glacis. In its middle course, the Cañas River adopts the general level of the plain, becoming vast swamps.

Towns and hamlets settled from the colonial period in this sector are on vast alluvial fans/glacis from the Upper Quaternary, and not in the depressed level of the often-flooded plain. This is the case for Nicoya, Santa Cruz, Belen, San Blas, Sardinal, urban centers, etc. The streets always follow the route of the Spanish lattice plan; however, they are influenced by the sometimes great unevenness of the ground. The sector where these alluvial fans/glacis are more developed is situated mainly between Belen and Santa Cruz and Cartagena villages; this due to the important surrounding reliefs, witnesses to the climatic crisis, which have provided the needed material for the towns' construction.

The Cañas and Andamojo rivers run in opposite directions; however, they occupy a unique flat-bottomed valley following a tectonic alignment (Stone, 1979 in Sprechman, 1982). Indeed, the thalwegs dig a NE–SW fault alignment. This wide valley of flat bottom allows an easy contact between the Tempisque River plain sector and the Nicoya central Pacific coast. In this sector, the Nicoya Complex is characterized by intrusive rocky outcrops (mostly much altered gabbros) (Dengo, 1962b) that are visible on the Nicoya Peninsula northern Pacific facade. These outcrops remain continuous until Bahía Culebra Bay (Kuyjpers, 1980).

From Cabo Velas Cape, the rectilinear layout of the coast changes, to give way to a series of bay wide open to the northwest, such as Conchal, Potrero, El Coco, and Playa Hermosa beaches. It is at Potrero Beach where we discover the "Potrero level" (Battistini and Bergoeing, 1983a,b). It is an old erosion level (Riss–Illinoian/Würm–Wisconsinian, 160,000 to 85,000 years old), which is at 50 ft above the current sea level. This same level is located in the Chocoyas Islands and Conchal Beach. Not far from this beach, on the road to the hamlet of Villareal, we find a sea-level fluvial terrace of 50 ft on the road slope. Villareal terrace is located 300 ft north of the intersection of the track that leads to Villareal hamlet. The profile displays on the base 20 ft of fluviomarine boulders, partly well worn, associated with clasts or angular rocks, all mixed with yellow alteration clay. The upper part of the profile is covered by a red clay soil. It might be an ancient pre-Flandrian beach.

The Nicoya Complex northern hills in contact with the Liberia plain were not formed before Hermosa Beach in the Culebra Bay sector. Here, the hills are covered by a gray ignimbrite blanket of variable thickness and can reach 16–33 ft in some places. Cretaceous–Tertiary columnar basalts of this sector are present but altered in surface (see Figs. 8.7 and 8.8).

Tectonics are also present in this part of the contact, with numerous parallel faults in the NE–SW direction; note also the great Cacao Volcano accident that constitutes a major fault escarpment continuing to Sardinal Beach. This first set of tectonic faults overlaps a second one, which presents cuts at right angles along a NW–SE direction. These shorter path failures are important because they are the basis for rock material preparation, before been cleaned by erosion. Here we see the extent of ignimbrite spillovers coming from the volcanic Guanacaste Mountain Range during the Quaternary period. Indeed, we find them on the tops of some Nicoya Complex isolated hills between Filadelfia town

FIGURE 8.7 Flamingo Beach, Guanacaste. Cretaceous red-brown radiolarite sedimentary outcrops, very altered and faulted. *Photo Bergoeing (2011a).*

FIGURE 8.8 From Flamingo Beach one can observe the promontory of Punta Sabana Cape, where various marine erosion levels are registered. *Photo Bergoeing (2011a).*

and Belen village, as well as in the sector of Culebra, which represents 43 miles as the crow flies.

2. TEMPISQUE RIVER TECTONIC DEPRESSION

A vast tectonic depression should emerge during the Pliocene–Quaternary orogeny. This graben is the logical continuation of the ancient subduction zone coast of Costa Rica, failed and filled by various Quaternary volcanic and sedimentary spills. At the beginning of the Tertiary was a sunken area that formed a vast gulf, ancestor of the current Nicoya Gulf. To the south appear the Barra Honda shale paleoreefs (Oligocene–Miocene), which are irrefutable proof of the orogeny. The depressed sector should emerge following the general movement that affected, during the Plio–Quaternary orogeny, neighboring units, thus becoming a vast marshy area of brackish water. It is in this landscape where the Tempisque River appeared and began to make its way, thanks to its important tributaries created by the progressive and continuous lifting of the volcanic Guanacaste Mountain Range.

The plain starts in San Luis, where the Tempisque River leaves the Liberia ignimbrite plateau that begins after crossing a stretch of river gorge of tectonic origin (Cacao volcano fault). In San Luis, we are in the presence of a narrow valley through which the river runs, leaving a fluvial terrace of 33 ft on its right bank, formed of fine ignimbrite material of silt and cobbles. This terrace is partly fossilized by a colluvial glacis. On the left bank, the river notches the ignimbrite Liberia plateau, leaving to appear an unimportant escarpment. The escarpment has fluvial–tectonic articulations where was found a lacustrine and fluvial series of ancient ignimbrite river deposits as well as rolled pebbles of the same origin.

From Hacienda San Miguel and downstream of Guardia hamlet, a vast alluvial fan/glacis of the Tempisque River is staggered. Here the plain begins taking amplitude. The western contact of the plain is made through major sedimentary deposits of the Quaternary period that gave rise to vast flattened alluvial fans, similar to those of the Santa Cruz sector. Outside of the Hacienda Porvenir estate area, the only depressed plain sector, it is covered by vast repositories of outfall fans. It is only from Filadelfia that we actually find the floodplain. Here the Tempisque has left traces of frequent overflows to the southeast, and exceptional floods have enabled these overflows in the plain of the "Monte Ucote" site.

Farther south, between Filadelfia and the northern contact with Nicoya Peninsula reliefs, the plain is dominated by one of the few western tributaries of the Tempisque River, the Cañas River, forming meanders and vast swamps and flooding the more depressed parts of the plain. The flooded areas are located in the Madre Vieja and Caballos riverbeds and Lake Laguna Chapernal.

Tempisque's plain could be divided into two well-defined sectors:

- **The filled plain** stretches between Guardia and Bolson hamlets. It is a plain that was filled during the Quaternary period and where the Tempisque River overflows during exceptional floods, from Filadelfia.
- **The floodplain** is located between the hamlet of Bolson and the Tempisque River estuary. It is a strongly depressed region whose altitude does not exceed 20 ft (without considering the isolated surrounding hills). It is periodically flooded during the rainy season and includes many small natural lakes where mangrove forests and riparian vegetation thrive, as well as a rich local fauna.

2.1 Tempisque River Plain Eastern Contact

The interest of this contact is where the Liberia ignimbrite plateau dominates the plain by approximately 330 ft. The plateau rises directly from the plain by a tectonic fault scarp. The Tempisque River follows a parallel path to this alignment. The rivers descending from the volcanic Guanacaste Mountain Range sink into the ignimbrite plateau and their mouths have built alluvial fans evolving to vast flattened glacis. The Matapalo River, upstream of the town of Filadelfia, built the first important alluvial fan in this plain sector. After the Matapalo River, it is necessary to go to the Salto and Cañas rivers (not to be confused with the Cañas River west tributary of the Tempisque River) to find such alluvial fans evolving into glacis. Among the aforementioned rivers, the ignimbrite plateau's contact with the plain is brutal. Then from the Cañas River to the plain the ignimbrites disappear.

The landscape here is composed of a series of volcano-sedimentary hills, folded and faulted, often isolated, and living traces of ancient Pliocene volcanic spots (Cerro Chopo volcanic hill). Between the Cañas and the Abangares rivers, owing to the significant piedmont contributions, the plain is covered by important alluvial fans evolving water down onto glacis. These alluvial fans are currently dissected by the rivers that fit them. They are composed of Quaternary sedimentary input, testaments of repeated climatic crises during this period, which allowed the creation of these voluminous accumulations, as well as the union to the continent of ancient islets. The series of hills formed by Cerros Cerco de Piedra lost its insularity little by little to finally join the mainland (Rivier, 1983). This is the fate reserved, in some thousands of years, for Chira Island, given the considerable contribution of river sediments that the Tempisque River constantly deposits in the estuary at its mouth, becoming a delta little by little.

From Filadelfia, the Tempisque River is directed toward the east. Arriving in the Laguna Piñuela lake sector it describes meanders through little, high, isolated Tertiary hills. Between Laguna Piñuela and the Estero Lagarto River, the Tempisque River runs through a depressed area (local subsidence?), where it has much difficulty in evacuating its waters. During floods, this sector is then transformed into a large marsh. The tributaries of the Tempisque River, which empty into this sector, do not do more than increase the amplitude of this expanse of water.

2.2 Center and South of the Tempisque Plain

These marshes are on track to becoming dry in the future, thanks to human actions. They are indeed rich soils for agriculture. An example of the occupation of the plain by humans is the Moracia District of irrigation, which is located at the Bebedero River mouth (Corella, 1979 in Battistini and Bergoeing, 1982a). Near the river mouth remain vast marshy areas such as Mata Redonda, Corral de Piedra, Sonzapote, and Lava Maiz, which are the remains of a wild area that lasted until the end of the 20th century. In fact, before the construction of the Inter-American Highway, during the years 1940–1950, the route of entry into the

Tempisque plain was the Tempisque River, navigable up to the Puerto Humo fluvial port, where the effects of the tide are observable 13 miles into the estuary. On the marshy banks of the Tempisque remain some alligators that find shelter in the mangroves (see Fig. 8.9).

It is interesting also to note that between the estuary and Bolson lives a brackish-water mollusk of this swampy area, which is very appreciated by the locals. It is the clam *Nephronaias tempisquensis*, discovered by Pilsbry in 1920. The lands of the Tempisque plain are silty clayey or clayey. They are waterproof and heavy soils, called locally "sonzocuitles," and are dark brown to black. They crack easily during the dry season (5 months), forming long polygons, 60–100 ft in radius. They are home to mollusks and snails hidden inside of this clay mass until the next rainy season. They are rich in phosphates and can be transformed into agricultural soil with good drainage (Dondoli, 1965b).

The Flandrian transgression stretched probably to the sector of Bolson (20 ft high and 13 miles to the interior from the current mouth), a place where the sedimentary input of the Tempisque has not yet filled this sector of the plain. The mangrove is currently backing up, because of humans, particularly on the Tempisque riverbanks, where it was important in other times, between the Puerto Humo fluvial port and Boca Toro (old ferry crossing).

Although the Legislative Assembly of Costa Rica has ordered a Palo Verde protected area, this has not been an obstacle for the degradation of nature for the benefit of real estate speculation. Within probably 20 years, the mangrove will have disappeared completely from the Tempisque riverbanks and with it the natural wildlife that survives thanks to the mangrove. On the shores of Nicoya Gulf, the mangrove is also the victim of humans and their lack of foresight. From the pre-Columbian period, the Tempisque River plain was occupied for a long time by the Chorotega civilization (5000-year-old skulls in the caves of Barra Honda). Owing to the marked seasons and the moderate regime of rainfall (1600 mm per year), today it is one of the most important agricultural regions of Costa Rica for cattle, sugarcane, sorghum, corn, rice, cotton, etc.

FIGURE 8.9 Lower Tempisque River course filled with sediments in the tectonic depression next to its mouth. *Photo Bergoeing (2009b).*

2.3 Tempisque Plain's Geomorphological Features

- The Tempisque River plain is the product of a major subsidence, framed in two sectors: northeast, the Liberia ignimbrite plateau and southwest, the Nicoya Mountains, both in horst position. If we take into consideration the Gulf of Nicoya as the natural extension of this depression, then we can consider this set as overall subsiding. Traces of three cyclic levels appear in the island of San Lucas (33, 82, and 130–160 ft), but are missing on Chira Island farther northwest (Fig. 8.10).
- The plain has been fed by alluvial contributions of the Tempisque River and its numerous tributaries, after the general uplift during the Pliocene–Quaternary orogeny. At the beginning of the Quaternary period, this sector should be a vast marshland where the influence of the sea was long inside. Witnesses to this last fluvial–marine stage are the Bolson sandy deposits with seashells (Pectinidae family).
- Alluvial fans and glacis installed at the river's mouth with the plain in both sides, Nicoya Peninsula, and the volcanic Guanacaste Mountain Range bear witness to the drier climatic crisis, probably from the Upper

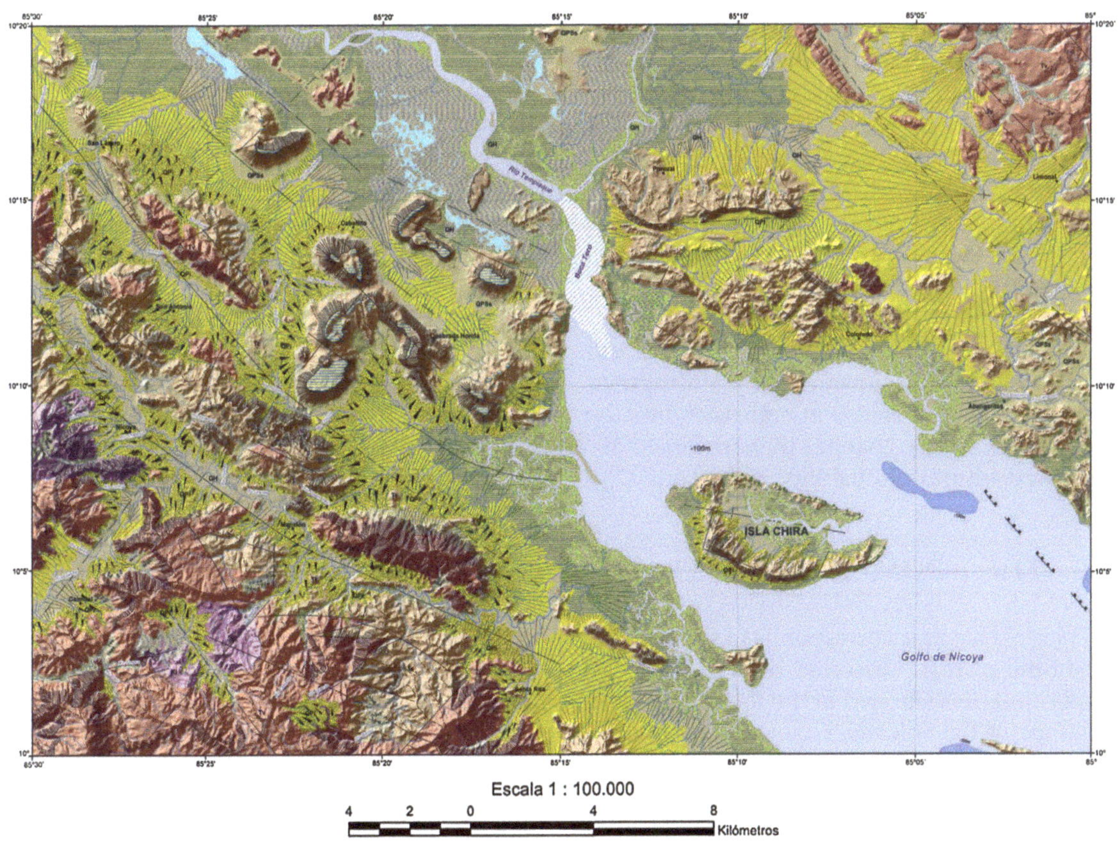

FIGURE 8.10 Tempisque River estuary and Chira Island. West, the isolated former Miocene coral reef hills today are under karstic domain. *Geomorphology Atlas of Costa Rica 2016. Bergoeing. Una-Costarica.*

Quaternary, given the youth of the material. We have also found the presence of a fluvial terrace level overlooking the Tempisque River plain 1 mile from Filadelfia City. It is an ancient terrace, which rises 50 ft from the current level formed by thick sand at the base and thinner at the top, resting on ignimbrite worn away by erosion. Close to Sardinal, a second cut shows us a fluvial terrace, but of torrential type, because over 16–23 ft of relative altitude we find rolled edges, 1.3 ft thick and medium mixed to a very sandy–clayey heterometric matrix (road from Sardinal to Comunidad).

- The Tempisque River is active only during the floods that occur in the rainy season; its volume decreases significantly in the dry season, and this has been increased recently because of inconsiderate deforestation in the high areas of the volcanic Guanacaste Mountain Range (the jungle plays the role of moderator of the river flows). It overflows, however, during particularly wet rainy seasons and spreads out over the plain.
- The Tempisque River plain, in its lower course, is a vast swamp in the process of drying since the Flandrian transgression. It is possible to see some isolated Cretaceous and Paleocene reliefs surrounded by a mangrove swamp; this last is also in regression because of anthropogenic changes that are aimed at the agricultural profitability of the sector.

3. THE IGNIMBRITE PLATEAU

This vast flat area corresponds to the top of an ignimbrite plateau, dissected by the rivers that deeply fit it. It is situated at the foot of the western slope of the volcanic Guanacaste Mountain Range. It is the consequence of the volcanic activity of this mountain range, because ignimbrite spills (Bagaces Formation) would come from an axial fissure zone (R. Weyl, 1960), where, later, various lava spills (mainly andesites) allowed the construction of the current stratovolcanoes (Middle to Upper Quaternary). This plain is interesting given its expansion, covering more than 62 miles in length and about 13 miles in width (Battistini and Bergoeing, 1983a,b). The oldest ignimbrites were issued almost continuously until the Upper Quaternary (Sapper, 1925). In fact, this type of acid effusive activity can play currently. The recent Arenal volcano Peléan eruption served to remind us, starting with a pyroclastic burning cloud eruption in 1968 followed by andesite lava flows.

The ignimbrite plateau can divided into three large sectors:

1. **Liberia plateau**
2. **Santa Rosa plateau**
3. **La Cruz plateau**

3.1 Liberia Plateau

The Liberia Plateau is located roughly between the northern Liberia City sector and the Tenorio River. Its uniformity and subequal altitude, dominating the plain by 330 ft, characterize it. The plateau's interior is a more depressed sector, which keeps the traces of its past lake (Hacienda Ciruelas, Pelon de La Bajura, etc.). K/Ar dates its oldest forms back to 4.87 MY (Pelon Quarry, Alto Mirador, Hacienda Tempisque) (Szymanski, 2009 in Alvarado and Gans, 2012, measurement). A gray tuff on the middle part of Bagaces Formation near the Pijije River gave an age of 3.8 MY (Mainieri, 1976). A deposit of pumice flow gave an age of 1.83 MY (Gillot et al., 1994 in Alvarado, 2000). Finally, Battistini and Bergoeing (1982a,b) dated a green tuff in Hacienda Ciruelas, giving 1.2 $\pm$ 0.7 MY. This helps us to understand that pyroclastic flows that gave rise to the first ignimbrite deposits on the folded structures of the Miocene originated in the Pliocene and continued during the Quaternary period until today. They are the products of an acid volcanism of the Guanacaste Mountain Range.

The external escarpment of the Liberia ignimbrite plateau is higher, which explains the internal lake forms: lacustrine terraces formed with sandy ignimbrite deposits. This escarpment dominates the Tempisque River plain from 330 to 500 ft in some places. It is a fault scarp result of regional tectonics, which places the ignimbrite plateau and the volcanic Guanacaste Mountain Range in horst position between the Great Northern Plain (Lake Nicaragua graben) and the Tempisque graben.

The plateau starts from the Tenorio River's right bank. Here is a vast erosion glacis that gradually descends toward the Tempisque River plain. This erosion glacis is widely attacked by the Paso Ancho, Jerónimo, Blanco, and even the same Tenorio river courses. In Aguacaliente de Cañas sector, we find a small fluvial terrace of 3 ft, resting on the ignimbrite, formed by well-rolled boulders and resting in a sandy–silty matrix. It is from the Piedras River. Heading toward the northwest, we find the first hints of ancient lacustrine depressions; however, the sector where these paleolake forms are better exposed is located on the side of Laguna Palenque.

Liberia plateau's northeast contact consists largely in the stunningly eroded volcanic cone/glacis from the Guayabo collapsed caldera, Liberia Formation (Granados, 1979, op. cit.). This erosion cone is crossed by numerous rivers (radial network), which have fitted it, thus giving two levels of erosion in the margins. These levels are 3–13 ft high in some places. This cone, affected by tectonic faults, is witness to the recent tectonic activity of the sector. Rivers that fit the ignimbrite plateau sink into the river canyons that evolved by linear erosion, forming ponds known as giant pots or kettles, as is the case of the Motañita River.

The Liberia plateau borders the Santa Rosa plateau in a contact area that is characterized, on one hand, by a fault escarpment, with NE–SW tectonic alignment, descending from Cacao Volcano and, on the other hand, by the Tempisque River canyon, which sinks before reaching the plain. The plateau in this sector gradually descends toward the hamlet of Guardia, forming an erosion glacis (subsidence effects by local neotectonics?).

3.2 Santa Rosa Plateau

This second sector is the widest and most solid or strong. It is situated north of the Tempisque River canyon and is bounded on the northwest by the Quebrada Aserradero River. The plateau is characterized by two main sectors:

- **A lacustrine depression**, which corresponds to the Tempisque River upper course. It is a vast north–south elongated extension. This river must have formed, during the Quaternary, a vast lake, before digging the current canyon and pouring its waters into the plain, drying the lake extension completely.
- **A higher plateau on the lacustrine depression**, which overlooks at 660 ft above sea level the Papagayo Gulf, giving rise to important active cliffs, whose base is form by basaltic to dacitic lava flows with an average age of 8 MY (Alvarado and Gans, 2012). The plateau has NW–SE and NE–SW faults in a grid system that gives a compartmental character. This is according to the rigidity of the ignimbrite and its low resistance to neotectonic effects (frequent earthquakes). Farther north, in the latitude of Santa Elena Peninsula (Azema and Tournon, 1980; Harrison, 1953; Dengo, 1969; Bergoeing, 1978b), the edge of the plateau dominates the serpentinized peridotite hill summits of the Nicoya Complex through a dissected erosion cornice. This western ignimbrite extension testifies to the importance of these volcanic effusions and the distances they reach during eruptions.

Down the road that leads from Santa Rosa to Cuajiniquil Bay, it is possible to observe the ignimbrite contact with Paleocene–Eocene sediment layers, separated by a small 3-ft-thick paleosoil. Farther north, from Santa Elena Peninsula, the Santa Rosa plateau thins progressively. The rivers here run southward and are tributaries of the Tempisque River (see Fig. 8.11).

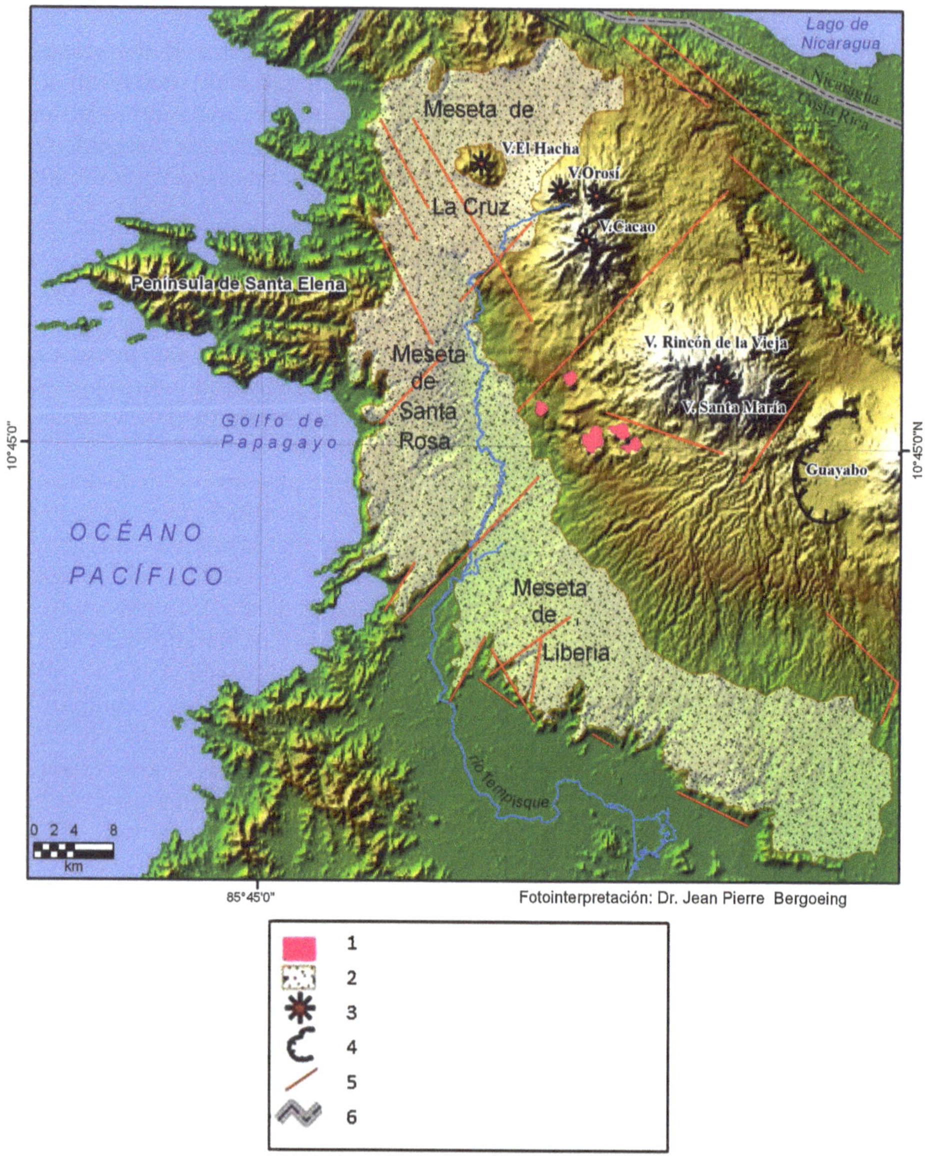

FIGURE 8.11 Ignimbrite plateau divided into three sectors (Bergoeing, 2014). (1) Dacitic domes. (2) Ignimbrite plateau. (3) Craters and volcanic cones. (4) Collapsed caldera. (5) Tectonic fault alignments. (6) International border with Nicaragua.

3.3 La Cruz Plateau

From Quebrada Aserradero, north of Santa Rosa plateau, rivers change direction and run northward through the plateau formed by ignimbrite much more altered. In some places, they change color from gray to light red. The volcanic flows are manifestly much older than those of Santa Rosa or Liberia, given the degree of alteration. The proof of their oldest age is provided by the northeast contact of the plateau of La Cruz: here the ignimbrite lies buried under El Hacha volcano lava flows (Lower Quaternary).

La Cruz plateau dominates, through a fault escarpment, Descartes Point and Salinas Bay. In this place, the ignimbrite is a powerful cornice 130–170 ft thick, which rests on a Lower Tertiary folded sedimentary series (Weyl, 1960). The cornice overlooking this escarpment takes on the appearance of a "pseudocuesta," whose base dominates Salinas Bay and ends on solifluction deposits, which meet the narrow coastal plain.

To the north, the plateau disappears, giving rise to sedimentary hills of a folded and failed landscape, which are the same ones that are southerly located but fossilized under the Santa Rosa and La Cruz ignimbrite. These hills are influenced strongly by tectonics (hydrographic network grid). The Sapoa River, flowing northward to the murky waters of Lake Nicaragua, is a natural boundary between sedimentary outliers of early Tertiary and volcanic basaltic–andesitic Pliocene–Quaternary lava flows located northeast of this river (Dengo, 1962b).

The ignimbrite plateau is characterized by a poor and xerophytic vegetation; this is due to the very nature of ignimbrite, a very porous rock, as well as to the existence of foehn winds that descend from the volcanic slopes, dehydrating the soils during the dry season. The twisted ignimbrite piedmont trees give testimony to the strength and direction of the wind and remaining shrubs are scarce. On La Cruz Plateau, rivers and especially the Sapoa River have dug erosion terraces into the ignimbrite. Three levels exist upstream of La Cruz Village. In the lower course, once the river has left the plateau, it forms two terrace river levels of 10 and 16 ft, respectively, formed by ignimbritic boulders, and sands associated with a silt clay matrix.

Different morphological episodes of the ignimbrite plateau are important because they constitute a link to establish a chronology of the various climatic events that occurred in Costa Rica during the Quaternary. Indeed, rexistasic periods (dry phases) have undermined the plateau, giving canyon forms, fluvial erosion terraces, and flat glacis on the Tempisque plain contact. This is due to the material delivered to the volcanic slopes, which served as a powerful abrasive. Obviously, we should take into account the volcanic eruptions, which could play a local rexistasic role for short periods. Biostasic periods (humid phases) have prolonged rain phases, which alternate with interpluvial periods, creating long periods of relative calm, during which the vegetation could grow everywhere, especially on the volcanic slopes, and as a result stabilized them shortly. During these wet phases, the ignimbrite plateau met with a saturation and overflow of water, and hence a greater aptitude for lake retention.

In conclusion, we can say that the ignimbrite plateau is the result of the following chronological phases:

- **In the Miocene** the region consisted of low sedimentary Paleocene–Eocene hills that began to emerge from the Pacific Ocean, forming small islands.
- **During the Pliocene**, the region had almost totally emerged. Volcanic activity, which began probably before, intensified, giving the first spill of ignimbrite as well as basaltic lava flows. Everything deposited discordantly on the sedimentary hills' erosion surface.
- **The Low Quaternary** was characterized by a strong volcanic activity, with a new cast of

basaltic–andesitic lava flows, which would form the basis for the construction of the volcanic Guanacaste Mountain Range. On the Pacific slope of this mountain range in construction, ignimbrite spills continued depositing. In the western slope of the Rincon de la Vieja Volcano appeared small dacitic domes like San Roque Dome (1 MY), whereas in the Liberia and Santa Rosa plateaus, natural damming formed vast lake expanses. It is possible that the Tempisque River formed a first mouth in Bahía Culebra before being definitely diverted by the undermining of the current canyon toward the Tempisque Plain. The Miravalles volcano sector shows how in this period an old volcanic source fell and gave rise to the Guayabo collapsed caldera and leaked white ignimbrite flows (Liberia Formation), progressively constructing a volcanic ignimbrite cone/glacis.

- **The Middle Quaternary** is marked in this region by a new acid volcanic phase, returning to depositing ignimbritic flows on the plateau. This period marks the beginning of the construction of the stratovolcanoes of the volcanic Guanacaste Mountain Range
- **The Upper Quaternary** it is also marked by new ignimbrite spills deposited in the plateau, especially between Santa Rosa and Bagaces. The neotectonics active since the Pliocene split the ignimbrite, giving important fault escarpments, such as Cacao Volcano or the faults fitting the volcanic cone/glacis of Guayabo caldera. Uplift tectonics helped rivers to undermine deep thalwegs following fault line crushing.
- **The Holocene** maybe was a rest period for ignimbrite volcanic activity, because spills that young are currently unknown. However, the Arenal volcano 1965 eruption with pyroclastic clouds is a warning that we should consider. The appearance of humans on the plateau has resulted in deforestation of the xerophytic vegetation that covered it, allowing a new phase of mechanical erosion by rivers, but especially of chemical dissolution (emergence of dissolution depressions in the ignimbrite).

CHAPTER

9

Coastal Geomorphology of Costa Rica

OUTLINE

Geomorphology and Volcanology of Costa Rica
http://dx.doi.org/10.1016/B978-0-12-812067-5.00009-5

1. THE PACIFIC COAST

1.1 The Role of Neotectonics

The Costa Rican Pacific coast stretches about 280 miles as the crow flies, but its development makes it much longer because of major accidents such as the Nicoya and Osa peninsulas as well as the gulfs enclosed by these peninsulas, Nicoya Gulf and Golfo Dulce, not to mention Santa Elena and Burica points (see Fig. 9.1). There is a great diversity of coastal landscapes: raised and cropped, or massive coasts with rectilinear sectors and very high tectonic cliffs, low coasts with large sandy arrows and vast mangroves. However, there are no important dunes and coral reefs are insignificant. Rainfall increases from north to south: 1600 mm in the extreme northwest in Puntarenas, between 2500 and

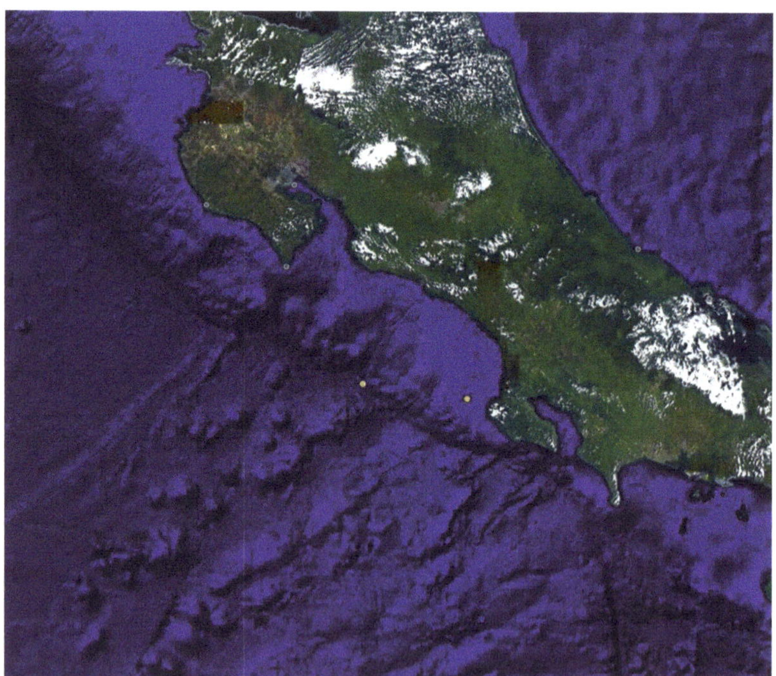

FIGURE 9.1 Seabed of the Pacific coast of Costa Rica, where it is possible to observe the continental shelf, the continental slope affected by landslides, and the contact area between the Cocos and the Caribbean tectonic plates as well as the area sliding against the Nazca plate. The Cocos plate with its numerous submarine volcanic cones can also be seen. *Image Google Earth (2010).*

3000 mm in Quepos, and 4500 mm in Golfito. The northwest is also marked by a dry season, December to May. The tides, in contrast to the Caribbean littoral, which are only 2 ft, can reach 10–13 ft high on the Pacific coast.

A fundamental tectonic direction, NW–SE, determines the disposition of the major units of the whole country (see Fig. 9.1). To this direction of fracturing belongs the Costa Rica Reverse Great Fault, a former subduction area. This tectonic fault is always very active and causes numerous earthquakes, especially in the southern part of the Pacific façade (Setumi Miyamura, 1980). It is also this tectonic fault that originated in 1914 the Orotina town earthquake in the central part of the Pacific coast of Costa Rica. This NW–SE direction determines a marquetry of faulty blocks, on a scale of miles, and then giving way to orthogonal salient coasts, as for example in the sectors between Puntarenas and Punta Judas Point.

There is also an east–west tectonic direction that is less well represented. On the Pacific coast of Costa Rica, there are large areas of low coasts with Flandrian sandy arrows, as for example Puntarenas Arrow, and mangroves (Coronado Bay background, Quepos, Nicoya Gulf coastal plain). However, rocky coastal landscapes are much more common. Thus, most of Nicoya Peninsula is a hill relief surrounded by the sea, which gives an aspect of detail very trimmed to the coast: a succession of steep spikes and small bays of micro mangrove forests and islands and numerous islets (see Fig. 9.2).

In addition to this are high rocky coast sectors looking solid or rectilinear, where it is possible to see the influence of a faulty tectonics, as for example in the Osa and Burica peninsulas. From the neotectonics point of view, we can distinguish four cases among these sectors:

- **Sectors of flooded morphology** but without terrace levels (sea or river)
- **Sectors with large amounts of lower "levels,"** witnesses to a moderate uplift
- **Sectors with a large number of staggered "levels,"** who are present in an important coastal mass, witnessing a fast rising
- **Sectors of large tectonic coastal cliffs**

By "levels" we understand the flattening of cyclical erosion, perfectly horizontal, often of a notorious sharpness and a great local continuity, sometimes truly "rasas" in the altered rock. But unfortunately, none of the places where we can see them have correlative marine deposits. Despite the absence of correlative marine deposits, it is clear that cyclical forms of erosion are related to ancient marine stands that have fitted the rock in situ. They can pass to boulders in fluvial terraces, for example, the Grande de Terraba river mouth.

1.2 Sectors of Flooded Morphology but Without Sea or River Terrace Levels

This is the first and most northern sector of the country, and it includes Santa Elena Peninsula and the Papagayo Gulf. Santa Elena presents a very high, much-cut cross-type coast, and hard rocks alternate with low-resistance rocks, with a sea submergence morphology without traces of cyclic levels. We explain this with a subsiding regime, limited in the south by a major NE–SW fault system well visible in the Landsat and SEASAT radar images. This fault system passes between the Cacao and Rincon de la Vieja volcanoes and plunges into the sea. This fracture address also determines the northeast coast of the large block of the Nicoya Peninsula and a subdivision of the Liberia ignimbrite plateau (see Fig. 9.3).

A second area is located on the central coast in the south of Costa Rica. From Parrita and for 93 miles to the south is a scarp landscape totally controlled by the coastal fault straight NW–SE from the great Costa Rica tectonic fault, by which the Coastal Mountain Range dominates the sea. The Coastal Range (Fila Costeña) was lifted up during the Plio–Quaternary orogenesis. These mountains exceed 3300 ft above sea level (the 3757-ft high summit is located just 3 miles from the sea). To the south, they culminate at 5600 ft. Regardless, all that is immediately before the escarpment is subsiding, that is, the Parrita–Quepos littoral plain, the Quepos Promontory group of hills, and coastal submergence reliefs,

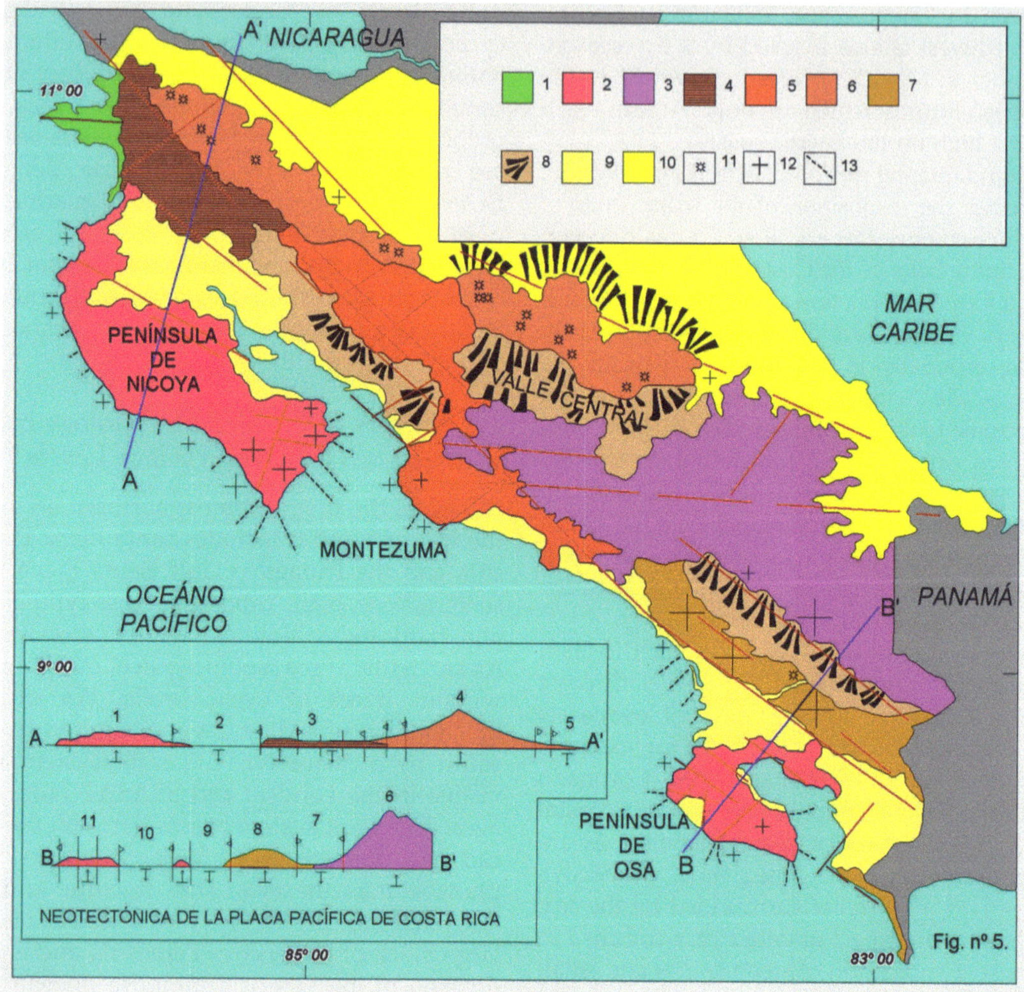

FIGURE 9.2 Neotectonics of the Costa Rica Pacific watershed (Bergoeing et al., 2012d). Big relief units: (1) Santa Elena Peninsula (Cretaceous–Tertiary ophiolitic series). (2) Nicoya and Osa peninsulas (Cretaceous–Tertiary). (3) Talamanca Mountain Range (Tertiary volcanic, sedimentary, and intrusive). (4) Liberia ignimbritic plateau (Plio–Quaternary). (5) Tilaran Mountain Range (Tertiary: Pliocene). (6) Guanacaste and Central volcanic mountain ranges (Quaternary). (7) Coast Mountain Range (Tertiary: Pliocene). (8) Big alluvial fans of the Quaternary period. (9) Tempisque, Diquís, and Coto Colorado tectonic depressions. (10) Nicaragua graben. (11) Montezuma and Puerto Jimenez sites (estimated 2–4 m per millennium rising). (12) Identified raised levels. (13) Subsidence tendency. *Profile A–A'*: Nicoya Peninsula, Tempisque depression, Liberia ignimbritic plateau, Guanacaste Mountain Range, Nicaragua graben.

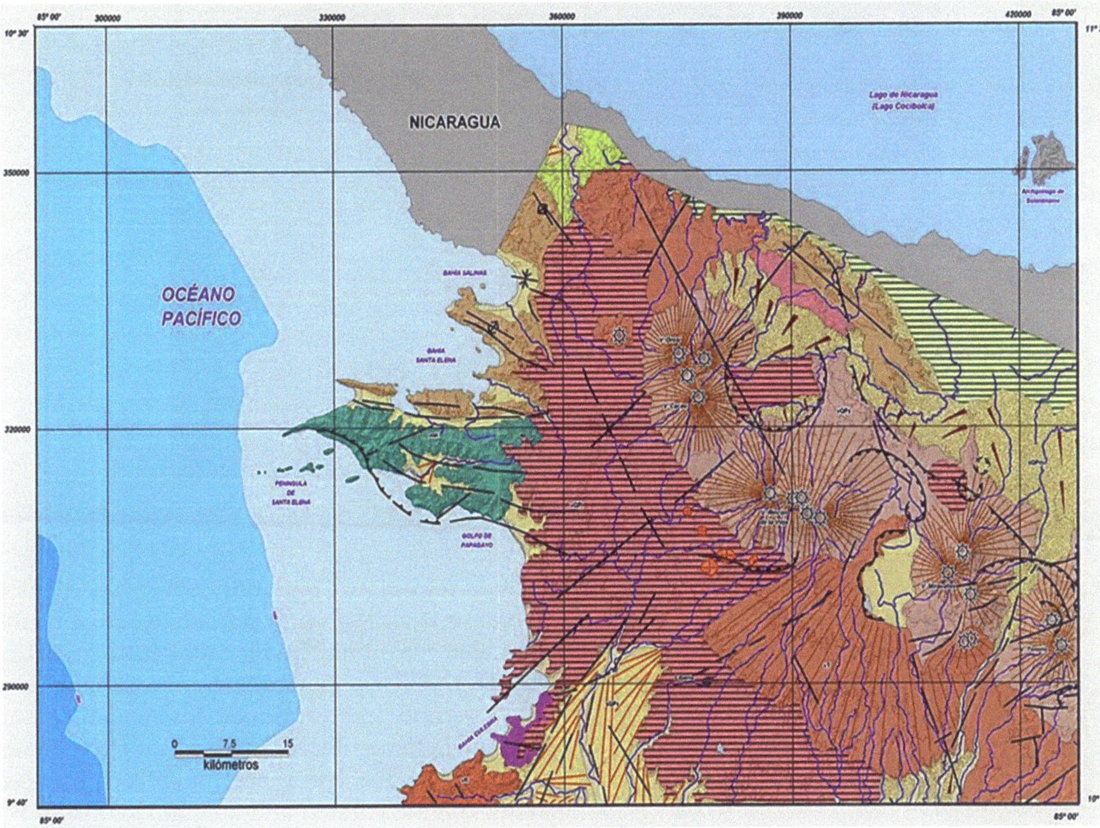

FIGURE 9.3 Costa Rica north Pacific geomorphology, highlighting the ophiolitic Santa Elena Complex, the Liberia ignimbrite plateau, and Orosí, Cacao, Rincon de la Vieja, Miravalles, and Tenorio volcanoes comprising the volcanic Guanacaste Mountain Range, as well as the Pliocene dacitic dome of Rincon de la Vieja. *Taken from the Costa Rica Geomorphological Atlas, 1:350.000 scale, Bergoeing et al., Costa Rican Institute of Electricity (ICE, 2010).*

giving numerous islands where we do not found remains of cyclic levels.

1.3 Sectors With Large Amounts of Lower "Levels," Witnesses to a Moderate Uplift

The study of the cyclic levels found south of the NE–SW Cacao Volcano fault is witness to a moderate trend in upheavals in the middle of the Nicoya Peninsula (Potrero level or "rasa": 33–50 ft above sea level) (Battistini and Bergoeing, 1984).

The sector between Puntarenas and Parrita provides one of the most beautiful examples of a coast of orthogonal tectonic ledges in a faulty grid structure. There is no proof of fast rising, at least between the Barranca and Tarcoles river mouths (see Fig. 9.4). We know, on the other hand, that the grids of the fault have not returned to play but an insufficient role since the Orotina piedmont was formed, whose age is at least from the Middle Pleistocene. This sector seems to belong to the Nicoya Gulf–Tempisque plain graben with a slightly positive

FIGURE 9.4 Puntarenas caldera, central coast of Costa Rica with a moderate coastal uplift. *Image Google Earth (2010).*

or slightly negative neotectonics depending on the site.

1.4 Sectors With Large Numbers of Staggered "Levels" Present in an Important Coastal Mass, Witnesses to Fast Rising

The southern Nicoya Peninsula rises much more quickly. Along the southeast coast, whose general design is commanded by the NE–SW fault system, the levels are multiplied and at least six or maybe eight can be identified up to 660 ft in altitude. At the southern end of Nicoya Peninsula, some high and straight sections of coastline correspond to fault escarpments. In one of these very faulted sectors lies the anglers' hamlet of Montezuma, built on a poorly resistant marine subhorizontal sandstone series, rich in shells, that was carried up to 26 ft above sea level. The shells (porcelains) were dated in France by the ^{14}C method, giving an age of 6620 ± 150 years (GIF 5480). These Holocene marine deposits are a support against the 263-ft high rectilinear dead cliff that we interpret as a fault escarpment. This dating provides an estimate of the lifting on the order of 3 ft per millennium. This value, of course, may not apply to the whole region, cut by faults, but it has a value because it allows us to provide a range of the possible maximum lifting order in this part of southern Nicoya Peninsula.

A second fast-rising area fits in the Coastal Mountain Range littoral (Fila Costaña) from Punta Mala. Here the levels are multiplied; between Punta Uvita, Palmar, and around Coronado, we can identify 8 to 10 levels. The lowest one starts at 82 ft of elevation. It is possible that this is an Eemian level, unless you estimate the Eemian level farther up, for example, at the levels of 130–165 ft or 245–295 ft, which are particularly well drawn. In the absence of correlative deposits, the problem cannot be solved. On both sides of Palmar village, at about 31 miles, the fault escarpment is at the same time a dead cliff, found in some places at 9 miles from the sea, and which continues to present staggered multiple steps that are easily identifiable up to about 820–980 ft above sea level.

The third sector here is represented by the Osa and Burica peninsulas and belongs to the same orographic complex (Nicoya's Cretaceous ophiolitic complex), an intensely folded series consisting mainly of basalt pillow lavas and the Nicoya Peninsula morphotectonics.

The grid failures of Osa Peninsula determine the detail mapping of the coastline and the disposition of the relief masses. The inner part of the peninsula is made up of Pleistocene deposits (Armuelles Formation), defined in the Burica Peninsula, and lifted up to several hundred feet above sea level. This formation rests on Cretaceous and Miocene marine formations. In Puerto Jimenez (Madrigal, 1978; Bergoeing, 1978a), it is possible to observe seven Flandrian beach crests, of which some reach 26 ft of altitude, which gives to this sector a value comparable to that obtained in the Montezuma rising (approximately 3 ft per millennium).

1.5 Sectors of Large Tectonic Coastal Cliffs

The southern coast of Burica Peninsula is an example of a tectonic faulting coast with high cliffs 800–1650 ft above sea level, making them megacliffs in the sense of Guilcher (1966). It is possible to find 650-ft isobaths between 0.6 and 1.2 miles from the coast. Flandrian marine erosion has not been able to rejuvenate the active cliff escarpment; but there is no doubt that in these 13 miles, we are in the presence of a large fossil sea cliff, which coincides exactly with the fault escarpment. The study of this sector of inaccessible terrain should done.

2. PACIFIC LITTORAL MORPHOLOGY

As we can see, because of neotectonics, the Costa Rica Pacific coast is an very trimmed area and therefore rich and varied in forms: islands, peninsulas, bays, active cliffs, points. For a better approach to the problems of the coast, we divide it, from north to south, into several sectors:

2.1 Between the Nicaragua Border and the Santa Elena Peninsula

This is a highly trimmed coastline area, with bays bordered by Flandrian coastal cords. A mighty scarp that dominates this sector is but the edge of the La Cruz ignimbrite plateau. It rises 330–650 ft and rests abnormally over the Cretaceous–Paleocene sandstone and graywacke of the Rivas and Sabana Grande formations. On some sides, the ignimbrites form a ledge 165 ft thick, with fossilized folded Tertiary sedimentary topography. Following the Cuajiniquil Port road, the scarp shows the contact between the ignimbrite and the poorly resistant sedimentary sandstone and graywacke. This contact is made through a thick red paleosol of 7 ft. Contact with the plain is made through a concave glacis presenting solifluction phenomena.

The most notorious accident on this coast is Descartes Point, whose southern coast follows a fracture. It separates Salinas Bay from Cuajiniquil Bay. Descartes is an east–west axis anticline whose southern flank emerges north of Santa Elena Peninsula. Decartes Point is formed by the Rivas and Sabana Grande formations sedimentary series giving active cliffs that can reach 330 ft high, with a small rocky abrasion platform ("estran") (see Fig. 9.5). At the southern tip of Junquillal Bay, there is the only example of dunes that exist in Costa Rica. They are linked to northeast winds and rise to 65–82 ft altitude.

2.2 Santa Elena Peninsula

This littoral feature is characterized by a high, much-cut coast. The axis of the peninsula consists of the Santa Elena ophiolitic complex (serpentinized peridotites from the Upper Cretaceous), following an east–west direction. The base of the peninsula is composed of a "native" series (Azema and Tournon, 1980), consisting essentially of radiolarite, graywacke, and basalt (pillow lava), over which has slipped from the north a peridotite oceanic crust. To the north, the Rivas and Sabana Grande formations sedimentary series relies on the complex.

The north coast is much notched on the back of the hogback (Playa Blanca row). This reverse constitutes the southern part of the Santa Elena Bay syncline, climbing farther north to the

FIGURE 9.5 Bahia Salinas Bay at the extreme north of Costa Rica near the border with Nicaragua. The coastline is formed by bays and Tertiary sedimentary cliffs. *Photo courtesy of Luis Carlos Paniagua (2010).*

Descartes Point anticline. This structural reverse plunges into the Gulf of Santa Elena without forming a real cliff. The numerous notches correspond to a submergence of the river cut morphology. The hogback is cut in the middle where the sea has penetrated into the subsequent depression, forming a bay surrounded by mangroves that rely on debris cones. South, the Varas–Naranjo Beach sector shows a succession of small tectonic notches due to a NE–SW tectonic fault that has recently become active contrary to the peninsula's east–west orientation. In Santa Elena's massif center, peridotites are deeply notched by the Potrero Grande River following an east–west tectonic fracture that allows observing, in the river thalweg, the volcano-sedimentary base (see Fig. 9.6).

2.3 Naranjo Beach and Culebra Bay Sector

Here the ignimbrite plateau reaches the sea. Between Naranjo Beach and Cabuya, the ignimbrites rest on the Nicoya complex formed of Paleocene basalt and sandstone. This area shows a great coastline cliff of 650–850 ft overhanging the Pacific Ocean. On the plateau, a dense tectonic faulting network shows preferential directions of NNW–SSE and NNE–SSW that have a decisive influence on the layout of this part of the coast. The buried base of the cliff is under colluvial cones formed by basalt, sandstone, and even ignimbrite fragments. The sea slowly clears this filling without actually attacking the cliff, even during major storms (see Fig. 9.7).

3. NICOYA PENINSULA LITTORAL

Nicoya Peninsula is oriented following a NW–SE direction determined by the tectonics. Three major fracturing directions are present, northwest–southeast, northeast–southwest, and north–south, playing a very important morphologic role because they order the hydrographic network, which allows the release of

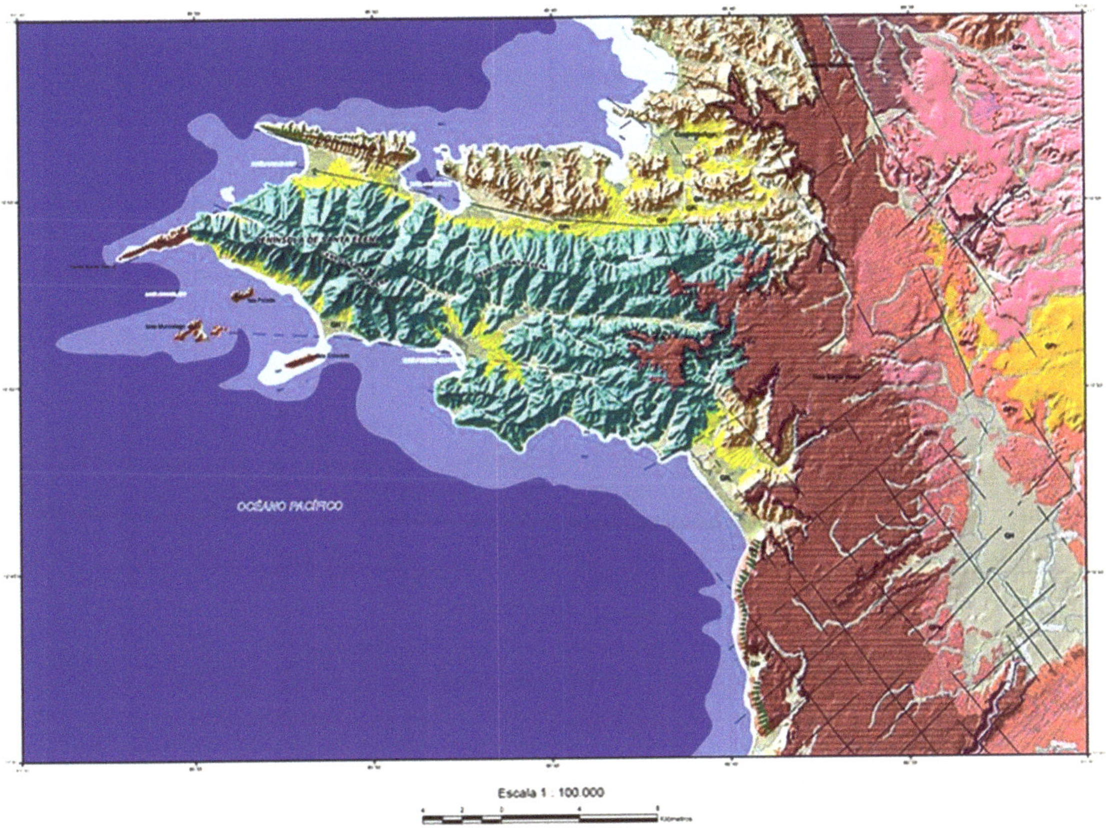

FIGURE 9.6 Upper Cretaceous serpentinized peridotites forming Santa Elena Peninsula in the Pacific Ocean, Wright ignimbrite Liberia plateau (Bergoeing and Brenes, 2016). Geomorphology Atlas of Costa Rica. UNA.

FIGURE 9.7 Santa Elena Peninsula, seen from Junquillal Bay. The sector shows a Tertiary folded series that tilts toward the sea. At the center emerge the Cretaceous serpentinized peridotite formations. *Photo Bergoeing (1982b).*

land mass by fluvial erosion and the hill disposition of the whole. These hills often dominate the shore by filling it and transforming it into small coastal plains at the edge of the sea. Submergence morphology closely linked to the Flandrian Transgression is characterized by narrow sandy cords on backs of the bays, sometimes with small mangroves. The coastline is marked also by numerous and abrupt islands. Basaltic outliers are the most frequent, providing not sand but a bit of rubble. The cliffs are carved in decomposed red clay, whereas basalt appears on littoral platforms and is prey to corrosion rather than marine abrasion.

It is only in some isolated littoral areas of the peninsula where Tertiary sandstone outcrops appear. These sectors are then provided with sandy beaches. The transit of materials from one cove to another is sometimes interrupted by the trimming of the coast, which gives the impression of a belly in the regulation of the coast. It evolves by the collapse of bays and the retreat of points, which are transformed into islands. The following are the various sectors of the Nicoya Peninsula.

3.1 Between Culebra Bay and Tamarindo Bay

Thirty-one miles long as the crow flies, this sector is characterized by a succession of steep points separated by sandy coves. The most important of these bays is Tamarindo Bay, drawing great arc of 3 miles. We observed a cyclical level at 50 ft in Punta Potrero (Battistini and Bergoeing, 1983a,b). We gave this level, fitted in the basalts, the name of "Potrero Level." It is also present in the Chocoyas Islands and along the Guachipelines Point coast. Behind Tamarindo Bay, following the road to Villareal, we find a terrace 40 ft above sea level overlooking the Flandrian filling and which could correspond to a cyclical episode. The cut in the road shows, resting on yellowish alteration clay (very altered rocks), boulders of a thickness of 12 ft, of which some are well carved, mixed with badly rolled elements, with a subhorizontal stratification appearance, all covered by a red clay soil. Although this material being hard to identify, we think that it belongs to an ancient beach.

3.2 From Tamarindo Bay to Punta Guiones

This is a 28-mile-long sector that is characterized by a succession of sandy beaches, little arched, which rely on red clay decomposition hills or rocky outcrops evolving into igneous origin platforms (Nicoya Complex). Here the basalts evolve rather by corrosion, whereas the clays are gradually undermined by the sea, thus regularizing the coastline.

3.3 From Punta Guiones to Bejuco Beach

This 19-mile-long sector is very trimmed. It essentially corresponds to sandstone outcrops of the Rivas Formation. Littoral points are fitted in large cliffs up to about 330 ft, enclosing horseshoe bays (Garza, Barco Quebrado, Buenavista, Samara, Carrillo, and Camaronal beaches), forming the most beautiful sandy beaches (Figs. 9.8 and 9.9).

Sandstones in this area have a marked dip of 45 degrees. They behave a little like resistant rock, making impressive active cliffs, as well as marine abrasion platforms. The Samara marine platform is more than 1650 ft wide at Samara Point. Sandy material comes mainly from sandstones and some scraps of corals and calcareous algae, but there are currently no real coral reefs.

Two erosion levels are always present following this littoral: a first, which is situated around 50 ft, and a second around 82 ft high. They are always abrasion levels in the rock altered in situ. Correlative deposits have not been discovered. The lower level has two steps: one to 20–33 ft and another to 40–50 ft. These cyclical elements can be observed not only along the coast but also in the small basal plains level

FIGURE 9.8 Culebra Bay and Hermosa Beach, where the Nicoya Peninsula begins. The northern part is covered by a discordant ignimbrite over the folded Cretaceous and Tertiary formations, forming a topographic plateau. *Aerial oblique image, School of Geography, University of Costa Rica (2008).*

inside the bays (Flandrian fillings). They give a little high hilly landscape of subequally reduced summits imposed on the Flandrian level.

3.3.1 *Playa Bejuco Bay to Cabo Blanco Cape*

This is a great rectilinear sector of sandy beaches (Bejuco Beach, 2.5 miles; Coyote Beach, 5.6 miles; Santa Teresa Beach, 6.2 miles) with the exception of Cabo Blanco Cliff. Bejuco and Coyote points serve as support points for the aforementioned beaches. Behind them grows some small mangroves. A fault escarpment of about 330 ft high characterizes the plain of Santa Teresa. This escarpment is a pre-Flandrian inactive, or dead, cliff.

3.3.2 *Between Cabo Blanco Cape and Ballena Bay*

This is a 12.5-mile-long sector. The coast is closely related to the NE–SW tectonic alignments, which translates into a 330-ft inactive cliff, straight and very continuous, dominating the small Montezuma hamlet. On the base of this inactive cliff, without a coastal platform, colluvial deposits are partially attacked by the sea. The cliff is at the same time a NE–SW fault escarpment. The cliff rests in gray sandstone of the Nicoya Complex, which, in turn, rests in this same basalt complex. The top of the cliff is spun in fossil shale from the late Miocene (Montezuma Formation). This sector is very interesting because it has a cyclic morphology with several very clear levels between 0 and 850 ft, but mostly from 330 ft (Cobano level: 525 ft) (see Fig. 9.10)

In the fishing village of Montezuma, between 0 and 23 ft, a lumachelle outcrop emerges on an ongoing base from the beach, where it rests on Cretaceous sandstone and rises to 23–26 ft above the current sea level. It is a formation composed of thin sandstone beds, well stratified, a result of the consolidation of thin grayish sand with a rich fauna: turritellas, olives, cerites, and various lamellibranchs. Some beds are of coarser sands, even thick, alternating with the finest sandy beds. This gives the appearance of a shallow marine deposit and not of a beach

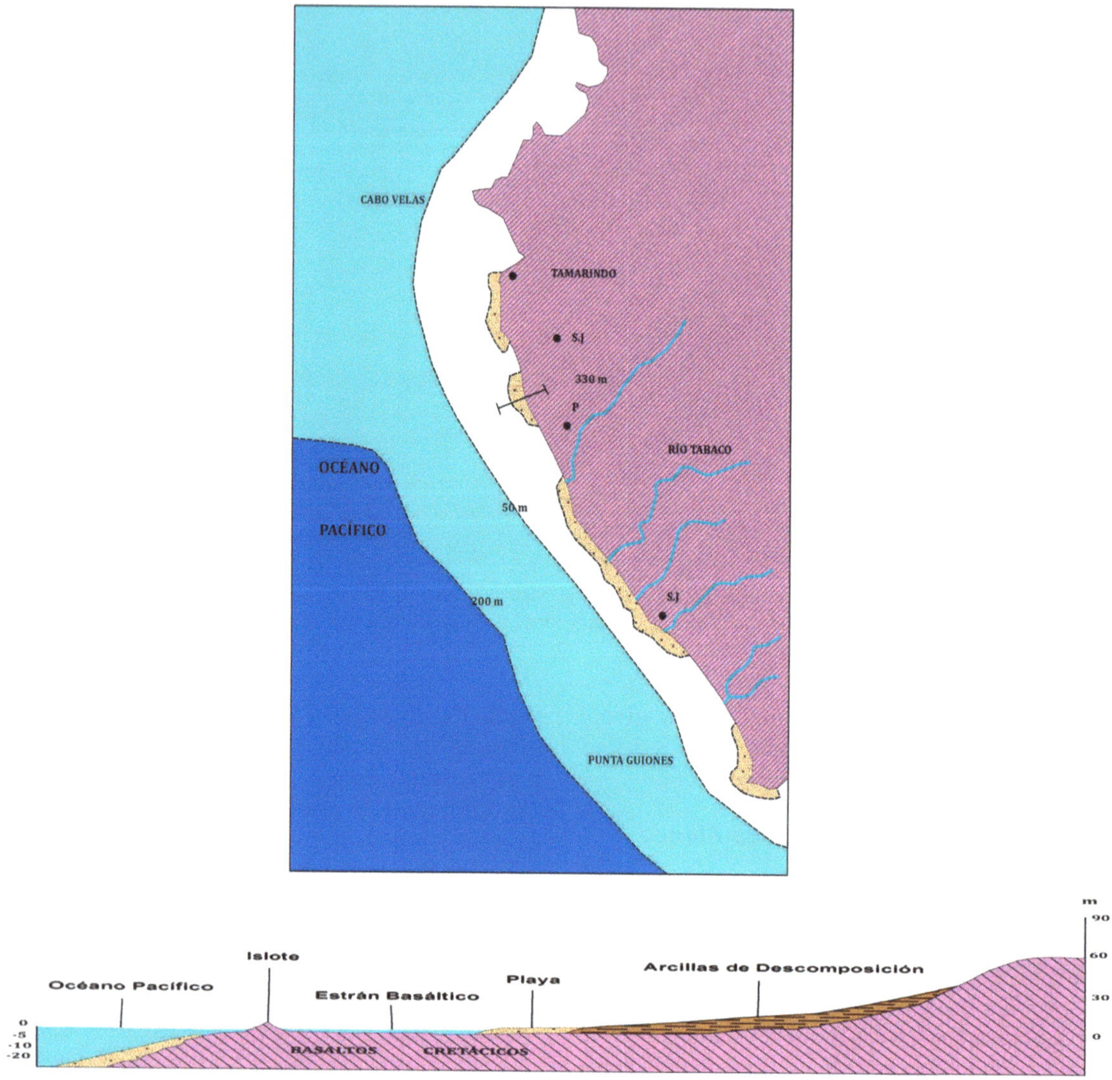

FIGURE 9.9 Nicoya Peninsula rectilinear coastline sector between Tamarindo and Punta Guiones. Bajo Grande beach profile with basalt platform (Bergoeing et al., 2010d).

because of the subhorizontal stratification and not the rolled shells. It was this formation of shells that built the angler's hamlet of Montezuma into what today is a tourist's resort. Radiometric dating of this lumachelle fossil deposit was carried out by G. Delibrias using the ^{14}C method in Gif-sur-Yvette (Gif 5480), France (see Fig. 9.11). The age obtained was 6620 ± 150 years BP, which corresponds to the beginning of the arrival of the Flandrian Transgression (13 ft) from the current sea level. This is a neotectonic vigor test in this sector because that deposit raised the seabed 23 ft in altitude, which could reach a total lift of 40 ft (6.5 ft per millennium). Obviously, this value is only local, given that the sector is very densely fail. Exhumed from

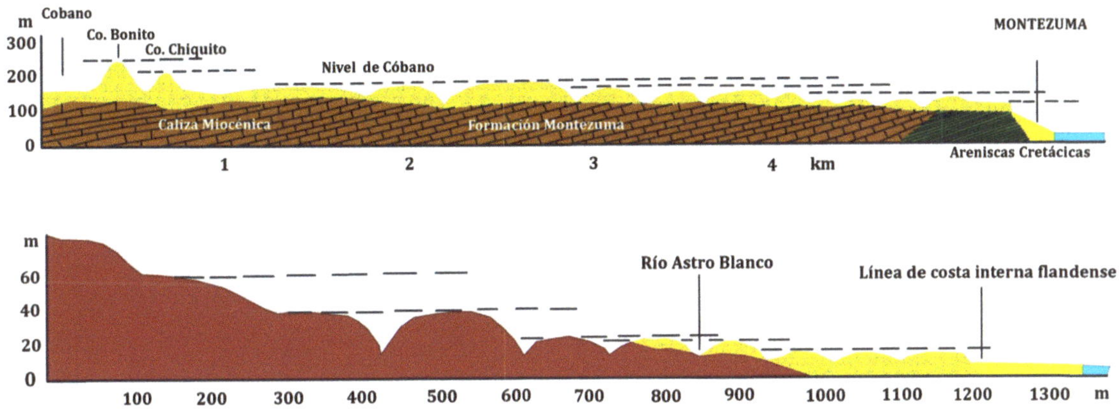

FIGURE 9.10 Cobano erosion levels (Bergoeing, 2007).

the sea and close to the escarpment deposit, the tectonic fault has been able to raise the floor locally in a brutal way. At the summit of the inactive cliff begins a first level at 262 ft and then a second that fits at 328 ft, a third widespread level is that of Cobano, 525 ft, and then Cerro Chiquito, 656 ft, and finally Cerro Bonito, at 853 ft. These levels can be related to uplift across the peninsula points. Its conservation in large areas is in contradiction to the tectonic faults that appear to have been active until a very recent period (Montezuma).

3.3.3 *Between Ballena Bay and San Lucas Island*

This sector covers 16 miles as the crow flies and is characterized by a very jagged coastline of deep bays and numerous islands (Tortuga, Negritos, Cedros, ...). It is a sector dominated essentially by neotectonics [succession of anticlines and synclines, faulty and folded in sandstone of the Rivas and Sabana Grande formations (Cretaceous) and Las Palmas Formation (Paleocene)]. The role of plate tectonics has allowed some cliffs to reach up to 660 ft (Punta, Tambor Point, Curú, and Punta Quesera). The bays are home to mangrove swamps, away from the coastal cord, unlike those that are directly exposed to the swell of the Pacific. At the other bays open to Nicoya Gulf, mangroves progress toward the sea, and this part of the coast is characterized by net cyclic levels: levels of erosion in sandstone of the Rivas Formation, filling levels carved in small hills of subequal altitudes.

The sector of Hacienda San Nicolas, Tambor Bay, shows the following succession:

- **Current cord**: 650 ft wide, with crests of net beaches on the outer side and covered with colluvium inward.
- **A small continuous uneven ground** 10–13 ft in altitude, with the appearance of a small inactive cliff, which is notched at 26–33 ft altitude. This level consists of a filling of small blocks taken in yellow clay, which includes waste from travertine sandstone, which appear to be identical to those of Montezuma. In this place lumachelles have not been seen. This level would be Flandrian if there were an analogy between lumachelle sandstone debris found there and the Montezuma lumachelles.
- **Hills of subequal altitudes and flat peaks**, dominating the previous unit. They result from a level dissected between 50 and 65 ft altitude, which in some places contains altered rock in place and elsewhere shows a filling of small blocks trapped in a yellow clay, identical to the level just described.

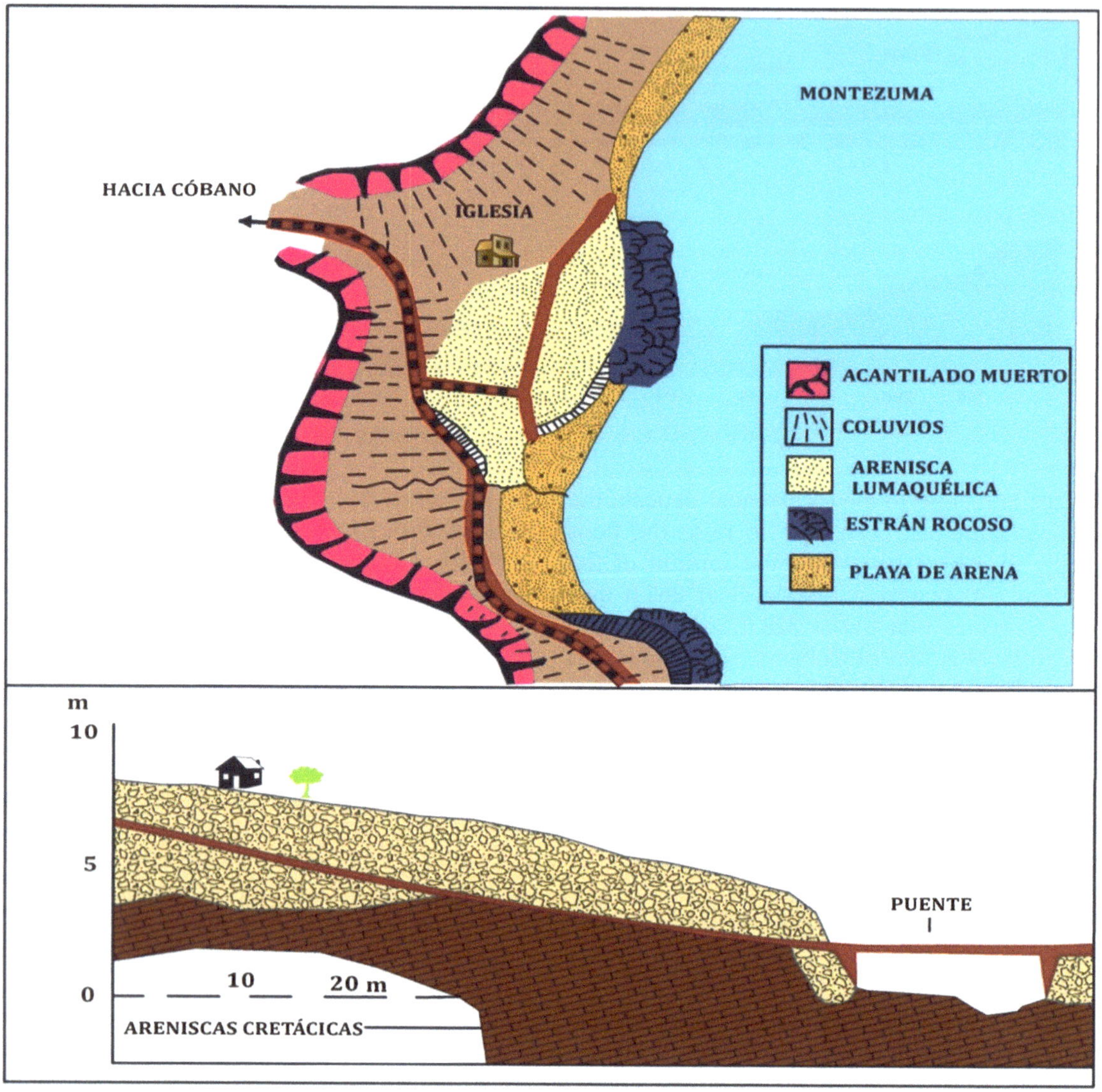

FIGURE 9.11 Location of Montezuma, where stands a Flandrian level of 23 ft above sea level. The deposit was dated to 6620 years BP (Bergoeing et al., 2010d).

- **Finally, hills that result from a dissected level of 115–130 ft altitude**, entirely carved in more or less altered rock in situ. Above, it is possible to recognize the remains of several other higher levels but which are less well preserved.

4. NICOYA GULF SECTOR

The several islands near San Lucas, as well as the surroundings of Punta Naranjo, are also witnesses to cyclic levels. Here are three net sectors with levels of 26–33, 80, and 130–164 ft.

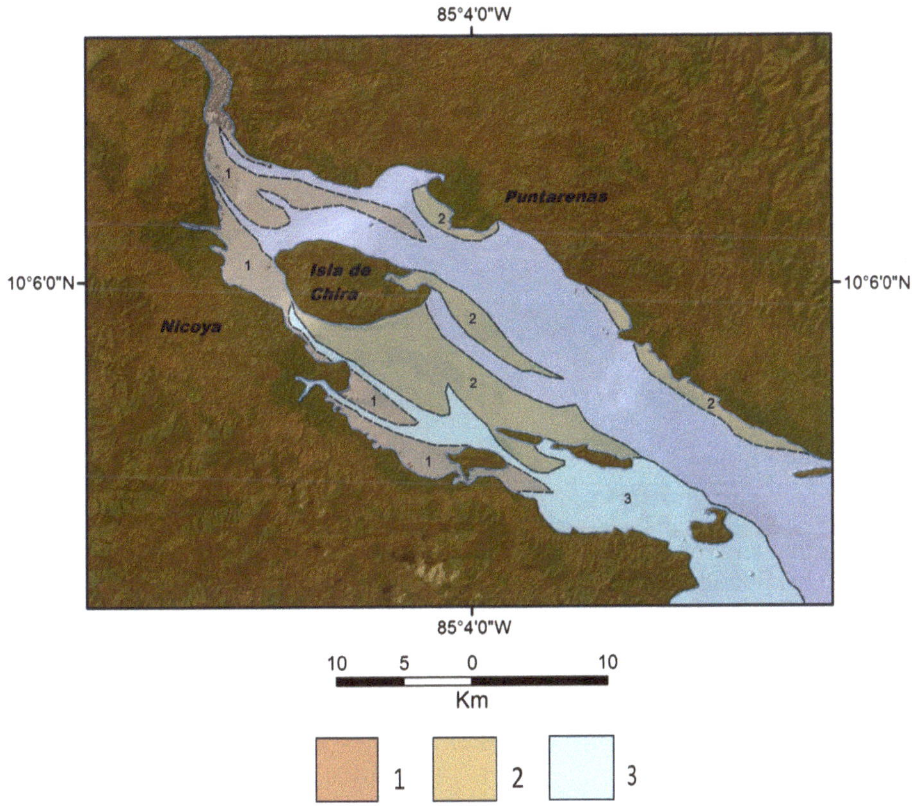

FIGURE 9.12 Gulf of Nicoya bordered by mangroves in a shallow marine environment of low depth (Bergoeing et al., 2010d). (1) Dense fluvial sediments in suspension. (2) Sediments in suspension. (3) Diluted fluviomarine sediments.

From Punta Naranjo the Gulf of Nicoya coastline is a plain that expands progressively toward the Tempisque river mouth and is occupied by a vast mangrove coastal cordless. The coastal plain is the result of fluvial filling, which translates into vast coalescent flattened alluvial fans evolving into deltaic cones. Contacts between the coastal plain and the interior are constituted through multiconvex-shaped hills under a forest cover of xerophytes.

Between Naranjo Beach and the town of Nicoya, the road cuts allow you to see an old glacis filling, fitted 33–50 ft high with microhills, which dominate the lower glacis. These glacis show a paleoclimate change and correspond to drier climate or more in contrast to the current one, with disappearance of a part of the forest cover. The last dry phase would be recent, Holocene (Battistini and Bergoeing, 1984). The Nicoya Gulf can be divided into two sectors: an internal sector between the Tempisque river mouth and a line joining the Puntarenas arrow to San Lucas Island, and a southern external sector beyond this imaginary line (see Fig. 9.12).

The internal Nicoya Gulf sector is characterized by a bottom usually 7 to 13 ft deep that can reach in some places 65 ft deep, creating banks of sand and clays, especially near the

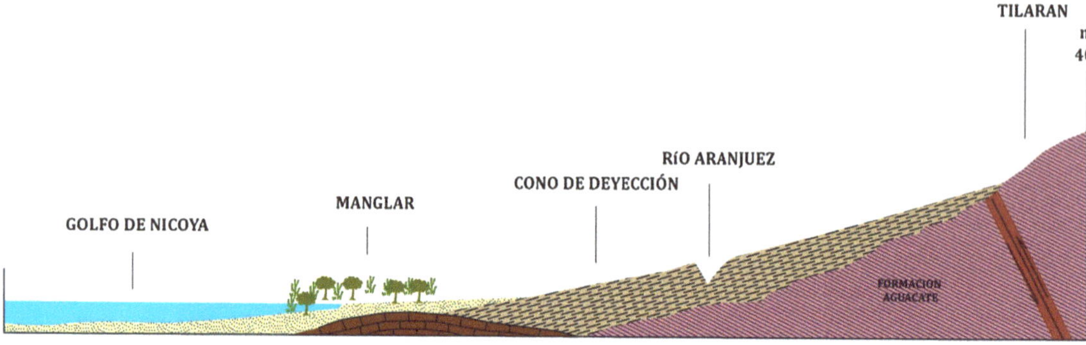

FIGURE 9.13 Nicoya Gulf mangroves and geological profile of the Aranjuez River sector with Tertiary tilted and folded sedimentary series cut by the Great Reverse Fault of Costa Rica (Bergoeing et al., 2010d).

Tempisque river mouth and Chira Island, the largest of the Gulf of Nicoya, and others like Bejuco and Venado islands, making a barrier effect. These islands are formed of volcano-sedimentary folded and faulted outliers, ranging from the Cretaceous to the Middle Tertiary. At the Tempisque river mouth, the tide can felt up to13 miles inland (Puerto Humo Port). This inner part of Nicoya Gulf is also characterized by the mangrove, *Rizophora mangle*, without coastal cords (see Fig. 9.12).

Chira Island is a synclinal depression keeping in its center a vast mangrove swamp. Some islets, like Berrugate or Punta Piedra, are surrounded by river mud that is fossilizing them inexorably, uniting them, little by little, with the continent. Here, mangrove has the role of setting the new muddy contributions. Vaster extensions are in the Estero Morote and Estero Letras estuaries, on the right bank of the Tempisque River. Farther south, in Puntarenas, another vast mangrove swamp also thrives.

5. PUNTARENAS SEA ARROW

This is perhaps the most beautiful section of the Costa Rican Pacific coast. Nine miles long, it is essentially formed of sand. It has two parts: a modern one, represented by the arrow where Puntarenas Port stands, and an old part formed by a stretched peduncle of beach crests, fossilized in part by the Barranca River delta cone. The Barranca River also has built fluvial terraces near its mouth, which we have already described (see Fig. 9.13).

The Puntarenas sea arrow is almost completely formed of sediments carried by the Barranca River, which descends toward the sea almost as a torrent because of its deep slopes. These contributions have, on one hand, allowed the construction of a powerful cone more than a delta and, on the other hand, carried thin elements for the arrow construction itself. If the arrow still exists, it is thanks to the construction at the end of the 19th century of the railroad line to Puntarenas, as the railway ballast was use as an embankment against coastal erosion (Fig. 9.14).

Thanks to the presence of this arrow, a vast mangrove swamp progresses in a nude inner zone, north of Puntarenas. The main species are *R. mangle*, *Avicennia nitida*, and *Avicennia bicolor*.

6. THE CENTRAL PACIFIC LITTORAL OF COSTA RICA

Costa Rica's central Pacific littoral is characterized by the large role played by tectonic accidents oriented NE–SW and NW–SE and combined with them the presence of the tectonic

FIGURE 9.14 Puntarenas sea arrow. There is a vast mangrove area to the left of the sea arrow peduncle, in the background. Maximum lift is only 10 ft above sea level, and thus the habitat is threatened by tsunamis and the imminent rise in the sea level during the 21st century. *Aerial photo, courtesy of Jean Mercier (2007).*

Great Reverse Fault of Costa Rica. This part of the coastline comprises several sectors.

6.1 Between the Barranca and the Tarcoles Rivers

The Aguacate Mountain Range piedmont is situated between the Barranca and the Tarcoles rivers, crossed by a grid of tectonic faults that still follow the predominant directions listed earlier. This piedmont evolved during the Quaternary period as a plateau of tectonic compartments. Each block has played an independent role thanks to neotectonics. The plateau stretches over 154 square miles between the internal contact (Costa Rica reverse tectonic fault) and the coast. The canyons of the Surubres, Jesús María, Machuca, and Tarcoles rivers carve it. Described geological profiles of the Orotina piedmont (Tilaran–Aguacate mountain ranges) briefly give the following distribution: at the summit, a series of 160–200 ft of clayey piedmont deposits resting on a thick series of ignimbrites (330 ft) with intercalated basalt lava flow passages. The whole rests on a tilted plateau structure formed by Miocene sandstone and shale (Punta Carballo Formation) (Fig. 9.15).

The Orotina piedmont ignimbrite basement, dated by the K/Ar method, gives an age of 1.28 MY. If we take the Punta Carballo cliff, which rises up to 500–660 ft above the current sea level, it gives us an idea of the lifting of this part of the coast from the Lower Quaternary, with a value on the order of 3 ft per millennium (taking into account local tectonic fluctuations) (see Fig. 9.21). The coastline follows as a series of cliffs with littoral points notched in the Miocene sedimentary series. This is a problem for the new Caldera Port, which is supported by a major fault escarpment of Alto de Las Mesas plateau. In the lower course of the Tarcoles River, as we described for the Barranca River, boxed fluvial terraces are present, giving fluvial levels of 130, 33, and 16 ft (see Fig. 9.16).

The predominant fracturing directions of NW–SE and NE–SW have determined in all the piedmonts the hydrographic network, but they have also been reactivated, because the present

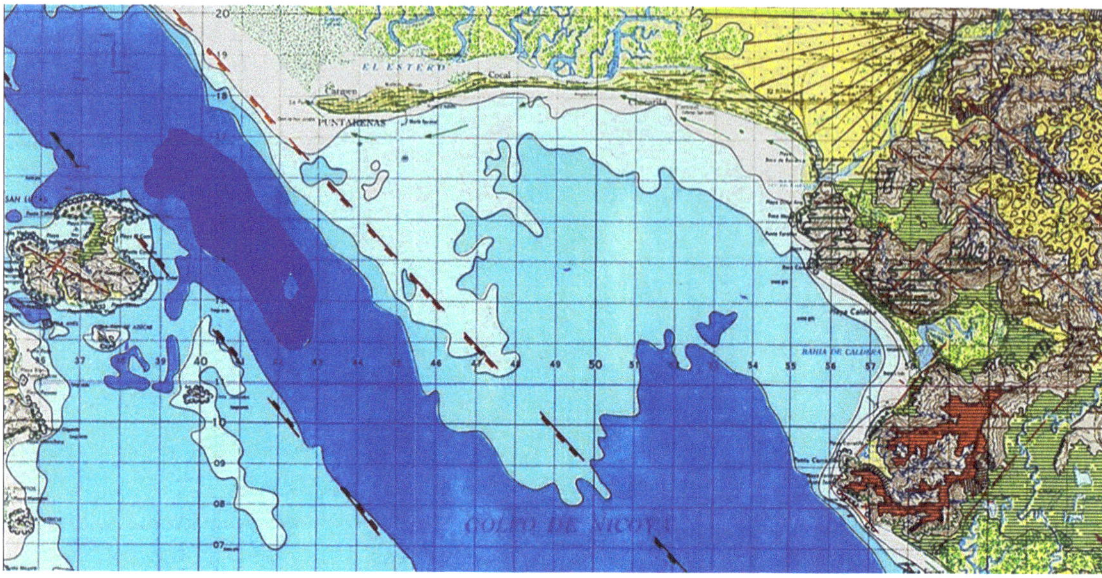

FIGURE 9.15 Punta Arenas sea arrow, fed by the Barranca River forming a vast alluvial–deltaic fan. At its mouth, the outcropping rocks include Tertiary sandstone and shale (Punta Carballo and Barranca formations). A tectonic depression on the bay is a prolongation of the Tempisque River graben. *Taken from North Pacific geomorphological map of Bergoeing et al. (1982b).*

130-ft Tarcoles River level implies a very recent neotectonics. Near the Tarcoles River, the sector is characterized as a volcanic-sedimentary horst, where the faults are predominantly NW–SE. The Cretaceous and Tertiary formations that emerge give a coastline of active cliffs in the Nicoya Complex basalt with small abrasion platforms. The reticule of faults is once again what determines the layout of the coast, giving a littoral shape of active cliffs, separated by sandy horseshoe bays like Herradura Bay and inlets, with an altitude of 330 ft.

Conejo, Bocana, and Guapinol points do not allow us to discover traces of cyclic morphology; however, farther south, behind Playa Hermosa, the Tusubres River has contributed in part to the construction of two fluvial levels.

There are two lower levels of 16–20 ft and 50–60 ft, fluvial but with small boulders, which dominate a low current clay terrace (stratified ochre yellow silt). Upstream are seen three levels ranging up to 490 ft above sea level. They are erosion levels of a Tertiary sedimentary series (sandstone and shale of the Punta Carballo Formation) (see Fig. 9.17).

6.2 Parrita–Palmar Sector

This sector is 74.5 miles long. The internal contact is closely linked to the tectonic inverse fault of Costa Rica. It is a beautiful coastal plain, which progressively thins toward the southeast to disappear completely after Dominical Beach. Beyond Dominical, the littoral is a longitudinal faulty coast. Outside of the accident of Quepos Promontory, the sector presents as a low plain of sandy Flandrian cords. The sector can be separated into two parts because of the Quepos Promontory:

- Parrita northwest sector
- Quepos–Dominical southeast sector

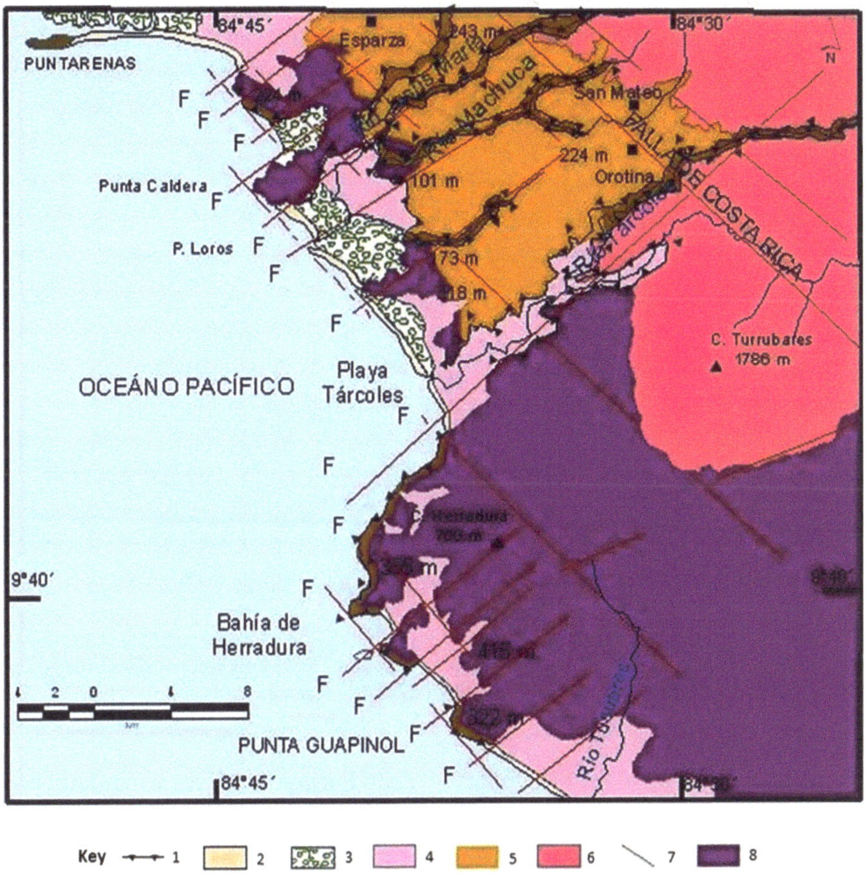

FIGURE 9.16 Littoral geomorphology between Puntarenas and the Tarcoles River Bergoeing and Brenes (2013). (1) Coastal of active cliffs. (2) Flandrian littoral cord. (3) Mangrove. (4) Flandrian fed littoral. (5) Orotina piedmont deposits from Aguacate Mountain Range. (6) Pliocene volcanic area (Aguacate Formation). (7) Principal tectonic alignments. (8) Cretaceous Nicoya Complex, volcanic and sedimentary outcrops and Tertiary sedimentary rocks (Punta Carballo Formation).

6.2.1 Parrita Northwest Sector

The Parrita northwest sector is a plain that reaches 6.2 miles wide formed by sandy cords and mangroves, covered by many rivers, of which the most important is the Parrita River, which has built a powerful Flandrian alluvial fan. The Paquita River, on the other hand, has built a significant delta where the lower course is formed by divergent meanders. The Flandrian plain is characterized by the presence of remains of ancient alluvial fans inside, staggered to 660 ft ahead of the Tertiary sedimentary hills forming banks of pebbles. We think that there are ancient mudflows such as are found in the north in the Orotina piedmont and Tilaran mountains. These alluvial fans then have a net paleoclimatic significance. Of these alluvial fans only the roots exist, preserved thanks to neotectonics.

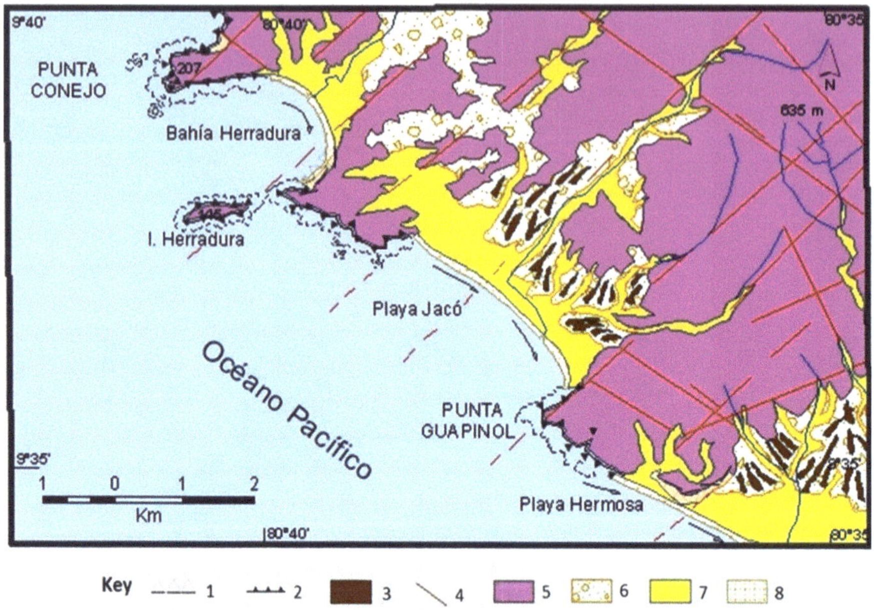

FIGURE 9.17 Costa Rica Pacific Ocean littoral geomorphology between Punta Conejo and Playa Hermosa (Bergoeing et al., 2010d). (1) Active cliffs. (2) Alluvial paleodelta fans. (3) Tectonic faults. (4) Nicoya ophiolitic complex outcrops and sedimentary Cretaceous rocks. (5) Pleistocene sediments. (6) Flandrian plain. (7) Littoral sandy cords.

From Vueltas sector to the center of the plain, there are sandstone hills where one can discern three not correlative deposit levels (260, 200, and 115–130 ft). Below, those hills are surrounded by two river levels of 50 and 16 ft formed by puddingstone and clay beds.

6.2.2 Quepos–Dominical Southeast Sector

The Quepos Promontory that separates this great plain is a set of Tertiary hills, with multiconvex shape, under an ombrophilic forest cover disappearing because of anthropic factors (see Fig. 9.18).

The Tertiary sedimentary series rests on a pillow lava base attributed to the Nicoya Complex (Azema et al., 1981). The sector is cut with a NW–SE tectonic fault system predominance (Punta Serrucho) (Bergoeing, 1978c). The coast presents itself as a very jagged coastline with cliffs carved in limestone and sandstone of the Brito Formation. Here we did not observe terrace levels (see Fig. 9.18).

Southeast of Quepos Promontory the coastal plain is higher in the internal part, made up of powerful alluvial fans at the mouths of the rivers, which below evolve to glacis. In these glacis rivers form vagrant meanders. The presence of the Great Reverse Fault of Costa Rica is felt, because these alluvial fans are tilted, forcing rivers whose bottoms are NE–SW to describe a curve to the south.

The coast is low and sandy with long coastal cords. The littoral drift takes an eastern direction, allowing mangroves to flourish in this sector protected by coastal arrows. The coastal plain disappears in at the Baru river mouth. Here the coast is bordered by a rectilinear inactive cliff of 330–490 ft, being at the same time the escarpment of the tectonic inverse fault of Costa Rica.

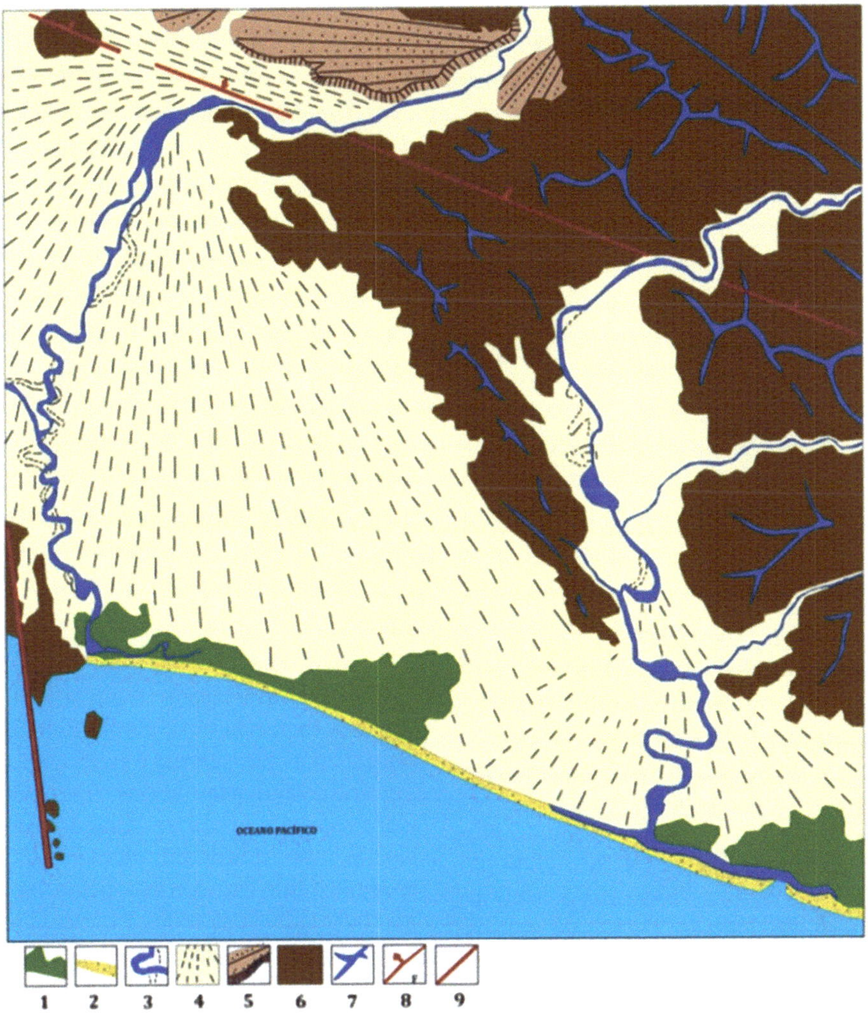

FIGURE 9.18 Naranjo–Savegre River geomorphology, altered by neotectonics by southeast tilting of the lower Naranjo River course (Bergoeing et al., 2010d). (1) Reduced mangrove in some sectors due to anthropic action. (2) Sea arrows and Flandrian littoral cords. (3) Rivers with abandoned meanders. (4) Alluvial fans tilted by neotectonics. (5) Alluvial fan terrace level of Naranjo River. (6) Tertiary sedimentary outcrops (sandstone and shale with volcanic intercalations. (7) Watershed divide line. (8) Costa Rica Reverse Tectonic Fault. (9) Normal tectonic faults.

7. BARU RIVER–CORONADO ELEVATED COAST SECTOR

This is a high coast that stretches 22 miles, guided by the Great Reverse Fault of Costa Rica and which can be classified as a longitudinal faulty coast without significant notches. Outside of the small plains drowned by the Higueron and Uvita rivers (Punta Uvita Tombolo) (see Fig. 9.19), the cliff is attacked directly by the sea until the Tortuga Abajo area. From this point, and to the Terraba river mouth in Coronado,

FIGURE 9.19 Punta Uvita Tombolo built by the encounter of two littoral converging drift currents forming two coastal cords due to the presence of a reef in the sea. *Oblique aerial photo courtesy of Alonso Badilla (2013).*

the escarpment turns into a 660-ft-high inactive cliff. This sector presents a general erosion level situated toward 82 ft altitude with no correlative deposits. Apart from the Holocene base small alluvial fan/glacis, rewaisted from 7 to 10 ft, there are no other cyclic intermediate levels.

Above 80 ft, we discover other levels near Coronado, which stand at 130–165, 250–300, and near 820–985 ft. The levels have trimmed steep hill slopes (35 degrees) that have evolved under forest cover.

8. THE PACIFIC SOUTHEAST COAST SECTOR OF COSTA RICA

This sector is characterized by vast coastal plains created by powerful rivers such as the Rio Grande de Terraba and reliefs such the Osa and Burica peninsulas, raised during the Quaternary period. The low coast is usually a dense mangrove forest, whereas high sectors present active cliffs, usually carved in basalt, with a fundamental role of tectonics.

8.1 Coronado Sector and Diquís Plain

This sector is largely the result of the Terraba River, which has built a vast delta protected by mangroves. No swampy parts are enhanced by agroindustrial plantations (bananas). Between Coronado and Palmar village, crossing Puerto Cortes, the dominant element is always the tectonic fault escarpment fitted as a Flandrian inactive cliff. It is from Puerto Cortes that we see a series of cyclic levels encountered along 43.5 miles of the southern Inter-American Highway.

The cut of this sector gives the following levels:

- **Coronado**: Carved at 6.5–10 ft, it is more recent than the Flandrian transgression (6000–7000 BP).
- **Alluvial fans/glacis**: These are flat and extensive, formed of badly worn boulders as well as small blocks. They are carved from 16 to 20 ft (example: Benjamin river mouth in Palmar Norte). Also, there are small embryonic alluvial fans similar to those described before.

- **Remains of a terrace of large boulders**: They present as well rolled and are 26–33 ft in relative height. It is this same terrace that can be found south of the Palmar Bridge and classified as an upper terrace of the Terraba River. The terrace rests on the altered bedrock and is 10–13 ft thick. It consists of boulders of volcanic origin, with altered surface edges.
- **Multiple levels of erosion:** They are characterized by a notorious horizontality. Razing the hills at their contact with the coastal plain and situated at 65–82, 164, and 250–260 ft, here are discovered correlative deposits. A vast mangrove forest protected by 8 miles of coastal cords occupies the Terraba delta. The shallows of Coronado Gulf partly explain the strong progression of this deltaic area, favored by major sedimentary deposits of the strong Terraba River. The current Terraba river mouth drains to the north, because of a general tilting of the sector by tectonic reactivation.

8.2 Osa Peninsula

The Osa Peninsula draws a NW–SE-oriented rectangle, completely dominated by tectonics. The whole is tilted toward the northwest. The interior of the peninsula is occupied by a rain forest, where gold diggers venture, which makes it doubly dangerous. Outliers in general are basalts giving tabular structures, especially to the northwest of the peninsula. The Corcovado National Park plain today is the result of a collapsed block (graben) occupied by a vast swamp (Laguna de Corcovado) (see Fig. 9.20).

In the Pacific sector of the peninsula, the coast is presented as a cliff at the foot of which are stuck long sandy cords, such the plain of Corcovado. The most interesting sector is undoubtedly the southeast coast between Puerto Jimenez Port and Sombrero Point (Madrigal, 1978). Here we see a tectonic fault scarp of 660 ft altitude that notches the Charco Azul Formation sedimentary series. At the foot of this fault escarpment, the plain is made up of a series of coastal terraces (16 levels) that lie between 0 and 80 ft above sea level.

The lower levels correspond to a Flandrian stepped crest, and the highest levels rise more than 26 ft, being present at 0.6 mile south of Puerto Jimenez (Bergoeing, 1978b; Battistini and Bergoeing, 1983a,b). The whole of this system, which cannot be but Flandrian by the well-preserved forms, gives testimony to the value of the uprising in the southern sector of the peninsula (7–13 ft per millennium).

8.3 Golfo Dulce Bay

Golfo Dulce is a deep bay (660 ft deep inside of the bay), 38 miles long and 12.5 miles wide. The tectonic faults of Golfito Port and the inner coast of the Osa Peninsula, oriented NW–SE and parallel to one another, give an undoubted tectonic origin. The gulf is bordered, on the northwest, by a string of hills much diminished by erosion, submerged during the Flandrian Transgression, with steep slopes, especially in the bathyal zone. They belong to the rocks of the Nicoya Complex.

The fact that this gulf, by it position, is protected from the strong swells of the Pacific, arriving much diminished, has made it so these hills do not present active cliffs. At the foot of the tectonic fault scarp, 1000–1300 ft above sea level, overhanging Golfito Port, is a bay of great beauty enclosed by a small Flandrian coastal arrow that relies on an isolated relief that emerges as a former island. Here the mangrove develops, protected from the sea by the Coastal Mountain Range (see photo 89). We found two fluvial terraces along the road (6 miles from Golfito). These two levels are at 33 and 164 ft relative altitude, formed by limestone boulders in large part.

FIGURE 9.20 Geomorphology of Osa Peninsula from the Sierpe river mouth. Observe the Cretaceous–Tertiary plateau landforms of Osa Peninsula and the tilted bottom of the gulf. Atlas of Costa Rica, 1:100.000 scale, Bergoeing and Brenes (2016). UNA–Costa Rica.

8.4 Burica Peninsula

This is a long stretched land, sinking into the Pacific, following a north–south orientation. It marks the international border with Panama. This stretched peninsula is composed in its northern part by Nicoya Complex igneous rocks, alternating in the southern part with the sedimentary, monotonous, much-raised sedimentary series of the Puerto Armuelles Formation (Lower Pleistocene). It is of recent tectonic origin, because here is where three tectonic plates converge: the Cocos and Caribbean plates and the Nazca plate that slides north with the Cocos plate. This explains the strong seismicity of the region (Miyamura, 1980).

The western coastline of Burica presents as a megacliff more than 1650 ft above sea level and about 6 miles long, which is nothing more than a tectonic scarp fault retouched by the sea. On the east coast, on the other hand, it is gently immersed in the sea (sector of Puerto Armuelles Port) (see Fig. 9.21).

9. BASALT AND SANDSTONE LITTORAL PLATFORM MORPHOLOGY

Basalt and sandstone are the most common rocks on the Pacific coast of Costa Rica. Basalt generally corresponds to Cretaceous marine volcanic deposits (pillow lava of the Nicoya Complex), whereas the sandstone is rather Tertiary and is attached to the old marine volcanic structures.

9.1 Example of the Basaltic Littoral Platform of San Francisco Point

This is located south of Tamarindo Bay (Nicoya Peninsula). It is a very faltering platform but does not present a pillow lava structure. From the cliff are recognized the following elements:

- **On the cliff base,** 3–7 ft of pierced rock of the small plug tray sponge type
- **In the middle part of the cliff,** healthy or little altered basalt
- **In the upper part of the cliff,** red laterite from basalt alteration
- **The vast horizontal corrosion platform.** It can reach 100 ft wide. It is formed of perforated rocks of 0.3 to 0.6 ft and small accidents that dominate the platform.
- **An external wall** (width: 33–100 ft) with a chaotic topography in basalt. The internal corrosion platform exceeds 60–100 ft. It is made up of small pieces of corrosion platform. It is directly attacked by the swell of the Pacific, which is a nonchemical but mechanical erosion, as is the case for the interior sector.

Many basaltic platforms of the Costa Rica Pacific coast do not have all this zoning. Only less fractured sectors are able to become flattened structures, so marine corrosion could form elements of stepped platforms (see Fig. 9.22).

9.2 Sandstone Sea Platforms

The powerful sandstone banks of the Pacific coast are often rich in limestone, with spherical forms that present as an "onion peel." Some feet from the cliffs, as in Punta Morales Point or Playa Naranjo Beach in the southeast of the Nicoya Peninsula, compact sandstone spherical forms are featured, but they are full of sowing in the form of a honeycomb (can reach up to 0.6 ft diameter) (see Fig. 9.23).

9.3 Samara Sandy Beach Platform

- **Active cliff at 115 ft** altitude in the sandstone fine laminated stratification with a slope of 45 degrees toward the northwest
- **Flandrian horizontal platform**, more or less at the level of the tide average, perfectly leveling the rock at 1000–1300 ft; this platform is characterized by:
    - sandstone intertidal hardening
    - cellular structure determined by the direction of three fissures
    - microcracks in projections from the presence of calcite

This platform is the product of differential erosion via the loss of hardened cliff sandstone and the sandstone's resistance to the platform.

FIGURE 9.21 Punta Burica sedimentary littoral point, the border with Panama, formed by sandstone from the Tertiary (Armuelles Formation). Here the Cocos and Nazca tectonic plates slide (Bergoeing and Brenes, 2016). *Geomorphologic Atlas of Costa Rica, 1:100.000 scale, UNA-Costa Rica.*

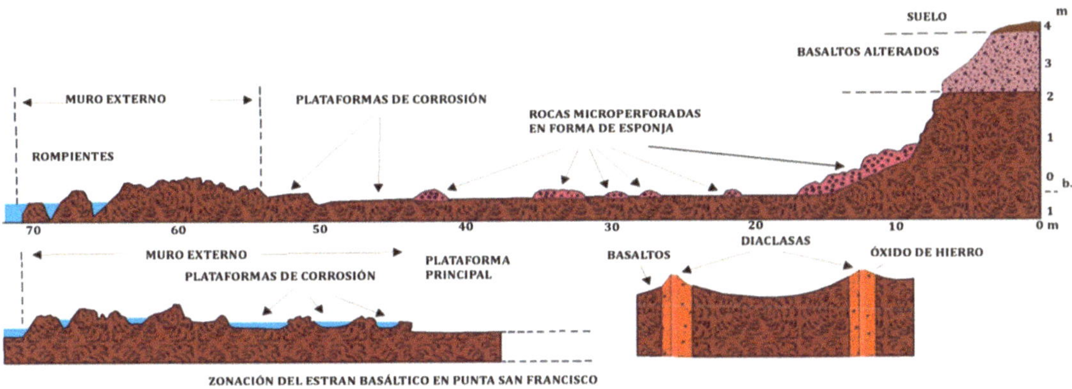

FIGURE 9.22 Basaltic sea platform of San Francisco Point northwest of Nicoya Peninsula (Bergoeing et al., 2010d).

10. CYCLIC ORIGIN LEVELS

Various staggered cyclic levels are found along the Costa Rican coast, and, in particular, to the southeast of the Nicoya Peninsula, they pose a problem because they have no correlative deposits. The fitted rock in situ is very altered. These levels, if they are perfectly horizontal, do not have regional continuity and this is due to a very active neotectonics.

In Ballena Bay, around Tambor Beach, these staggered erosion levels show their most important development. They are behind the low filling of the Flandrian plains (filling of the bays' foundations). Because of this and because of their sheltered position, there cannot be older

FIGURE 9.23 Punta Leona faulted and uplifted sandstone platform. In the background are active cliffs on a Tertiary sedimentary series. *Photo Bergoeing (2011b).*

marine abrasion platforms. Today are not present but on the ropes of San Francisco and Samara Points. These levels cannot be of continental origin either, because the flat surfaces observed are fillers. The hypothesis here retained (Battistini and Bergoeing, 1983a,b) is an alteration of rock in situ up to the coastal water table level. Pushed sheds free exposing rock and the altered level.

This process should have settled for quite long periods, which would have allowed fluctuations in the sea level and therefore in the coastal water table (Eemian high levels in the inter-Riss–Würm–Sangamonian = 140,000 to 80,000 BP). In short, it would be a process of differential continental erosion, because erosion would clean the altered area, leaving the unchanged rock. This liberated area would correspond to the level of the coastal mantle at a given time of the corresponding sea level.

11. THE CARIBBEAN COAST

The Caribbean coast of Costa Rica is located between 11°00′ and 9°39′ north latitude and extends over 124 miles between the border with Nicaragua to the north and the Sixaola river mouth, which serves as the border with Panama.

This coast divides into two sectors:

- **A Northern sector** of 75 miles; formed by a low, sandy, rectilinear coast
- **A Southern sector**, from Puerto Limon, with ribs, which gives a high rocky coast

The tide level is low, 2.4 ft between high tide, with higher levels observed, and low tide, compared with the increase in the sea level on the Pacific side (13 ft in Puntarenas) (see Fig. 9.24).

Climatologic data are those of a humid tropical coastline, with no real dry season. The extreme northern part of Barra del Colorado is the rainiest, with 5451.5 mm of annual rainfall against 4686.7 mm in Puerto Limon (Instituto Meteorológico Nacional de Costa Rica, 1972).

Frequent winds are from the north (11 days per month with forces of 9 knots), then the northeast (11/10) and the southwest (19/5). The winds from the east and southeast, though less frequent, are generally strong (8–9 knots). The Costa Rican Caribbean coast is, by its geographical position, away from the path of hurricanes, and strong surges are exceptional. The Seawater has a temperature of 83.3°F (1972 US Coastal Geodetic Survey—IGN-CR, 1972). The sea temperature never drops below 71.6°F, which allows coral life, but this is embryonic and thrives only in the sector of Punta Cahuita reef.

11.1 Northern Sector

The northern sector is part of the Nicaragua graben, dominated to the west and to the south by the volcanic mountain ranges of Guanacaste and Central, in horst position. To the north is the San Juan River border with Nicaragua, and the small alkaline volcanic effusions of Tortuguero and Cerro Coronel characterize this part of the country (Tournon, 1972). Puerto Limon is located in a middle area between Lake Nicaragua to the north and the raised southern littoral: the division of the coast into two large morphologically distinct sectors seems to correspond to one major tectonic reality (Battistini and Bergoeing, 1982a,b) (see Fig. 9.25).

In this sector, the Flandrian parallel construction of coastal cords has been important. It includes a coastal band of 6–10 miles in some sectors, which has allowed the rise and consolidation of the vast level of base plains, covered with rainforest. The importance of coastal sedimentation is explained by copious deposits of sediment provided by the rivers that descend from the volcanic mountain ranges of Guanacaste, Tilaran, and Central, also draining the piedmonts at the foot of these mountains. The uniformly flat plain is marked by some isolated volcanic reliefs of 330–660 ft altitude that correspond to Plio–Quaternary volcanic activity.

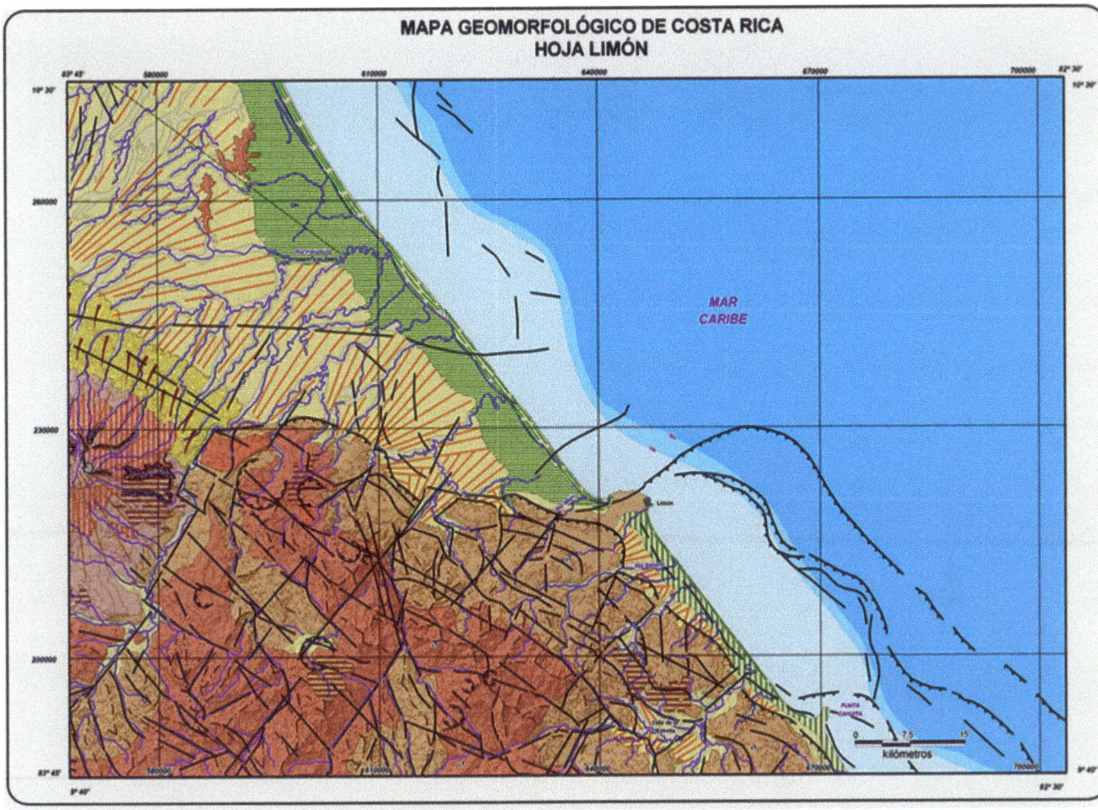

FIGURE 9.24 Costa Rica's Caribbean coast geomorphology. In green, the Flandrian Transgression maximum reach. The Talamanca Mountain Range Caribbean watershed formed by Tertiary volcanic and sedimentary outcrops as well as NW–SE orientation tectonic faults are shown. *Taken from the Geomorphology Atlas of Costa Rica, 1:350.000 scale, Bergoeing et al. Costa Rican Institute of Electricity, 2010.*

The Tortuguero Strombolian volcanic cone is the only witness to a calc-alkaline eruption on the Flandrian waterfront that served to support the development of this point. Tortuguero Volcano is a small cone of poorly welded pyroclastic rocks, which the summit of the cone rises 985 ft of altitude. It is possible that this is a recently formed Holocene volcano, which emerged first in the sea and then became attached to the post-Flandrian coastal cords (6000 years BP). The volcano is typically oceanic, made up of compact blocks of healthy gray basalt with olivine phenocrystals and augite and interstitial alkali feldspars (Tournon, 1972) (see Fig. 9.26).

Farther north, the Colorado volcanic cone with basanite outflows has been dated at 1.20 MY (Bellon and Tournon, 1978). In Nicaragua, beyond the border of the San Juan River, these volcanic reliefs had a most important development at the beginning of the Quaternary. Inside the Flandrian coastal cords developed vast lagoons parallel to the coastline. The San Juan River is a much-flattened delta at its mouth.

The Colorado River is the southern branch of the San Juan River delta. To the north of the Colorado extends a system of lagoons (four parallel draws, 660–985 ft wide and over

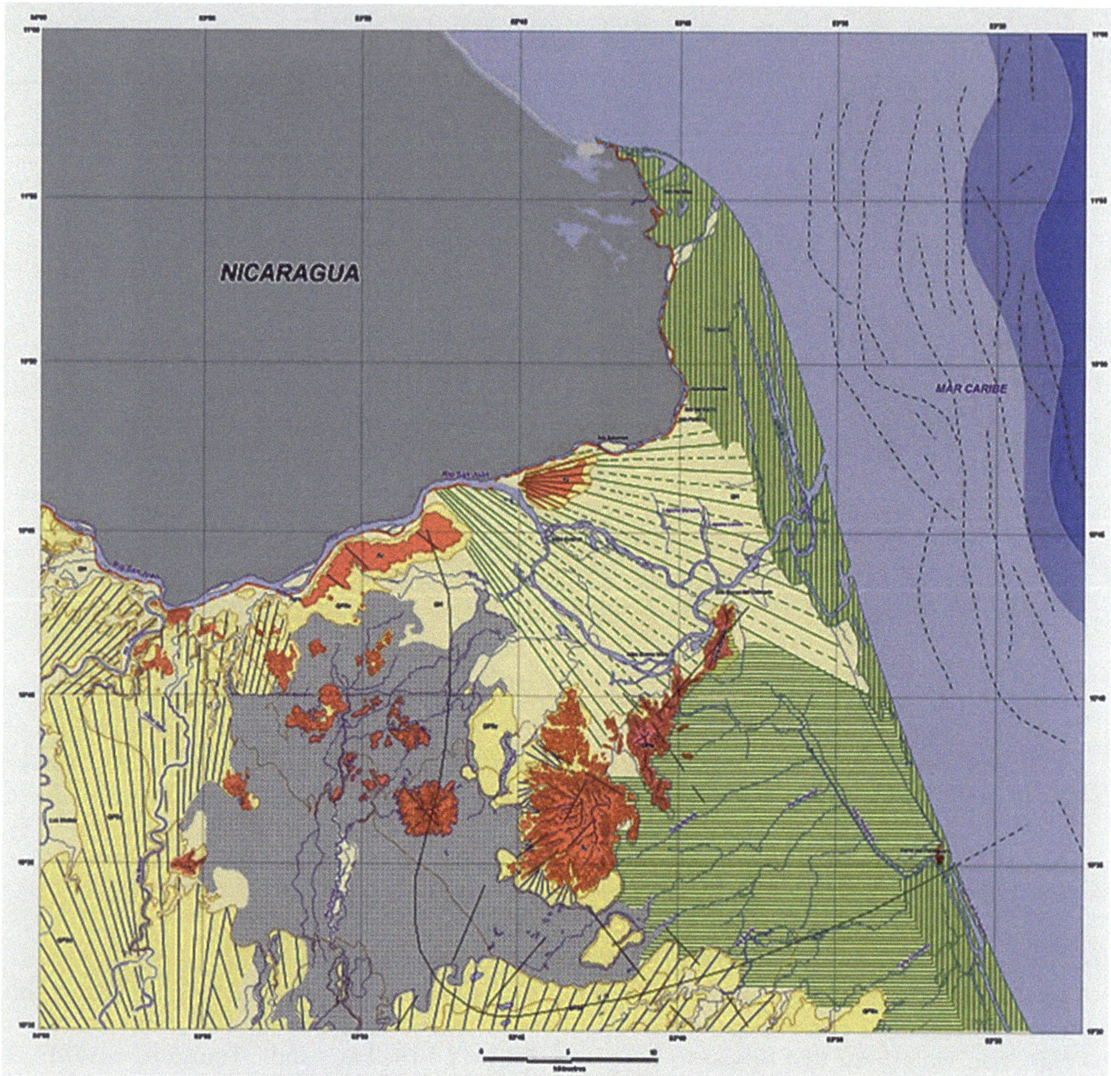

FIGURE 9.25 San Juan River delta geomorphology and Tertiary volcanic outcrops. This is a post-Flandrian sedimentation area, where the costal drift disposes fluvial sediments. The Colorado River in the delta is the main course of the San Juan River by neotectonic tilting effects associated with the Hess marine scarp. *From Bergoeing et al., 2010 Caribbean Geomorphologic Atlas—Costa Rica. 1:100.000 scale.*

1.25–1.9 miles long) that make a plain that can be compared to a very vast and very flat Flandrian delta, in which little marked surges contain vast marshes. The system of parallel strings of cords separated by lagoons is visible only in a wide strip of 3 miles along the shore.

South of the San Juan River delta, the rivers follow parallel coastal cord depressions before draining in the lagoon system and then arriving at the sea. Direct drains to the sea are few in number: the mouths of the Colorado, Tortuguero, Parismina, and Pacuare rivers and, close

FIGURE 9.26 Tortuguero pyroclastic cone resting on Holocene basalts. *Aerial photo Bergoeing et al. (2010d).*

to Limon Port, the Matina river mouth. All evolved in the same way, under the action of the south–north littoral drift, where the coastal cord construction tends to push the drain northward (see Fig 9.27).

Weak depths border this coast (−33 ft isobaths are located 1.5 miles from the mouth of the Colorado River, whereas the −66 ft isobaths lie between 2.5 and 3.7 miles from the coast). Approximately from Limon Port Promontory the isobaths progressively approach the coastline.

11.2 Southern Sector

This extends over 50 miles between Limon Port Promontory and the Sixaola River. It is much more varied and of greater morphological wealth. Between Limon and Puerto Viejo, the coastal plain starts wide at 2.5–3 miles, becoming more and more stretched toward the south, reaching the Estrella river mouth and Cahuita coral reef (averaging 0.6–1.25 miles wide). The tops of Fila Carbon row (average altitude 820 ft) dominates directly this narrow coastal plain,

FIGURE 9.27 Post-Flandrian coastal cords forming long channels but altered by human presence. *Aerial photo Bergoeing et al. (2010d).*

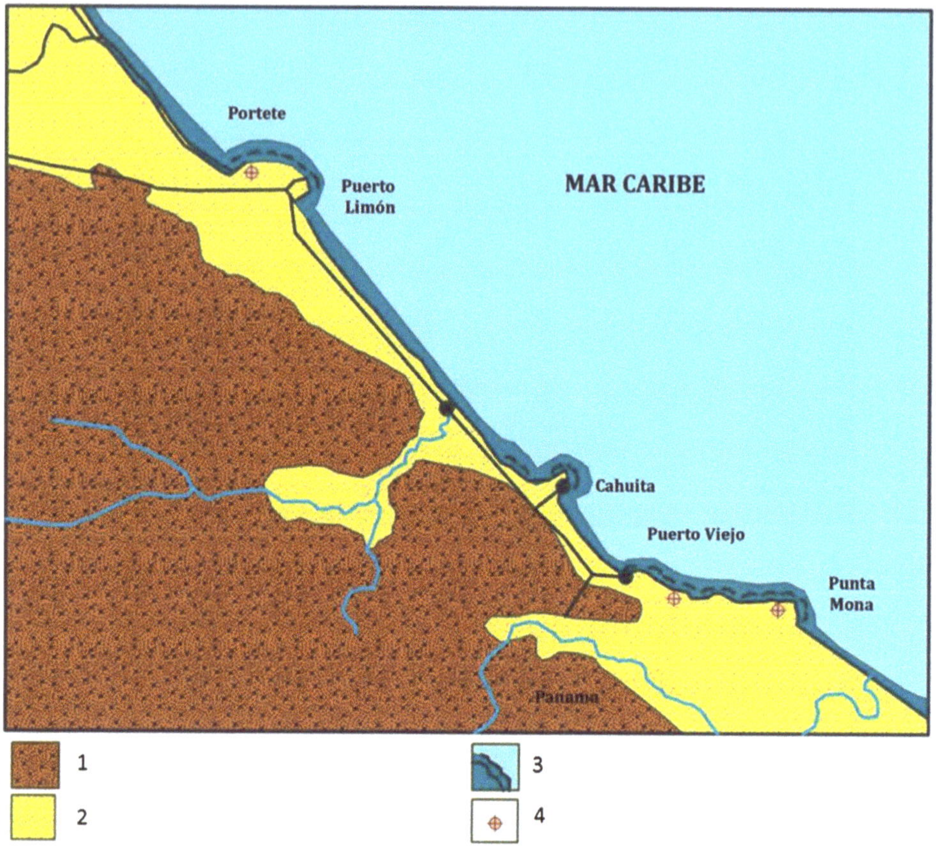

FIGURE 9.28 Costa Rica's southeast Caribbean littoral area (Bergoeing et al., 2010d). (1) Mountain shapes of Talamanca. (2) Coastal plain. (3) Puerto Viejo Formation outcrops area. (4) Portete Formation outcrops area.

sometimes also presenting the appearance of an old very eroded inactive cliff (see Fig. 9.28).

Between Puerto Viejo and Manzanillo Point, the coastal plain exists everywhere, but in the form of a narrow ribbon of less than 0.6 mile wide. Apart from Manzanillo Point, the coastal plain stretches out, benefiting the Sixaola River delta mouth, which continues onto Panama territory until Almirante Bay in Bocas del Toro.

Here spurs separate the rocky sectors, in a succession that is a characteristic, between long sandy beaches. The rocky outliers of Limon, Cahuita Point, Puerto Viejo, El Bluff, and Punta Mona allow us to outline the evolution of this coastline from the Pleistocene to the present day (Battistini and Bergoeing, 1980).

The existence of a Tertiary marine formation series, attributed to the Miocene in the Limon and Punta Mona promontories, was put in evidence through previous geological studies (Taylor, 1975) (see Fig. 9.29).

11.2.1 The Lower Pleistocene

We recognized an old coral complex that we consider Lower Pleistocene and that the former geological survey did not differentiate from the

FIGURE 9.29 Coral reef strip of Puerto Viejo made up of old dead corals of the Upper Pleistocene. *Photo Bergoeing (2008).*

Miocene Complex. It is an ancient massive coral reef in growth position, associated with coral scars and fine-grained limestone (ancient sediment beach or lagoon), which lead up to an altitude of 50–66 ft and form a very different topographic and geomorphologic shape from the Miocene Complex. This ancient reef is preserved to the north of Puerto Limon, also in Punta Bluf (south of Puerto Viejo), and finally at Punta Mona. We gave it the name of Portete Formation (Battistini and Bergoeing, 1980), from the name of the cove where it is exposed to the north of Puerto Limon. The Portete Formation differs from the Miocene coral reef not only by its topography but also by its less advanced alteration.

11.2.2 The Middle Pleistocene

A most recent coral reef complex, of Middle Pleistocene age, to which we gave the name Puerto Viejo Formation (Battistini and Bergoeing, 1980), gives very numerous rocky outcrops to Limon Promontory, Cahuita Point, Puerto Viejo Bay, and Punta Mona. They are clumps of coral growth, often large, alternating with gaps in the coral and with ancient sandstone beach beds. This complex is therefore very similar to the Portete Formation. However, it is distinguished from it very easily: whereas Portete Formation corals are always entirely calcite recrystallized, dense and resistant to the blows of a hammer, the aragonite structure remains predominant.

Whereas the Portete Formation is much altered, forming red decomposition clays with crypto-karst, sometimes released in large residual sinkholes, the Puerto Viejo Formation has no altered red soils and is witness to a little-advanced pedologic evolution. Finally, Puerto Viejo Formation corals rise just above the level of the highest current tides, in Puerto Viejo and Cahuita, and 10 ft high in the Puerto Limon sector. On the ground, these two formations are visibly discordant at numerous points.

11.2.3 The Flandrian

Rocky coastline outliers, consisting of these two ancient coral complexes, constitute the points of support of the Flandrian construction. It's also in this rocky material, mostly of Puerto Viejo Formation, where the elements of a low tide platform and the numerous breaks of the shore line have been formed. **Portete Cove**, North of Puerto Limon, shows a double notch corrosion system, with old platform at low tide that dominates the current 3.2ft-6.6ft levels, and provide testimony of the Flandrian sea level variations.

11.2.4 Current

Coral life, without being non-existent, has not reached the construction of true coral reefs, except in **Punta Cahuita Point**. They are isolated corals colonies, which have taken advantage of the hard substrate constituted by the Pleistocene slabs.

11.2.5 Littoral Evolution and Current Forms

It is difficult to get an idea of the layout of the coast, in The Low Pleistocene, during the emergence of Portete Formation. In Puerto Limon sector must have existed a coral reef strip that was attach to the Miocene coastal relief. Southern, in the sectors of Puerto Viejo-Punta Mona, the old coral reef should have been taken off, of the coastline, forming, a Coral Barrier reef with width channel.

The Middle Pleistocene Coral reef it's easier to reconstitute because is a Strip reef, lush, and highly developed throughout, both in Limon area and in Cahuita and Puerto Viejo sectors. It also covered Punta Bluf and Punta Punta Mona "Islands", and should come off significantly from Punta Uva Bank to the South (see general sketch of the southern sector of the Caribbean coast Fig. 9.28).

It can be assumed that the Islands Rosary and little high islets that form in Panama, the great **Bocas del Toro Gulf** and **Chiriqui**, is the Southern logical follow-up of the Pleistocene coral complex: the aligned arrangement of these islets of coral origin paralleling the coast indeed suggests such explanation.

In the Limon region, the Middle Pleistocene Strip Reef of Puerto Viejo Formation is a still well preserved shape in many places in front of inactive cliffs forming a rocky bench of 6.6ft-10ft of free fall. The effects of the marine erosion tend to reduce in all parts of this bench, giving a small cliff, with lapiaz forms and basal corrosion carvings (Portete, Playa Bonita Beach, Piuta, ICE area).

Sandy Flandrian construction (coastal cords) does not present anywhere the amplitude observed more in the North, especially in what refers to the border with Nicaragua sector. This is no doubt due to the lack of important fluvial sedimentary contributions (the Sixaola River contributes very little to the power of this sector since the littoral drift goes mostly to the South). There is however, an important Flandrian arched coastal cord between Cahuita Point and Limon Promontory, which is a progradation course, as well as between Cahuita Point and Puerto Viejo. From Puerto Viejo to Punta Mona Point, the impression is of a general fast rewind of the coast (coconut excavated or staved roots on the beach). Finally, the most interesting aspect of Limon concerns the current morphology of the Sandy-limestone sea platforms that comprise the following elements:

- **A slightly emerging platform** at low tide and that equals a stepped platform; It's perfect but just few feets wide (sector of ICE in Limon), except may locally exceed the 164ft wide (Piuta, Playa Bonita). The outer side presents a vertical front churning surf. This side evolves by crushing in large blocks with a shoulder that hinders the tide and dominates the platform at low tide on the outer side. The limestone at the top, is often much eroded in grooves or lapiaz erosion. This morphology is sometimes locally, complicated by the presence of a double undermining of corrosion.

- **A little step** that hinders the tide and dominates the platform at low tide on the outer side. The limestone at the top is often much eroded forming lapiaz. This morphology is sometimes locally, complicated by the presence of a double corrosion undermining.

CHAPTER

10

Cocos Island Geomorphology

OUTLINE

Cocos Island belongs to Costa Rica and is located in the Pacific Ocean between 5°29′52″ and 5°33′50″ north latitude and 87°01′44″ and 87°06′23″ west longitude. It is 267.9 miles from Cabo Blanco Cape on the Pacific coast of Costa Rica. In the political and administrative division of the country it is the 10th district of Puntarenas Province. On June 22, 1978, by Executive Decree 8748-A, the island was established as a national park, and on December 4, 1997, it was declared a Natural Patrimony of Humanity by UNESCO. Finally, in 1998 it was incorporated as an International Important Wetland (RAMSAR site). The island covers an area of 9.26 square miles and covers an absolute protection navy surface of 375.2 square miles. To the southwest lies Yglesias Hill, which rises to 1886 ft above sea level; it is the highest point on the island (see Fig. 10.1).

The island is characterized by a humid tropical climate with 7000 mm annually of almost permanent rainfall. The average temperature is 80.6°F. This climate has acted on the basaltic island surface, creating an important water supply in a small area, which includes the Yglesias River, which has eroded much of the Yglesias volcanic cone, describing jumps and an important waterfall before flowing into the sea south of the island. Other important rivers that are worth mentioning are that of the Quebrada Alfaro Gorge and the Gema River, which flows northward where stands the housing facilities of the national park in Wafer Bay. In general, the hydrographic system crosses the island, creating V-shaped valleys, rushing to the sea often by waterfalls from river ruptures on the active cliffs, results of the continuous sea erosion. Cocos Island was discovered in 1526 by the Spanish captain Juan Cabezas and we have testimonials from visitors to the island during the 18th and 19th centuries, such as Wafer in 1699, Duret in 1720, Colnett in 1788, Morell in 1832, Coulter in 1836, and Belcher in 1836 (Foreign Office, 1919). They described on Cocos Island, about 9 miles from the Genio river mouth in Wafer Bay, the existence of a circular lake, whose volcanic formation was undoubtedly powered

Geomorphology and Volcanology of Costa Rica
http://dx.doi.org/10.1016/B978-0-12-812067-5.00010-1

FIGURE 10.1 Cocos Island topographical map IGN-CR (1975). It highlights Yglesias Hill, in the southwest at 1887 ft altitude, and more to the south Jesús Jimenez Hill at 1411 ft, eroded by the Yglesias River. (Reduction of the topographical map; original scale 1:25.000.)

by small waterfalls, according to the description of Wafer in 1699, falling vertically on columnar basalts, but without clear drain, though it was probably underground:

> The Rocky Sides of the Hill being more than perpendicular, and hanging over the Plain beneath, the Water pours down in a Cataract, as our[sic] of a Bucket, so as to leave a Space dry under the spour [sic], and form a kind of Arch of Water …

In 1720 Duret adds: "… c'est une grande abondance d'eaux fort claires & fort douces, qui descendant par plusieurs endroits du haut de la montagne, se réunissent dans un profond & large bassin, de figure ronde qui est dans le roc; cette eau n'ayant point de canal pour sortir du bassin, se répand au dehors & forme des cascades qui jaillissent & tombent dans la plaine, ce qui joint à la beauté des vues & des perspectives, a la quantité d'arbres de cocos, & a la fraicheur de l'eau ...."

The lake descriptions continue until 1919, so its disappearance may be due to natural causes (rain and river erosion) or seismic causes. In fact, the island was sporadically a penal colony (1879–81) under the rule of Costa Rica's President Tomás Guardia. Later it was an agricultural colony authorized by the government of Rafael Yglesias and formed by German emigrants under the command of August Gissler (1894–1905), who was looking for hidden treasure on the island. The colony was a total failure, and in 1905, Gissler was the last to leave the island.

1. AN EMINENTLY VOLCANIC ISLAND

Cocos Island is of volcanic origin and corresponds to an emission of the Galapagos hot spot, so its lava flows are eminently basaltic. The Galapagos Islands, as well as the aseismic ridge system of Carnegie, Cocos, and Malpelo today, are two separate tectonic plates. The eminent Canadian Geophysicist J. Tuzo Wilson (1908–93) enunciated the hot spot theory in 1963. Cocos Island is the emerged point of the tectonic Cocos plate, which collided at the end of the Pliocene with the Caribbean plate, creating the Central American Isthmus. A ridge, or underwater volcanic chain, that consists of a series of volcanic cones characterizes the seabed between Cocos and Costa Rica (see Fig. 10.2).

According to Hey (Hey et al., 1977), Cocos Island would have an age between 10 and 15 million years, but Tournon dating (in Saénz, 1982 in Rojas and Alvarado, 1982) gives ages ranging from 2.0 to 1.2 million years, which means it would be a Lower to Middle Pleistocene island. The same is given by the Dalrymple and Cox (1968) dating process (in Dengo and Levy, 1970). So these dates significantly rejuvenate the island. The lava flows feature trachyte, basalt with olivine, and hornblende in smaller amounts.

The island is characterized by two large volcanic episodes. The Mount Yglesias volcanic cone is the main emission focus. It is an eroded crater that opens toward the south and is drained by the river of the same name; its walls are eroded, forming vertical erosion scarps. West, Mount Yglesias (1886 ft), which is the youngest of the island volcanoes, belongs to the end of the Middle Pleistocene (1.2 million years ago) and presents two very eroded craters and three volcanic domes, of which the highlights are the Yglesias summit itself and the Jesús Jimenez summit (1411 ft). Cerro Pelon (1640 ft) is part of the crater or rim of the volcano structure. To the east, and part of the Yglesias volcano structure, extends a structural plateau known as Llanos de Palo de Hierro, whose average altitude is 984 ft (see Figs. 10.2 and 10.3).

Farther east is the oldest volcanic structure, which would date from the Lower to Middle Pleistocene, characterized by three very eroded volcanic domes, which are called Scorpion Dome (919 ft), to the south, and Venado Dome (787 ft), to the northeast, drained by the rivers with the same names, and Chattam Crater, next to the park's housing sector. Finally, Weston Bay, enclosing Pajara Island, could be just the submerged structure of a much eroded crater. This second area of the island is older than the first one described and corresponds to the Lower Pleistocene, about 2 million years. All are based on radiometric dating carried out by Dalrymple and Cox in 1968 and by Azéma and Tournon in 1982.

The island is also characterized by active cliffs, eroded by the sea and reaching 660 ft of freefall with basaltic sea platforms in some parts of its base; islands; and smaller islets that accompany it (Fig. 10.4). In some sectors, debris cones cover the cliffs. The cliffs contain a large number of underwater caves, expressions of basaltic volcanism exploited at its weaker parts by both

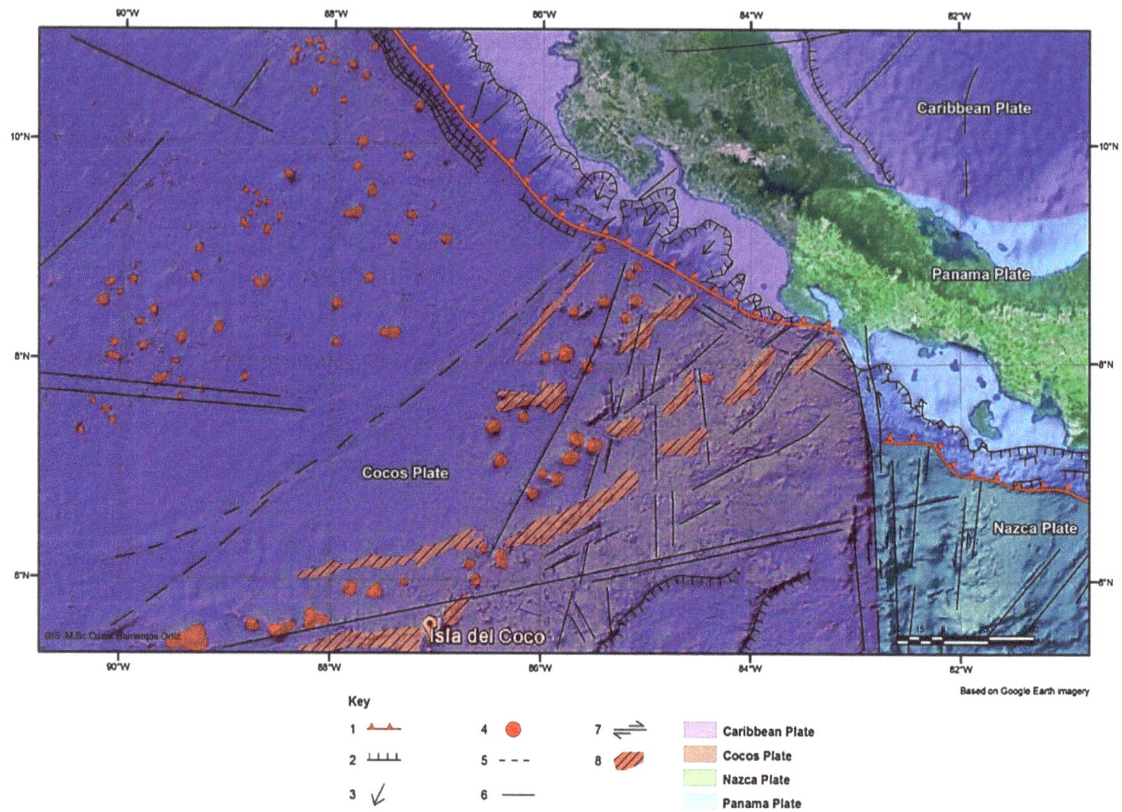

FIGURE 10.2 Cocos plate submarine geomorphology by Jean Pierre Bergoeing. (1) Subduction area. (2) End of continental sea platform. (3) Submarine landslides. (4) Submarine volcanoes. (5) Probable tectonic faults. (6) Tectonic faults. (7) Sliding faults. (8) Submarine volcanic range. *Drawn by Oscar Barrientos (2016).*

fluvial and marine erosion. This is because the island has not had volcanic activity, at least during the Holocene, which has allowed the sea to erode its flanks by not feeding with new lava spills. Alteration of the basalt, particularly in the northern area, reaches important thicknesses that translate into landslides that do not reach great proportions, thanks to the prevailing vegetation that holds the soil. The rivers in general erode the island, forming V-shaped valleys. The northern Gema River has created a vast river depression with major escarpments on its edges. It is an older volcanic area and therefore more altered than Mount Yglesias (see Figs. 10.5 and 10.6).

Here the river has formed a broad valley with a NW–SE direction. Two important tectonic fault accidents section the island in this direction. The first dissects Mount Yglesias and the second the Gema River. Other SW–NE tectonic structures are also present. The first sections the Gema River obliquely, the second is parallel to the first, sectioning the Scorpion Dome. Tracing the Genio River, this is born on the Scorpion Dome's northern slope, where are also born the Scorpion River (to the east) and the El Gato, El Ancla, and Quebrada Alfaro rivers, who adopt a radial fluvial pattern, typical of a volcanic structure. It is very likely that where the Genio

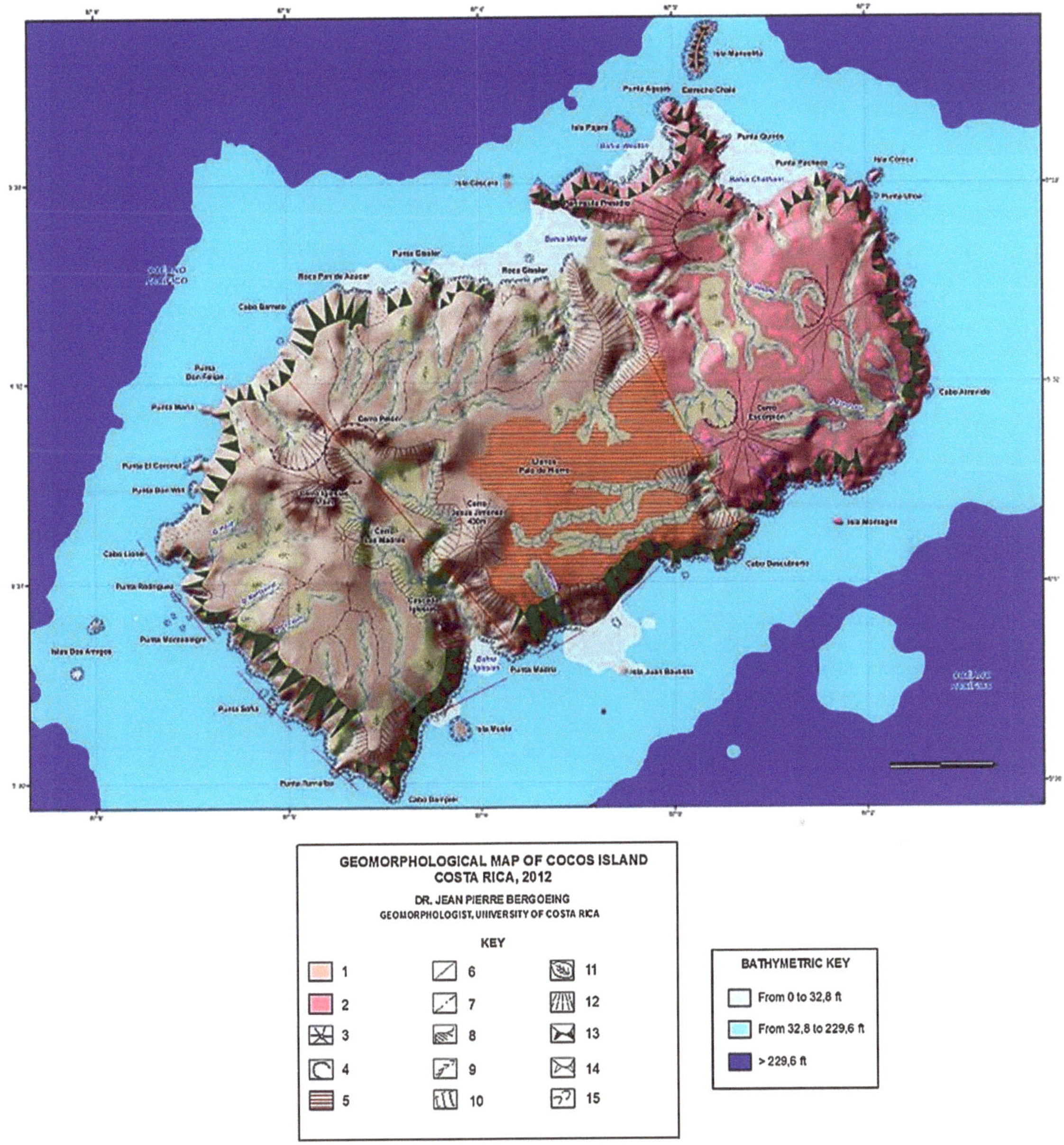

FIGURE 10.3 Cocos Island geomorphology by Jean Pierre Bergoeing, 2012, using an IGN-CR map at 1:40.000 scale. (1) Middle to Upper Pleistocene basalts. (2) Lower to Middle Pleistocene basalts. (3) Volcanic cones. (4) Craters. (5) Structural plateau. (6) Tectonic alignments. (7) Water divide line. (8) Erosion talus. (9) V valleys. (10) Flat-bottom valleys. (11) Landslides. (12) Scree cones. (13) Active cliffs. (14) Inactive cliffs. (15) Coral reefs.

FIGURE 10.4 Inclined columnar basaltic lava flow forming an active cliff. *Photo Barbara Sperl (2008). Courtesy of Michel Montoya.*

FIGURE 10.5 Pajara Island in Weston Bay, formed by a series of tilted basalts indicating the sense of a volcanic wash that once united Pajara Island to the big island. It is the oldest island volcanism. The island base has a sea-level erosion concavity. *Photo Pucci (2006). Courtesy of Michel Montoya.*

River is born could be the place of the circular lagoon described by multiple navigators who visited the island in past centuries, because the topography seems to indicate it. It would then be an old Scorpion Dome volcanic crater that housed the lake until the beginning of the 20th century. It was described by many sailors who visited the island and were amazed by its splendor. It is very likely that the lake disappeared because of natural causes, such as just

FIGURE 10.6 Tubular basalt of an active cliff 330 ft high. A stratification of marine sandstone is attached to the cliff. Cliff debris and a basalt pebbles beach form the base. *Photo Silvia Gudino (2008). Courtesy of Michel Montoya.*

FIGURE 10.7 Mount Yglesias volcanic summit showing an opened crater to the south. In the foreground are island tubular basaltic active cliffs. Tropical vegetation thrives everywhere but some sectors show landslides, victims of gravity. *Photo Expedition Phoenix (2004). Courtesy of Michel Montoya.*

rain erosion of the volcanic walls that held it (see Fig. 10.6).

It can be summarized that Cocos Island is an eminently volcanic basalt-type structure, a product of the Galapagos hot spot and its subsequent drift of tectonic plates in the direction of Central America. It is characterized by two Quaternary volcanic events that formed the island as it is currently known. The high rainfall produced by the hyperhumid tropical climate is an element that has helped to stabilize its slopes, and because it was declared a national park, the anthropic erosion is almost zero. Vertical active cliffs are proof that the island has not had volcanic activity, at least during the Holocene, adopting a typical erosive coast profile of the Pacific Ocean islands without volcanic activity. The real treasure of Cocos Island is its well-preserved terrestrial and marine environments (Fig. 10.7).

CHAPTER

11

Costa Rica Quaternary Chronology Proposal

OUTLINE

Costa Rica is a humid tropical country where meteoric alteration processes are very advanced, which sometimes prevents the establishment of age correlations by means of radiometric measurements through valid samples, such as the K/Ar method. This is equally true for organic matter using the ^{14}C method. This method allowed us to date the Flandrian Montezuma Beach terrace, but failed with coral samples of the Portete and Puerto Viejo Caribbean formations. In addition, we had the difficulty of fragmented or nonexistent geological information for some sectors, which forced us to take on the painstaking work of bibliographical research and development of a geological synthesis, which culminated in the publication by the Instituto Geográfico Nacional de Costa Rica of the geomorphological charts of the North Pacific, at 1:100.000 scale, and the Central Valley, at 1:50.000 scale. In 2010, we published the geomorphologic Costa Rica Atlas at 1:350.000 and the Caribbean Atlas of Costa Rica at 1:100.000 scale. In addition, in 2016, the Costa Rica Geomorphology Atlas at 1: 100.000 scale, the first one for a Central American country, was published.

Costa Rica is a country of a mountainous and flat relief, located in the isthmus sector of Central America. Above all, it is a humid tropical country, which shares its thermal and hygrometric graduation in altitudinal levels. In addition, it is a country

Geomorphology and Volcanology of Costa Rica
http://dx.doi.org/10.1016/B978-0-12-812067-5.00011-3

with two oceanic facades that expose it to the double swing of warm, moist air masses from the Pacific and especially from the Caribbean. This relief is of great importance because it allows the understanding that the climatic oscillations during the Quaternary that disturbed North America acted in decisive ways in Central America, because of the displacement of large anticyclonic centers.

The Mississippi Valley sector in North America is a climate corridor where cold polar air masses traveled toward the Gulf of Mexico. Born on the old continental, American–Canadian ice caps, they directly influenced Central America, creating pluvial and interpluvial periods during Quaternary ice ages and also permanent ice deposits in the high peaks of Guatemala (Altos Cuchumatanes), the Dominican Republic (Duarte Mountain Range), and Costa Rica (Talamanca Mountain Range).

There is a concordance of paleoclimatic events in North America, the West Indies, and Venezuela, published by C. Schubert (Schubert, 1984). Now let us see if such correlations can also be set for Central America, and in particular for Costa Rica. First, let us see what major morphological problems are posed for the Costa Rican landscape before starting a detailed chronology:

1. **Fluvial levels**
2. **Alluvial fans of the Tilaran Pacific Piedmont**
3. **Nicoya Peninsula low eustatic levels**
4. **Caribbean coast paleoreefs**

1. FLUVIAL LEVELS

Observations made in Costa Rica allow us to affirm that there are river levels in almost all the rivers of the country. Apart from the ignimbrite erosion terraces of La Cruz (Guanacaste), there are many deposits of boulders with well-polished edges that lie in a clay matrix whose origin is in the foothills. Two lower river levels (T1 and T2) are always present. A top river level (T3) is sometimes recognizable in some hydrographic cuts (Reventazon and Pacific Chirripó rivers, Tilaran Pacific foothill alluvial fans). This level is sometimes found deformed by tectonics and even fractured (Tarcoles River facing the South Coast Bridge). T3 would be the oldest river level (except for the San Pablo de Turrubares T4 level). It often arises with very cohesive decomposed clays, with much altered boulders (T3 level of the Reventazon River).

One may wonder if all these river levels are the result of large climatic fluctuations despite the observed tectonic deformations. We agree, and so we will rely on the case of the Reventazon River (see Fig. 11.1).

1.1 Reventazon River

The T1 level shows a high fluvial terrace 6.5–13 ft above the riverbed, which is embedded in the volcanic Cervantes lava flow (^{14}C, 13,800 years BP, Murata et al., 1966; $^{238}U/^{230}Th$, 23,000 years BP, Allegre and Condomines, 1976). The river terrace is formed of fresh andesitic boulders resting on a brown clayey matrix. This terrace is attached laterally to numerous bases of red fan slopes. This level is obviously newer than the already dated Cervantes volcanic flows. So this T1 river level should be placed in the Upper Pleistocene and could be associated with a final high-mountain cold-phase Würm III–Wisconsinian (the period between 40,000 and 20,000 years BP and which in Costa Rica can be translated as a rexistasic phase, with ice deposits on Chirripó Mountain and a large runoff in the intermountainous area) (see Fig. 11.2).

The T2 level is the best developed in the sector and is represented among the towns of Paraiso and Turrialba. On the right bank of the river lies the T2 fluvial level of Tucurrique village. This T2 level has a northeast flexure from the effects of neotectonics. It is constituted by a dense reddish soil of 3.3 ft thickness with healthy boulders but possessing a dark color. The Cervantes lava flow (23,000 to 13,800 years BP) fitted the rivers of this second level, which makes it younger than these rivers.

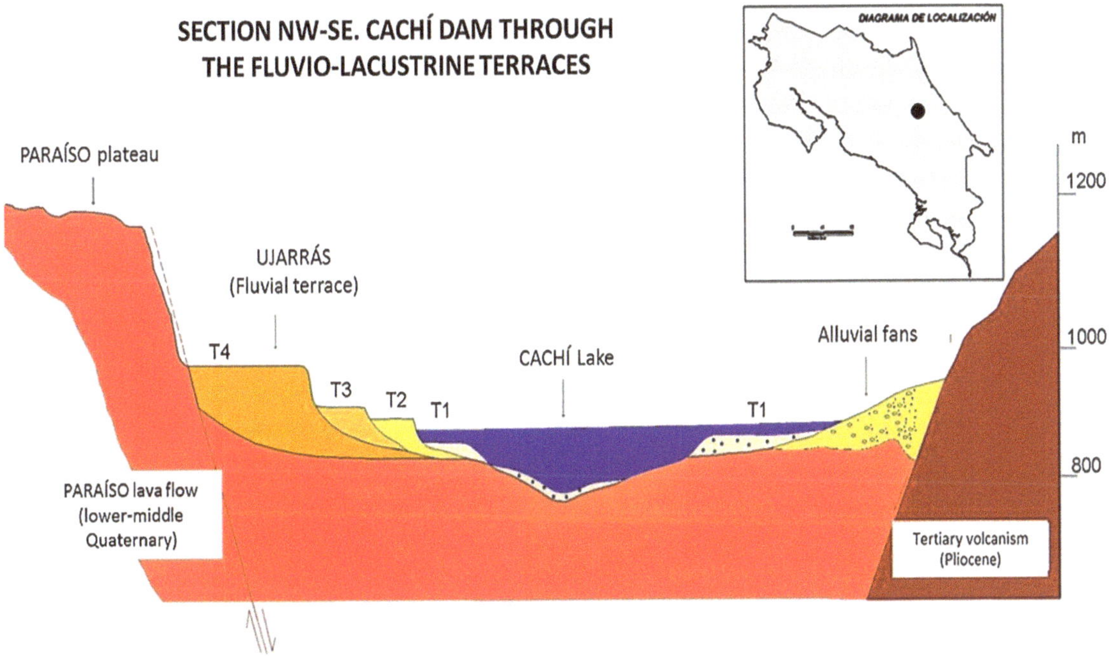

FIGURE 11.1 Fluvial–lacustrine terraces in Cachí Dam sector (Bergoeing, 2012a).

The T3 level is represented by the remains of a high terrace, also distorted by neotectonics, and is present in Infiernillo (between Paraiso and Turrialba). The T3 level consists of a 5-ft-thick red clay matrix enclosing old altered andesitic boulders. Downstream, this level joins the coalescent alluvial fans, suspended and of imposing dimensions (millions of cubic feet), that dominate the vast Northern Plain from the Turrialba and Irazu volcanoes' northern slopes.

The Reventazon River shows us, in these profiles, three levels of paleoclimatic origin (Riss–Illinoian and Würm–Wisconsinian ice ages) forcibly placed between the end of the Middle Pleistocene (date of last ignimbrite spills of the Central Valley, 0.33 MY) and the beginning of the Upper Pleistocene (Cervantes lava flow, 23.000 to 13.800 years BP).

Therefore, everything favors the climatic hypothesis despite subsequent tectonic deformations suffered by the T2 and T3 levels. These three levels are present in the middle of the Reventazon course and downstream merge perfectly with the alluvial fans, which ossify the Turrialba and Irazu volcanic bases of the northern slopes. Its composition is that of material deposited during drier phases (Riss–Illinoian and Würm–Wisconsinian rexistasic periods in Costa Rica), with torrential rains transporting lahar material from the volcanic slopes temporarily devoid of sufficient vegetation cover to keep them.

1.2 The General River Depression

The San Isidro de el General sector offers us an example of stratified alluvial fans, which are undoubtedly of climatic origin:

T1 level: This is a lower river terrace composed of heavy material (clay–sandy), which is the most common and is everywhere in Costa Rica.

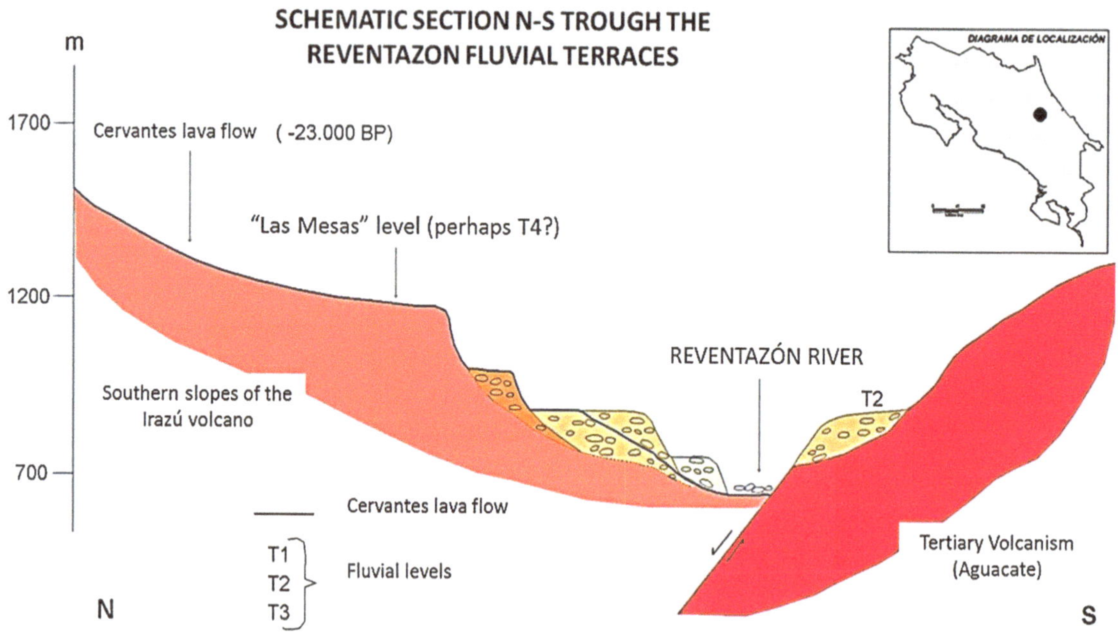

FIGURE 11.2 Reventazon river fluvial terraces (Bergoeing, 2012a).

T2 level: This is made up of large-dimension blocks (diameters greater than 1 m) and forms very steep alluvial fans.
T3 level: This is at the mountain terrace mouths and hanging alluvial fans composed of coarse and thick material. On the surface is denoted an alteration of 3 ft thickness (cf. Fig. 11.3, Pacific Chirripó River profile).
T4 level: This is the oldest level, formed by an old outfall alluvial fan constituted by an accumulation of clay, profoundly altered and tectonically folded, forming a syncline (General Valley Syncline). It has well-rolled boulders from 1.6 to 2 ft in diameter, but so altered that they present as "ghosts," because of an extreme decomposition. This level is credited with a Villafranchian age because it contains, upstream, blocks of lava dated to 2.6 MY (Beaudet et al., 1982a).

Upstream, the Pacific Chirripó River presents three fluvial terrace levels that combine, downstream, with the three lower levels of the alluvial fans of the General Valley. We think that the T4 alluvial fan level formed during the Villafranchian process in which Talamanca rose by effects of the orogeny, a moment in which the current rivers were not creating a river network big enough to create a fluvial level (or the fluvial level is gone because of erosion effects). Once again, here we find the three classical upper levels of Costa Rica, separated from the T4 level (Lower Pleistocene) by a large chronological interval (T4, 2 MY, and T3, T2, and T1, 0.3 to 0.05 MY).

Costa Rica is a highly volcanic country that was built after the Cretaceous period and particularly during the Quaternary (the large volcanic cones of the Guanacaste and Central Mountain Ranges were built up during the Quaternary orogeny), leaving signs in the landscape only of those episodes that the volcanic activity has allowed to conserve superficially (see Fig. 11.3).

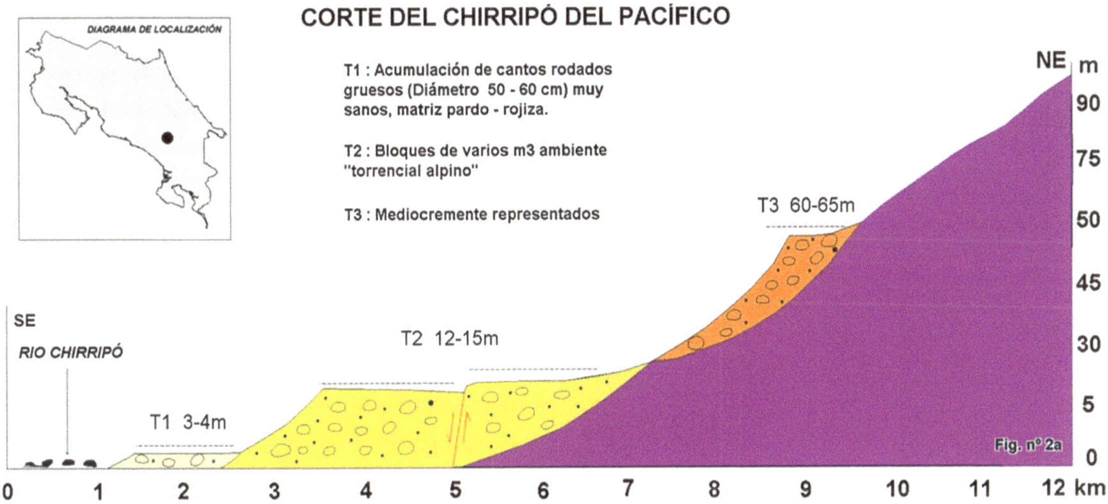

FIGURE 11.3 Fluvial terrace levels of the Pacific Chirripó River (Bergoeing, 2012a).

In the General Valley, alluvial fans in the T2 position pose a question because they present on their surface large lithic granodioritic blocks (tens of cubic feet) striated by rainwater dissolution, in the Talamanca Pacific slope, which were stopped at 4 miles from the slope rupture of the Volcan River. The transportation of these blocks was allowed by the sudden ice meltdown on the high peaks of the Chirripó, creating avalanche conditions (Battistini and Bergoeing, 1982a). As we see it, only under a drier climate, which corresponds to an ice age period in the high peaks, might this have happened, and therefore it indicates a southern shift of the global climatic mechanics in Central America (Bergoeing, 1977; Beaudet et al., 1982b) (see Figs. 11.4 and 11.5).

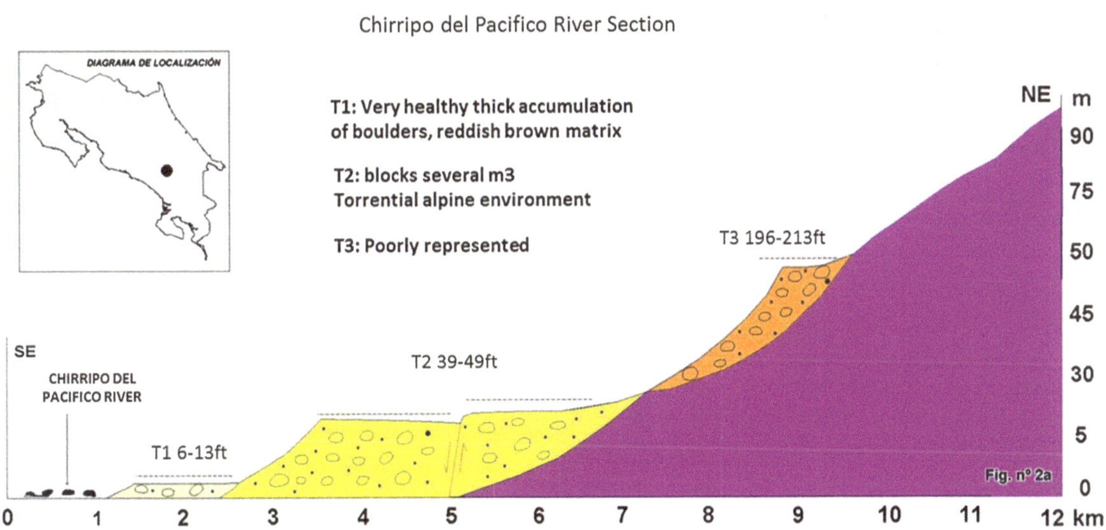

FIGURE 11.4 The General Valley profile between Santa Maria row and the Coastal Mountain Range (Bergoeing, 2012a).

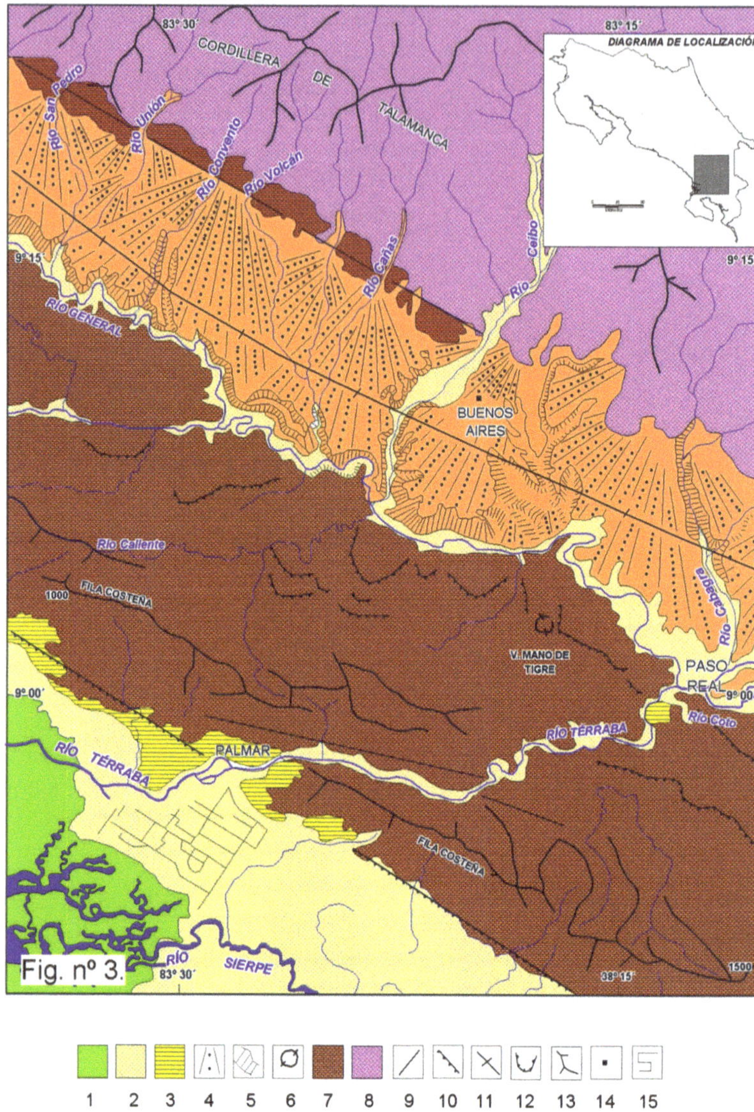

FIGURE 11.5 The General Valley alluvial fan geomorphology (Bergoeing, 2012a). (1) Mangrove. (2) Alluvial fills. (3) Fluvial terraces. (4) Large Quaternary alluvial fans. (5) Erosion talus. (6) Mano de Tigre Volcano. (7) Folded Tertiary sediments and tectonic faulted Coastal Mountain Range. (8) Tertiary intrusive and volcanic deposits of the Talamanca Mountain Range. (9) Tectonic faults. (10) Great Costa Rica Reverse Fault. (11) General Valley syncline. (12) Tilted blocks in chevron position. (13) Water divide lines. (14) Major urban areas. (15) Banana plantations.

2. TILARAN PACIFIC WATERSHED ALLUVIAL FANS

Southern Nicoya Peninsula hides many cyclical levels with no correlative deposits. These levels are located in the sector at 33, 82, and 130–164 ft of altitude; however, they are also present particularly in Parrita and Coronado. We have issued an explanatory hypothesis that they are of a climatic origin, that is to say, these rocky levels that fit in situ with profoundly altered rock would be in relation to coastal groundwater levels for old sea-level change episodes. This would partly explain why these levels have no correlative deposits (Battistini and Bergoeing, 1983a) (see Fig. 11.6).

We have discarded the ancient abrasion platforms hypothesis because of its position (the levels are not in the place of old littoral points, but in sheltered positions in the bay's interior). Nor do we think the erosion levels are of continental origin, because currently the unique flat surfaces found in Costa Rica are filling surfaces (for example, alluvial fans/glacis of the Tempisque River plain).

These erosion levels must therefore be placed in an age range that should oscillate between the Middle and the Upper Pleistocene–Holocene. They correspond to the Eemian (120,000 years BP) and Flandrian sea transgressions (8000 to 6000 years BP). This period is long enough to allow a profound alteration above the water table and subsequently an altered material clearance level, allowing the underlying rock surface to be discovered.

3. CARIBBEAN COAST PALEOCLIFFS

Costa Rica's Caribbean coast has two ancient coral cliff levels. The first and oldest is called the Portete Formation (Battistini and Bergoeing, 1984). It is an old strip reef, stuck on the Miocene coastline calcareous formations, but which was detached from the coastline southwest of Punta Mona, forming a small barrier reef.

The coral's calcite concentration in a growth position as well as the degree of erosion of this compact mass leads us to infer that it is an old Quaternary reef. In contrast, the Puerto Viejo Formation paleoreef (Battistini and Bergoeing, 1980) has the youngest and best-preserved forms. This makes us think it would be an Eemian Formation because of the position of the reef relative to the current coastline and the abrasion notches on the base, which can only be a Flandrian high marine parking, despite the strong neotectonics of the area (see Fig. 11.7).

In summary, the coastal Quaternary of Costa Rica allows us to establish four clear elements:

1. **The Eemian transgression** (120,000 years BP) is well represented on the coast. Higher erosion levels are seen on the southern Nicoya Peninsula and Terraba river mouth.
2. **This high-sea-level period** was followed by a period of low sea levels, which corresponds to the Würm–Wisconsinian great recession (sea level under 460 ft).
3. **The Flandrian sea transgression (13 ft)** is represented and attested to by the fossil deposits of Montezuma Beach, Nicoya (6000 years BP) (Battistini and Bergoeing, 1983a), by the coastal ranges of the two oceanic facades, as well as by the silting up of the interiors of bay and paleocliffs (coastal row).
4. **The current sea level** lies with minor modifications from 6000 years ago, taking into account a very active neotectonics (Limon earthquake, 1991).

It is with these land research elements and the radiocarbon (^{14}C) and K/Ar radiometric dating methods that we will try to summarize the Quaternary morphology of Costa Rica.

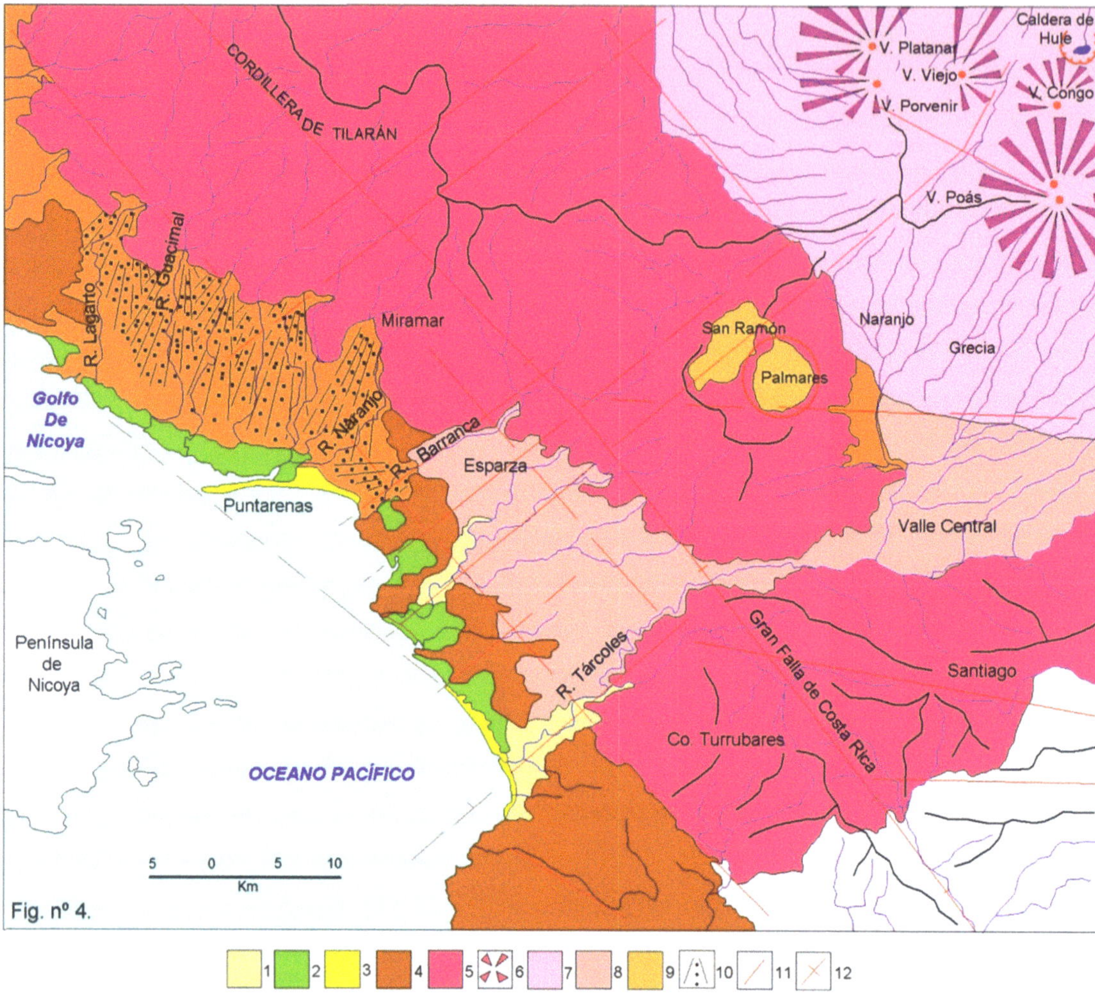

FIGURE 11.6 Tilaran Mountain Range geomorphology (Bergoeing, 2012a). (1) Recent alluvium. (2) Mangroves. (3) Flandrian sandy littoral cords. (4) Cretaceous–Tertiary sedimentary and partly volcanic areas. (5) Miocene–Pliocene volcanic areas, Aguacate Formation. (6) Quaternary volcanic cones (stratovolcanoes) of the Central Mountain Range. (7) Quaternary volcanic foothills of the Central Volcanic Mountain range. (8) Virilla–Tarcoles River canyon eroded in the Central Volcanic Plateau and in the Orotina ignimbrite piedmont. (9) Paleolake areas. (10) Coalescent Quaternary alluvial fans. (11) Main tectonic alignments. (12) Syncline.

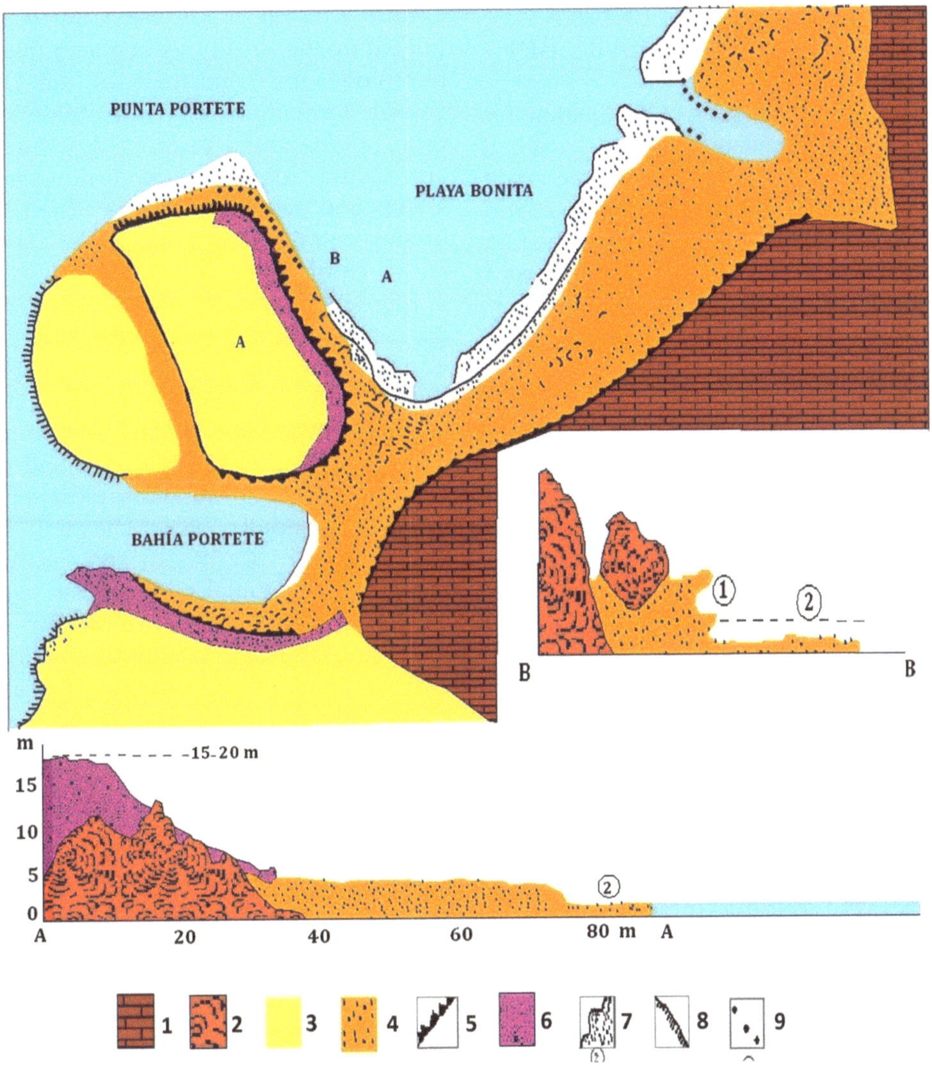

FIGURE 11.7 Quaternary coral paleocliffs. Geomorphology of Portete and Puerto Viejo formations (Bergoeing, 2012a). (1) Miocene–Pliocene outcrops. (2) Portete Formation with coral forms. (3) 15–20 m level. (4) Puerto Viejo Formation (inter-Riss–Würm or Eemian). Coral gaps and chunky thick coral in the growth position. (5) Fossil marine cliff (inter Riss–Würm). (6) Portete Formation altered red clays. (7) Current sea corrosion platform in Puerto Viejo Formation limestone, ②. (8) Current cliff. (9) Traces of a high Holocene marine surface, ①.

4. THE LOWER QUATERNARY (2,000,000 TO 300,000 YEARS BP)

Costa Rica at the beginning of the Quaternary was only a string of islands formed by small volcanic cones (Lloyd, 1963) stuck to the rising Talamanca and Tilaran massifs (Picher and Weyl, 1975). The most remote period of the Old Quaternary is represented only in the Talamanca piedmont and in particular in the General Valley (Beaudet et al., 1982b). It is here, during this period, that the powerful coalescing alluvial fans (T4 level) were constructed. Might they equally have existed buried at the base of the Tilaran Pacific foothills under the Middle Quaternary lava flows? In the Talamanca Caribbean slope, would the remains of high terraces (puddingstone) of the Telire River be witnesses to the same event?

In the General Valley, the old alluvial fans (T4), folded by plate tectonics (the General Valley Syncline) during the Quaternary period, flowed along the coast, because at that period the Coastal Mountain Range was just beginning to emerge and order the route of the General and Coto Brus rivers (Mora, 1979).

From the volcanic point of view, the Lower Quaternary in Costa Rica was a very active period (Tournon, 1984a). Later the Central and Guanacaste volcanic mountain ranges would grow, but at that time they were limited to small volcanic pockets, lined up on a NW–SE fissure. In addition, they corresponded roughly to the southern edge of the Nicaragua graben (Battistini and Bergoeing, 1982a). These foci issued fluidic lavas (basalts) that allowed the emerging continental isthmus to be better consolidated (Dengo, 1962a). Welding established in the Tertiary (Pliocene) in the Rivas Isthmus for the second time allowed the flora and fauna to migrate between North and South America (Schmidt-Effing, 1980; Coates and Obando, 1996; McNeill et al., 2000).

A minor volcanism emerged in what would later be La Carpintera Mountain Range in the Central Valley (2.25 MY; Bergoeing et al., 1980). It lifted up the already sedimentary folded series (limestone and sandstone) of the Coris and Patarra formations via fault strike–slip (Pliocene) of the Central Valley. This natural boundary would constitute the continental divide waters and determine therefore the river paths of the Central Valley in the western and eastern basins (Bergoeing et al., 1980).

The Tilaran Mountain Range started forming during the Miocene, and the Aguacate Mountain Range also experienced volcanic activity until the Lower Pleistocene (Cerro Anunciacion and Palmares volcanoes). The ancient Palmares volcano magma chamber exploded and created a collapsed caldera (acid and explosive magmatic activity) (Dondoli, 1951; Malavassi, 1965). At the base of the Caribbean and Pacific Tilaran sheds, first torrential alluvial fans began their slow construction. This was mainly due to a dry and contrasting climate that reigned at the time, and whose only witnesses are the T4 levels of the General Valley alluvial fans. Others, much more modern, have buried the old Tilaran alluvial fans, as we see today and of which we have described three generations (Battistini and Bergoeing, 1983b).

The Lower Quaternary in northern Guanacaste is also characterized by the extinction of small volcanic cones, like El Hacha Volcano (Dengo, 1962b), and by the continuous ignimbrite spills that buried the folded and eroded Paleocene reliefs, gradually forming a topographic plateau with a variable thickness of 165–330 ft. Ignimbrite volcanic activity was constant (Bagaces Formation) and the proof was obtained based on three radiometric K/Ar samples taken by the Costa Rican Institute of Electricity (ICE) giving ages of 7.5, 3.8, and 0.63 MY (ICE, 1976). It is on the ignimbrite plateau where we discovered an old lake complex with sandy deposits in Hacienda Ciruelas. K/Ar dating on some samples showed them to be 1.2 MY old (Battistini and Bergoeing, 1983b).

One major event was the Guayabo collapsed caldera (Granados, 1979), covering about 116 square miles, separating the modern Rincón de la Vieja and Miravalles volcanic cones. It is from this volcanic depression that tuffs and ignimbrite spills issued, creating the Liberia Formation at the end of the Lower Quaternary. A first dating of the volcanic collapse places it at 1.6 MY. The subsidence of the ancient Guayabo volcano occurred around 0.6 MY (Guillot et al., 1994 in Alvarado, 2000); however, this volcano underwent several stages of collapse and successive building that rejuvenate it to 0.4 MY. Subsequently, at 0.1 MY began the construction of the Miravalles stratovolcano (Alvarado, 2000). It is estimated that the Guayabo caldera issued a volume of more than 12 cubic miles of pyroclastic spills (Chiesa et al., 1992 in Denyer and Kussmaul, 2000).

The Caribbean Sea during the Lower Quaternary penetrated deeply to reach Lake Nicaragua. It was during the entire period of the Quaternary that the Northern Great Plain was gradually filled, despite a weak subsidence of the sector (tectonic graben). The gradual filling of the sector allowed it to become a plain that ends at the partially buried small calc-alkaline isolated volcanic foci of the sector (Tournon, 1972).

It is only at the end of the Lower Quaternary that the roots of the Guanacaste and Central Volcanic mountain ranges began to build and therefore rise up off the monotony of the marshy plains landscape. K/Ar dates are witnesses to this slow evolution. Basalt flows on the base of Turrialba Volcano date to 2.15 and 1.67 MY (Tournon, 1984a). Later, and due to continuous andesitic and basaltic lava spills, the Central Mountain Range built a structural plateau that would make up the bottom of the Central Valley. This plateau makes contact with Pliocene volcanic deposits (Aguacate Formation) of the northern end of the Talamanca Mountain Range. It is dated from the deposits of ignimbrite tuffs of San Pablo de Turrubares to 1.38 MY (K/Ar dating; Battistini and Bergoeing, 1983b). This allows us to say that the fluvial terraces that erode the ignimbrite, waters beneath the Central Plateau, cannot be later than this date and therefore should be attributed to the Middle or Upper Quaternary.

It is at the end of the Lower Quaternary that the volcanic activity of the Central Mountain Range became more acidic. This is translated by eruptions and ignimbrite spills (Madrigal and Malavassi, 1967) covering all the western sector of the Central Valley. In the eastern Central Valley, ignimbrite spills are present in Cartago but buried under more modern laharic deposits (Tournon, 1984a). At the Puente Mulas Bridge near San José, Jean Tournon (op. cit.) dated an Irazu lava flow (afiric andesite) topped by a pumice tuff that threw an absolute age of 1.0 MY. However, more recent dating by P. Gans rejuvenated it to 0.5 MY (Denyer and Kussmaul, 2000).

The Central Valley western sector ignimbrites are topographically joined through the Tarcoles river canyon with the Orotina foothills. The San Pablo de Turrubares fluviovolcanic terrace (1.38 MY) confirms it by its presence and is the logical topographic sequence between the Central piedmont and the Orotina foothills, whose base was dated to 1.28 MY (Battistini and Bergoeing, 1983b). We asked ourselves about the possible common ignimbrite origin of the Central Valley and the Orotina foothills. In the Central Valley, they cover an area of 290 square miles and have been dated by $^{40}Ar/^{39}Ar$ to 0.33 MY (Alvarado et al. in Denyer and Kussmaul, 2000). The ignimbrites of Orotina "would have been channeled by the Paleo-Tarcoles River to form distal facies" (Alvarado et al., in Denyer and Kussmaul, 2000).

At the end of the Pliocene the Guanacaste Mountain Range (Rincon de la Vieja volcano) presented a peculiar activity that translated it into a series of dacitic domes (Cerro San Roque) whose ages vary from 1.6 to 1.0 MY (Tournon, 1984b) and which activity was completed in the Lower Quaternary.

Finally, volcanic activity in the Miocene was manifested in the country's southeast (Coastal Mountain Range) by the presence of the Mano de Tigre aborted volcanism (Bergoeing et al., 1978b) and by volcanic deposits of the Panama Tisingal Volcano in the sector of San Vito de Java, according to K/Ar dates of 66 ± 0.49 and 1.18 ± 0.40 MY (Camacho, 1992).

If we take into account the strong volcanic activity in the Central and Guanacaste mountain ranges during the Lower Quaternary, these sectors are the actors that would produce the most significant changes in the Costa Rican morphology. Therefore, that climate should have been very disturbed locally, by ash-laden skies and flooding rains that had consequences. (Observations made in 1963 during the Irazu eruption showed that these climatic phenomena associated with eruptions, common in the past during intense volcanic activity, are the direct causes of mud wash "lahars" that destroyed Taras houses in the area west of Cartago.) This is an important fact, which should be taken into account and could be added to the explanation of deposits forming the powerful alluvial fans at the base of the volcanoes, mostly relating to the slopes of the Caribbean (see Fig. 11.8).

It was also during the Lower Quaternary, on the Caribbean coast, that a strip coral reef (Portete Formation) was built and attached to the old Miocene limestone cliffs (Battistini and Bergoeing, 1980). This reef is preserved even in the form of some witnesses between Portete Bay and Punta Mona. It is at this last point where the reef fringe took off from the coastal line to constitute a barrier reef that lasted probably until Admiral Bay in Panama. All these observations allow us to deduce that there were two alternating and different climatic periods during the Lower Quaternary:

1. **An old interpluvial phase (rexistasic)** was responsible for a drier climate, with torrential rains that led to the creation of the large alluvial fans in the General Valley in the T4 position, which are today folded, forming a leaky syncline.
2. **A pluvial phase (biostasic)** was the first successor and explains the long time that separated the T4 alluvial fan deposits of the General Valley from the T3 levels of the Pacific Chirripó River (Beaudet et al., 1982a).

5. THE MIDDLE QUATERNARY

It is very likely that in the Middle Quaternary, taking into account the continuous uplift of Talamanca, the summits reached an altitude near the current values and therefore they met with some ice age periods. However, we have no elements to say with certainty that during this period there was an ice age on Chirripó (Gunz period) and the surrounding summits, even if it is more than likely because the conditions, both climatic and altimetric, were favorable, at least for a snowy parking lot (Bergoeing, 1977). It is during this period that North America was suffering the Kansas ice age, which equaled Gunz in Europe (Imbrie and Imbrie, 1979). If this is accurate, Costa Rica was experiencing a pluvial climate consistent with that of the West Indies and Venezuela (Schubert, 1984, op. cit.).

During the Middle Quaternary period the central volcanic plateau ended its formation of lava flows, and its ignimbrites would be partially buried by laharic deposits and mud washes from the volcanic alluvial fans in the formation of the Central Volcanic Mountain Range, which started to rise in altitude (stratovolcanoes) (Alvarado, 2000). It was during this period that the T3 fluvial terraces of the Reventazon River were created, as well as the embedded alluvial fans of the Tilaran Mountain Range; this is judged by the similarity and significant degree of alteration of the materials that composed them in Tilaran and in the Central Valley (Battitistini and Bergoeing, 1982b).

We also think that it was during this period that the first large alluvial fans/glacis of the

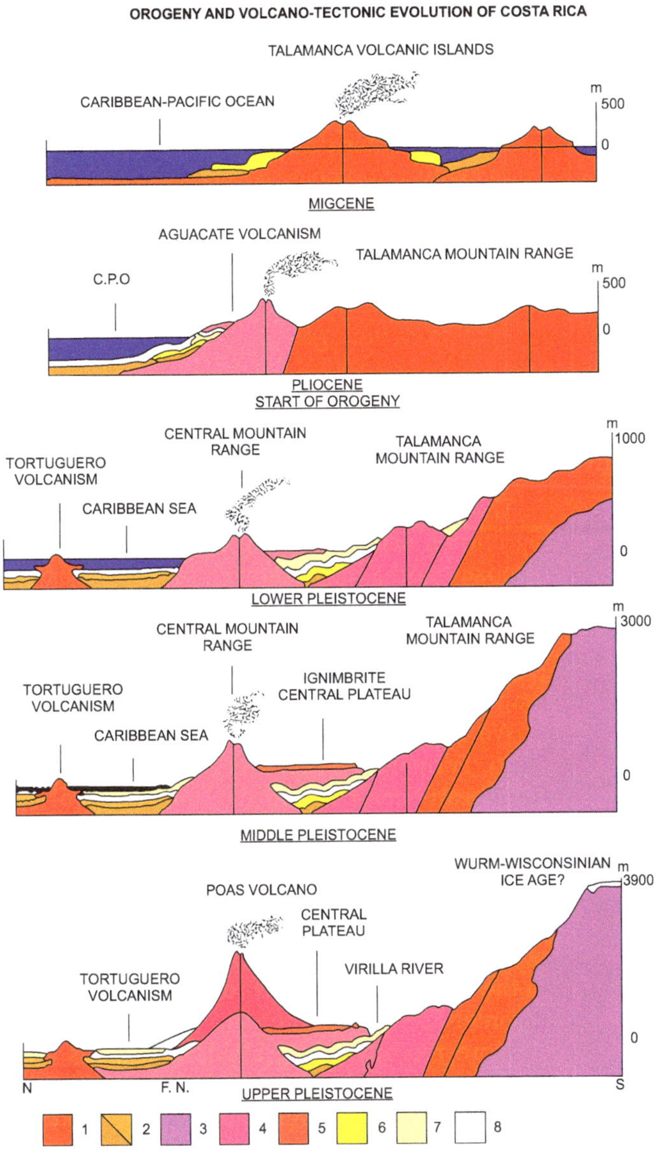

FIGURE 11.8 Orogeny and volcanic–tectonic evolution of Costa Rica (Bergoeing, 2012a). (1) Basal Pacific-type volcanism (basalts, andesites, dolerites); Cretaceous, but above all Tertiary and Quaternary formations. (2) Marine and continental sedimentary rocks. (3) Tertiary; Talamancan granodiorites. (4) Pliocene; Aguacate Formation volcanism. (5) Ignimbrites of the Central Valley (Middle Quaternary). (6) Subsident Northern Great Plain (Nicaragua graben). (7) Alluvial fans of the northern slopes of volcanic mountain ranges. (8) Detachment faults of Talamanca.

Tempisque plain were deposited; this is based on the observations made of the road slopes, 4 miles away from Nicoya Town, where an outfall fan cone composed of limestone debris is well exposed. The exposed part shows two embedded cones; the lower level consisting of an array of light red clays reveals an advanced degree of alteration of both clays and limestone debris.

The interpluvial period that occurred later is responsible for the former cold altitude pluvial period. We think that was the determining factor that gave rise to the higher river levels (T3) as well as to the old alluvial fans and glacis (if we leave aside the T4 level of the General Valley alluvial fans and the Telire River, which as we saw is attributed to the Lower Quaternary).

It is possible, therefore, to attribute to this period the higher cyclic levels that we have seen, in particular between Coronado and Palmar Sur. These levels indeed rise to 80, 130–160, and 250–300 ft and even up to 800–1000 ft, and cut the tectonic escarpment that dominates the Flandrian plain. It is true that here neotectonics play a very important role. If the upper levels are young, say Eemian, we will demonstrate that in this sector the positive neotectonics are very strong (Battitini and Bergoeing, 1983b) (Beaudet et al., 1982a). Therefore we can say that during the Middle Quaternary large coastal depressions began to be fed and constituted the first plains, and the first and powerful torrential deposits were built at the base of the volcanic mountain ranges (the eastern sector of Orosí Volcano and the sector of San Lorenzo to Tilaran are good examples).

Palmares caldera in the western Central Valley sector knew in this period a lacustrine episode. This depression retains ignimbrite deposits of the late Lower Quaternary that were deposited in an aqueous environment (clay balls), which proves there was still a lake. Was it compressed during the Middle Quaternary? One thing is certain, and that is that the Río Grande, which passes through the caldera, eroded a deep river canyon following an east–west fault alignment. It is equally possible that a revival of neighboring faults of the depression allowed the retained water to evacuate abruptly, or the lake simply disappeared after its silting, which transformed it into a swamp and then dried. The Middle Quaternary ended with a cold phase covering North America (Illinoian, known in Europe as Mindel). This period translates into Central America as a pluvial phase, and then a biostasic period would be set, particularly in Costa Rica, separating the T3 level from younger lower levels.

6. THE UPPER QUATERNARY

The beginning of this period is marked by two significant occurrences: a cold period (Riss–Illinoian) and a high marine park known as the Eemian. On the Pacific coast, particularly in the southern Nicoya Peninsula, high levels of erosion (115–130 ft) may be related to the high Eemian sea parking (Battistini and Bergoeing, 1983a).

We think that during this period the Puerto Viejo Formation strip reef appeared on the Caribbean coast (Battistini and Bergoeing, 1980). Indeed, those coral deposits are hardly or not at all recalcified with an aragonite domain structure. This reef is discordant with the Portete Formation complex reef. Although we cannot obtain radiometric measures, all evidence leads to the conclusion that this reef appeared during the inter-Riss–Würm period.

More recently (since about 6000 years ago) these old coral formations were assaulted by the Flandrian Transgression. Marine stands slightly higher than the current level have left their traces through observable fossil notches in the sector of Limon Promontory. This period is characterized by the following events in chronological order:

1. **Eemian transgression**
2. **Würm–Wisconsinian marine regression**
3. **Flandrian Transgression**
4. **The current sea level since 6000 years ago**

6.1 The Eemian Interglacial Period

The Eemian interglacial period (name taken from the Eem River in The Netherlands) was the penultimate warm period to which the land was subjected before the Holocene (current period). This period began 127,000 years ago and extended to 106,000 years ago. It is known thanks to the recognition of fossils and temperate fauna in studied sediments. It is estimated that the Eemian temperatures were 1–2° higher than the current ones (Kaspar, 2005). In England, the fossil studies show tropical fauna such as hippopotami, and this period has been called Ipswichian. In addition, the study of corals and foraminifera shows that the seas were 2–3° degrees warmer than the current temperature (Lea et al., 2000; Pelejero et al., 2010; Martrat et al., 2004). The Eemian sea level was 4–6 m higher than the current coastline, which probably was due to the melting of the Greenland ice sheet (Cuffey and Marshall, 2000). During this period the earth's orbit around the sun was more eccentric and its perihelion coincided with summer in the Northern Hemisphere (today it corresponds to the aphelion). The inclination of the earth was also greater than the current 23°27′. This led to a seasonality and the heat was greater, with colder winters, than at present, corroborated by the study of the former coral reefs (Felis et al., 2004). In Africa, the equatorial forest occupied a greater extension than currently and the Sahara was constituted by a savannah with transition to steppe vegetation. In addition, it was inhabited by lakes such as the Azawag between the Niger and Mali. The study of samples of sediments from Lake Ioannina in Greece suggests that the Eemian interglacial suffered no alternating periods of cooling and overheating and that it was similar to the current Holocene (Frogley et al., 1999). However, it appears that there could have been a notable Eemian cooling climate period, according to McCulloch's studies on the Bahama reefs (McCulloch et al., 1999) (Thompson and Goldstein, 2005), during which the sea level could have fallen 50 ft. In Germany (Eiffel region), there are traces of a remarkable cooling and arid climate toward the end of the Eemian interglacial due to the recoil of the Gulf Stream.

The Eemian is well represented in Costa Rica on the Pacific coast, and in particular in the south of Nicoya Peninsula. Levels of erosion without correlative deposits in Tambor, Paquera, and San Lucas Island between 130 and 164 ft are likely witnesses to this high marine parking (127,000 to 106,000 years BP), increased by strong positive neotectonics (Bergoeing, 1998).

6.2 The Big Würm–Wisconsinian Sea Regression

It slowly began about 106,000 years ago and extended to 12,000 years ago. The sea floor reached 394 ft below the current sea level. This period is characterized in Costa Rica by the death of the reef strip of Puerto Viejo on the Caribbean coast. As it was a cold period, the Talamanca summits experienced a real glacier parking at this time. The rest of the country (lower mountains and plains) was in a biostasic phase and therefore the rainforest dominated the landscape. This means that very few sedimentary deposits were made because of the protection exerted on the relief by the dense tropical vegetation (Bergoeing, 1998).

According to our observations in the Cerro de la Muerte and Chirripó summits, we note that a recent ice age (Würm; Chirripó II) occurred between 9850 and 12,470 ft above sea level, modeling small cirques, small glacier tongues, locks, and navels and moraines (Bergoeing, 1977). Moraines form three distinct groups that mark three glacial episodes during the last glacial phase. Investigations by Lachniet determined that glacier parking fell to 9974 ft altitude, covering an approximate area of

14 square miles (Lachniet and Seltzer, 2002). The dating of pollen and charcoal, using the ^{14}C method, gave an age of 10,400 years BP for these deposits, which indicates a climate change that heralded the end of the last ice age (Horn, 1993). Lower, between 9850 and 8500 ft altitude, traces of paleosnow levels are evident and are shown by small snow niches and material from the accumulation of melt water (yellow clays mixed with angular elements, products of the gel fraction, which can be observed on Cerro de la Muerte; Bergoeing, 1998). Unfortunately, our investigations have not allowed us to find remains of the oldest ice ages; however, this should happen at least during the Riss (Chirripó I), as Talamanca had already reached altitudes (orogeny) close to the current values. Otherwise, finding systematically in the rivers of Costa Rica two or three encased levels of fluvial terraces, formed by boulders, gives weight to the theory of climate change (T3 and T2 would be interglacial Riss–Illinoian and T1 postglacial Würm–Wisconsinian). Following this logic, we think that during the Upper Pleistocene the high peaks of Talamanca were subjected to two interrain periods alternating with two rain periods that correspond to the Riss–Illinoian and the Würm–Wisconsinian.

6.3 The Flandrian Transgression

The last great eustatic variation that modified the coastline, known as the Flandrian Transgression, stabilized the current sea level some 6000 years ago. This was due to the melting of the glaciers from the Würmian–Wisconsinian ice sheets, the Finland–Scandinavian (8000 years BP), the Laurentide (7000–6500 years BP), and the Siberian, which resulted in a rise in the sea level of about 420 ft above its previous level, with annual increases that reached in some cases 0.3 to1.6 ft (Delort and Walter, 2001).

During the Quaternary period, the sea level fluctuated dramatically, varying between ice ages and interglacial periods. During the Flandrian period (Holocene), the sea invaded the current coastal plains to about 10–13 ft above the current sea level throughout the entire world. This is known as the Flandrian Transgression, a name taken from Flanders in Europe, where three periods can be distinguished:

- **The Lower Flandrian (15,000 to 8000 years BP)**, to which is given the opening of the Calais pass between the North Sea and the channel of La Manche;
- **The Middle Flandrian (8000 to 5000 years BP)**, with low sea levels, transforming the Baltic Sea into an inland sea;
- **The Upper Flandrian**, to the beginning of the Christian era. It corresponds to the emersion of Flanders and the birth of the Zuiderzee.

Most of the coasts have undergone changes in the past 2500 years. It is worth remembering the change in landscape in the narrow coastline of Thermopylae, Greece, where Leonidas stopped the Persian army on a narrow coastal strip, which today is a plain of 1.56 miles in length opposite Euboea Island. We may also invoke the Picts Gulf in the west of France, today a plain, where Julius Caesar beat the Gauls on the western coast of Europe.

During the Quaternary period, there were important climatic changes that were translated by glacial and interglacial periods due to variations in the earth's orbit that resulted from insolation fluctuations of the planet (Milankovich theory), particularly during the Middle and Upper Pleistocene. It produced, in the equatorial and tropical latitudes, rexistasic periods, during which the vegetation receded because of glacier stands in the high peaks, and biostasic periods, in which, in contrast, the vegetation thrived. In the same way the ecosystems of the middle latitudes were modified by herbaceous vegetation during the ice age periods followed by a settlement of forests during the interglacial periods. The situation is analogous to the current (Holocene). In Europe the last interglacial, known as the Eemian, is one of the most studied by the scientific community and considered analogous to the current period.

The Serbo-Croatian Milutin Milankovich established three large climatic cycles. The first corresponds to the equinox's precedence of 21,000 years (great sidereal year). This affects 50% of climate change. The second cycle corresponds to periods of 41,000 years and is directly related to the inclination of the earth's axis, which varies from 22 to 25 degrees. Currently the axis has an inclination angle of 23°27′. Any variation in the earth's axis accounts for 25% of climate change (Bergoeing, 2002). The third factor corresponds to 100,000- to 400,000-year cycles and is related to the earth's elliptical travel around the sun, and that can reach up to 7%. Currently the value is close to 0 degrees. This factor influences 10% of the earth's global climate change.

One hundred twenty thousand years ago, in the Eemian period (last interglacial maximum), i.e., a period of natural global warming of the planet, the temperature exceeded 35.6°F at the current global temperature and the concentration of CO_2 in the atmosphere was 200 ppm. Currently this value is 370 ppm. However, the value of atmospheric CO_2 is not so important if it is associated with the value of the heat that the terrestrial globe receives (Hogg and Bernier, 2005).

In Costa Rica, the Eemian level is represented in levels that rise 328 ft above sea level in the northern mouth of the Terraba riverbank and on the Osa Peninsula, on sedimentary plateaus forming active cliffs, especially in Punta Carate Cliff. The positive neotectonics can explain only that altitude of the ancient sea level. Eemian levels have not been fully studied in Costa Rica and must be the subject of further investigations.

On the other hand, the Flandrian level has been studied in Costa Rica, but not in a systematic way, by the author of these lines. It is very well represented in the coastlines of the two oceanic facades of the country and, in this case, neotectonics have played a fundamental role, allowing the Flandrian Transgression to occupy coastal plains that are raised today to 33 ft above sea level.

On the Caribbean coast, at Playa Bonita (Limon sector), it is possible to see the traces of a Holocene marine park translated by the remains of an old platform of coral limestone that dominates the current platform of 6.6–10 ft (before the earthquake in Limon on April 22, 1991, which caused a vertical lift between 1.6 and 4.9 ft) (Denyer and Kusmaul, 2000). The old platform is also characterized by two erosion levels with shaped notches located at 4 and 6.6 ft from the current abrasion notch line. On the Caribbean coast of Costa Rica, the tidal swing is only 1.64 ft.

Farther to the north, between Moin and the Nicaraguan border (San Juan delta), old rectilinear littoral cords are also witnesses to the Flandrian Transgression that must have ventured up to 23–33 ft above sea level, to the interior of the grounds, despite the positive neotectonics particularly active in this sector of the country. In Barra de Tortuguero, coastal cords are supported on a small half-destroyed volcanic cone, which should be noticeably contemporary (Tournon, 1984a).

On the Pacific coast, the traces of the Flandrian Transgression are numerous: coastal cords and arrows, silting up of interior bays, and ancient swamps transformed into coastal plains, as in Parrita, for example, and inactive Flandrian cliffs (Montezuma in Nicoya and Coronado in the Diquis delta). We believe that the 33-ft level to the south of Nicoya Peninsula corresponds to this high marine parking (always taking into account the important positive neotectonics also in this sector of the country). In effect, the lumachelles (little rolled shells) deposit of Montezuma Beach that we dated to 6620 years BP (Battistini and Bergoeing, 1983b) is a beautiful example of the rising of the sector. The deposit was made at the Flandrian sea level and is currently at 6 or 7 m above sea level.

During the Flandrian, inland, strong erosive activity washed the slopes and in the main rivers formed fluvial terraces at the T1 level. The lower terrace of the Reventazon River is a good example, and the Cervantes lava flow dating back

23,000 years BP covered this fluvial level (Allegre and Condomines, 1976). For this river we have therefore a logical succession from the Middle Pleistocene to the Upper Pleistocene. In the Tarcoles River as well as the Barranca River (Pacific slope), the lower river levels run from 20 to 26 ft, and we see the slightly altered material of the boulders that make up the terraces with little or no tectonic deformation, which makes them contemporary with the T1 level of the Reventazon River. They are undoubtedly reservoirs of climate crisis origin (post-ice age landslides).

The great stratovolcano slopes of the Central and Guanacaste mountain ranges are characterized by volcanic activity that translates to large laharic flow deposits (lahars in top position, very well preserved, north of Cartago City) (Bergoeing, 2009a).

6.4 The Current Sea Level Since 6000 Years Ago

The coastal aquifer mantle has not varied much since 6000 years ago and is located near the current level that it would probably give (if the sea level come down again from the current level), a new coastline, comparable with erosion levels that we know and have described in the Nicoya Peninsula.

Since the last period, the coastal plains have been gaining ground near the sea, thanks to the fluvial sedimentary contributions that are increasingly important (in particular since the 20th century due to anthropogenic effects). This is combined with a tectonic hoist, which continues today. The Diquis and Coto Colorado plains are good examples in which marshy areas have disappeared or are in danger of extinction. A second area is the Tempisque River plain in its lower course, between Bolson and the estuary where every day coastal sedimentation wins more land facing Lake Areas or the sea.

Large Upper Quaternary climate changes probably took place in Costa Rica, in more or less strong oscillations. Researchers from Ohio University (Athens, OH, USA) have used ice drilling done at 19,700 ft altitude in the Peruvian Andes Mountain Range to study pollen trapped

QUATERNARY CHRONOLOGY OF COSTA RICA BY J.P.BERGOEING Ph.D.

Years in thousands	Period	Volcanism		High Mountain over 9800 ft	Piedmont and Piedmont deposits	Fluvial Terraces	Plains	Pacific coast	Caribbean coast
0-0,006	Recent	Arenal volcano	Strata-volcanoes	Moor Vegetation	Solifluction		Lake Tempisque areas	Mevorgna	Cahuita coral reef
0,006-0,012	Holocene	Tortuguero volcano Cot lahars			T1 Tilaran	T1 Reventazon	Santa Cruz Glacis1	Montezuma Flandrian Transgression	Playa Bonita Shaped Notches
0,012 0,085 0,0140	Upper Quaternary	Cervantes lava flow Fortuna de Bagaces lava flow		Wurm-Wisconsinian Ice Age Chirripoll			Pluvial? Inter-pluvial		
	Quaternary			Inter-glaciar	T2 Tilaran	T2 Reventazon	Santa Cruz Glacis 2	66ft-82ft Nicoya level without correlative deposits	Puerto Viejo Coral Reef. Eemian
0,1600 0,300	Middle	Central plateau lahars Las Nubes volcano? Zurqui volcanoes?		Riss-Illinolian Ice Age Chirripol	T2 Tilaran	T2 Reventazon	Pluvial? Inter-pluvial	130ft-150ft Nicoya level without correlative deposits	
0,6	Low	Bagaces Ignimbrites				T3 Telire River Liberia Lacustrine levels	Start filling of recent shorelines		
0,7		Central plateau ignimbrites							
0,88		Guayabo caldera ignimbrites							
1,2		Dacitic Domes Rincon de la Vieja							
1,3		Central volcanic plateau			Orotina Piedmont base	T4 San Pablo Turrubares			
1,8	Quaternary	Palmares caldera Small volcanoes			T4 El General Valley	T4 Tio Telire?	Caribbean Sea in contact with Nicaragua Lake	Tempisque depression and Nicoya uplift	Portete Coral Reef Villafranquian?
2,3	Pliocene	End of Guayabo							Rio Banano Formation
3		La Carpintera volcanisms							
		Middle Bagaces ignimbrites							

FIGURE 11.9 Quaternary synoptic overview of Costa Rica (Bergoeing, 2007).

in the ice at a depth of 535 ft (Thompson, 1980). The results were the following:

Drought periods

570 BC to 610 AD
1250–1310 AD
1710–1860 AD

Humid period

1500–1720 AD

A very wet period occurred between 1500 and 1720, which coincides with the end of the "Little Ice Age" of Europe. This result should be taken into account in Central America because of the relation to anticyclonic centers, in particular, the one in South America, bound to the Intertropical Convergence Front and therefore responsible for rains in the Pacific watershed of the Central America Isthmus, as well as to the El Niño climatic phenomena, which provide more and more evidence of their presence in the distant past (Fig. 11.9).

Conclusion

During the Quaternary period, Costa Rica was shaped by active tectonics, volcanism, and large periods of climatic change (pluvial and interpluvial). The Plio–Quaternary orogeny made Costa Rica a predominantly mountainous country, characterized by different types of models distributed at various altitudes. It is a predominantly tropical model of altitudes, and we can classify them in the following way:

12,467 to 9840 ft altitude: This range is dominated by paleoglacier modeling. Traces of the last Würm–Wisconsinian ice age are observed, described, and analyzed in the Chirripó and Kamuk summits.

9840 to 8530 ft altitude: These altitudes are dominated by paleosnow modeling. This modeling is a seasonal snow parking during the last ice age.

8530 to 3300 ft altitude: Here stands the multifaceted model. This is a typical shape that occurs in the humid tropics from the effects of rain erosion and a modeling that exposes parts of the underlying stone structure.

3300 to 164 ft altitude: This range is dominated by multiconvex modeling, which occurs where alluvial fans, fluvial terraces, and generalized solifluction prevail. It is where the deposited material decomposes in situ and takes the form of domes, or "half-oranges," as they are known in Brazil.

164 to 0 ft altitude: This is the domain of the floodplains such as the Great Northern Plain (San Carlos) that extends into Lake Nicaragua and the San Juan River (Nicaragua graben).

Various modeling forms in Costa Rica bear witness to climatic crises that occurred during the Quaternary period and which were translated by phases of rexistasis (drier) and biostasis (wettest). It was during the rexistasic phases that rain erosion built three levels, characteristics of the observed fluvial terraces and corresponding to the Middle Pleistocene (T3 level) and the Upper Pleistocene (levels T1 and T2). The oldest (Lower Pleistocene), the T4 level, is sporadically represented throughout the country and in particular by the more antique alluvial fans folded by the neotectonics of the General Valley, as analyzed in this work.

During the interpluvial, the coastal plains suffered marine transgressions, of which at least two left traces, the Eemian and the Flandrian.

Altitude is a witness to at least one recent ice age (Würm), in the peaks of the Talamanca Mountain Range and particularly in Cerro Chirripó (12,530 ft), which occurred during the last biostasic phase, establishing a balance and only local erosive effects due to specific causes (tectonics, volcanism).

Currently the country is building and shaping its landscape, this time coupled with anthropogenic actions—farming, for example, which has intensified the silting up of the Nicoya Gulf and the creation of submarine deltas such as at the mouth of the Tarcoles River. Natural positive tectonics also contribute to the great changes, with the ongoing uplift of the Caribbean coast, of the San Juan River middle course, and of the Nicoya Peninsula and the Coastal Mountain Range due to active neotectonics. Volcanism is still expressed in magmatic emissions, products of the permanent displacement and collision of the Cocos and Caribbean tectonic plates. Finally, the inexorable rise in the sea level tends to delete the arrows and current littoral cords, being a threat to the urban settlements of the coast (Fig. 11.10).

Glossary

A

Advance: This is used to describe preexisting river thalwegs when a relief is in formation; antecedent is also used.

Adventitious: This describes a parasitic cone on a volcano whose activity was very low.

Alkaline: This is a soil whose pH is higher than 7. They are usually calcareous soils.

Trade winds: These are the tropical zone northeast winds.

Alluvium: This is a sedimentary deposit formed of sand, silt, and clay with rolled edges deposited by fluvial action.

Amonite: This is a fossil mollusk that lived during the Jurassic and Cretaceous periods.

Amphibolite: This is a deep calcareous pelitic rock or basic eruptive rock transformed to a metamorphic rock.

Andesite: This is lava of intermediate acidity, which contains between 52% and 60% silica.

Anticline: This is a structural form folded in the shape of a roof.

B

Barrier reef: This is a coral reef far from the coastline sometimes leaving a lagoon between the coast and the reef. See fringe reef.

Basalt: This is basic volcanic rock. It contains less than 52% silica. Lava flows are fluid and extensive, typical of oceanic volcanoes like the Hawaiian Islands.

Batholith: This is a deep magmatic ascent, usually of granite or granodiorite and cooling in the interior of the earth's crust. A batholith could be the heart of a mountain range.

Biostasis: This is a wet period or cycle, which allows vegetation to grow and protects soils. It is a phase of biological stability, giving rise to alteration in situ.

Burning cloud: This is a thick volcanic, acid, gaseous emission, a product of violent eruptions with pyroclastics and abundant pumice. Burning clouds descend from the craters with temperatures above 1300°F.

C

Caldera: This is a usually circular depression formerly occupied by a volcanic focus. One speaks then of a collapsed caldera (by magma chamber collapse) or, if it remains part of the volcanic cone, it is usually an explosion caldera.

Cherts: These are volcanic rocks.

Climax: This is the optimal state of a relatively stable balance between vegetation and soil.

Colluvial: This refers to side accumulations of sediment in a valley whose travel and transport are reduced.

Coastal arrow: This is a cord of sand and boulders accumulated by littoral drift. Coastal cords are formed of the same mode that originates beaches.

Chipboard: This is a set of rolled, sedimentary rocks of small size united by a sedimentary matrix (clay, silica, sandstone, or limestone).

Cryoclastism: This refers to rocks broken by the effects of frost weathering.

Cuesta: This is a tilted monocline with concordant sedimentary layers and a cornice on the top front. If they are discordant layers, it is a *pseudocuesta*.

D

Dacite: This is a volcanic rock.

Diabase: This is a mineral element of volcanic rock.

Dissected: A dissected structure is eroded, cut, or undermined by physical–meteorological mechanisms.

Diorite: This is an igneous rock formed by plagioclase-associated minerals (biotite and hornblende).

Dip: This is a slope that has an inclination value. It must be combined with the address measured with a magnetic compass.

E

Eemian: This refers to the interglacial Riss–Würm period of overheating of the terrestrial globe that occurred between 130,000 and 90,000 years BP, characterized by a sharp rise in the sea level of 130–165 ft. (It is also called Sangamonian, Ipswichian, Mikulin, Kaydaky, Valdivia, or Riss–Würm.)

Escarpment: This is a steep rocky slope (higher than 45 degrees).

F

Fault: This is a tectonic fracture at the surface of the earth's crust due to tectonic plate movements or, locally, due to stresses on hard surfaces, causing the displacement of two compartments. If it is recent it is called a *failure mirror*. If it has been altered by erosion it is a *fault escarpment*. If the escarpment has been regressed by erosive effects it is a *fault line escarpment*. If a block that has been sunk by erosion effects remains in the upper position it is an *inverse fault escarpment*.

Foehn: This is a hot and dry wind descending the mountain slopes, due to subsidence, which makes sure the air is hot and loses its relative humidity. Its strength is due to the fact that it is channeled by valleys.

Flandrian: This is the period of the climate optimum reached 6500 years BP and that was characterized throughout the world by a 4-m sea-level rise due to widespread melting. It is also called the *Flandrian Transgression*.

Fringe reef: This is a coral reef stuck to the beach or coastline.

Fumaroles: These are posteruptive volcanic phenomena that translate to emissions of sulfurous vapors or other components with high temperatures (as on Poas Volcano).

G

Gabbro: This is an intrusive igneous rock. By its composition, it is similar to a basaltic rock.

Gasmaar: This is an explosive volcanic eruptive caused by an encounter between water and ascending magma (as in Lake Río Cuarto).

Geyser: This is source of hot water associated with volcanic activity that arises suddenly at high altitude.

Giant kettle: This is a circular cavity undermining the bedrock of a river or the coast, through the effects of a whirlwind conjugated with the abrasive effects of boulders.

Glacis: This is a vast sedimentary area composed of fine particles that extends to the feet of an alluvial fan and imperceptibly meets the plain.

Graben: This is a tectonic depression.

Graywacke: These are volcanic deposits characterized by black sands, lapilli, and pumice.

Graphite: These are minor scars of the dissolution form of karsts but more commonly are characterized by vertical slots.

H

Hypercalcareous: This refers to sediment oversaturated with calcium carbonate.

Hogback: This is a rectilinear crest in a folded sedimentary structure (such as the Appalachian Ridge).

Halophyte: This is a plant that lives in the middle of salty waters.

Heterometric: This refers to sedimentary material where coexist fine and coarse materials without a scale of values.

I

Igneous: This refers to something that comes from the earth's interior. Igneous rocks can be intrusive or extrusive.

Ignimbrite: This is an acid consolidated volcanic deposit. It is associated with stratovolcanoes and violent volcanic eruptions (burning clouds).

Ice caps: These are glaciers of convex shape that cover a large area in Antarctica or Greenland.

Isobath: This is an isoline indicating a depth determined in oceans, seas, or lakes.

J

Jasper: This is a metamorphic rock of sedimentary origin (clay).

Joints: These are cracks in rocks. Sedimentary rocks are perpendicular or oblique to the stratification planes. In crystalline rocks they are curved.

K

Karst: This is a dissolution process of rocks due to rainwater in limestone rich in calcium and magnesium carbonates.

L

Lahar: This refers to wash mud and heteroclite materials on volcanic slopes associated with a volcanic eruption. When it comes from the overflow of a crater lake it may be hot. In the case of the sudden melting of an ice cap the volcanic cone can be cold. A lahar flow can cause great destruction.

Lagoon: This is a semiopen seawater area surrounded by coral banks.

Lapilli: These are volcanic projection slag-type fragments of degassed lava.

Latite: A volcanic rock.

Lava, pillow: Basalt lava flows in contact with the sea that take the forms of pads are known as *pillow lava*.

Limestone: This is a sedimentary rock composed of the skeletons of marine microorganisms.

Lumachelles: These are mollusks, commonly known as "porcelain."

M

Madreporic: This refers to something of coral origin.

Meander: This is a sinuous curve formed by a river in a plain (lower course) because of lack of competition. Meandering is typical of low tropical coastlines.

Multiconvex modeling: This is the first form of humid tropical landscapes, where reliefs adopt round or dome forms due to the effects of erosion.

Multifaceted modeling: This is the second form of tropical landscapes. The relief adopts chevron forms due to the effects of erosion in direct concordance with the underlying structure.

Monocline: This is a tipped but not folded sedimentary structure. If the upper stratum is tough and consistent with respect to the bottom tier it gives rise to a *cuesta relief*.

N

Nuée ardente: This is the French expression for violent volcanic acid gaseous flows with pyroclastics and pumice. They descend from craters at high temperatures (more than 700°C).

O

Ophiolitic complex: This is a complex of volcanic and intrusive basic and ultrabasic rocks (diorite, dolerite, gabbros, and peridotite) from offshore funds (greenstones) typical of geosynclinal volcanism.

Olive: This is a small dark green mollusk of the "porcelain" type.

Ombrothermic: This refers to a dense tropical rainforest with elevated temperatures.

Orogenesis: Also tectogenesis, this is the process of uplifting of a relief by the collision of tectonic plates (such as in the Andes).

Ox bow: This is an abandoned lake formed by meanders.

P

Peridotite: This is a plutonic rock of green olive color composed of olivine.

Piedmont: This is a region that falls in contact with a mountain massif and is characterized by alluvial fans and landslides.

Puddingstones: These are strongly cohesive river deposits composed of boulders of different sizes and usually cemented by a clay matrix.

R

Rasa or sea platform level: This is the level of marine erosion on a rocky surface.

Radiolarites: These are sedimentary rocks composed of organic marine sediments, the radiolarians.

Rexistasis: This is a drier period associated with the evolution of a vegetative soil cover and modeling forms. It allows the description of sedimentary cycles: the opposite is the *biostasic period*.

Riss: This was the Illinoian ice age in North America. It was the second youngest ice age during the Pleistocene, 300,000 to 130,000 years ago.

Rhyolite: This is a very acidic volcanic rock.

S

Syncline: This is a tectonically folded canoe-shaped sedimentary form.

Superimposition: This describes river thalwegs that have subsequently fitted an ancient preexisting structural surface.

Subduction: This refers to an area of confrontation of two tectonic plates and where the first plunges into the mantle.

Subsidence: This is the progressive collapse of a basin or a tectonic pit from the effects of the weight of the deposited sedimentation.

Shale: This is a metamorphic rock of sedimentary origin (compacted clays), also called flysch.

Solifluction: This refers to landslides of clay and other materials due to massive water contributions (rains). It may originate from laminar to mass catastrophic landslides and mud flows.

T

Table: This is an aclinal area between sedimentary structural forms that gives rise to an elevated and isolated plateau.

Thalweg: This is the maximum depth line in the longitudinal profile of a river.

Tarn: This is a glacial lake.

Toleitic: This refers to an abyssal rock type composed of biotite, hornblende, andesite, mica, and sometimes garnet phenocrystals.

V

Volcano: This is a conical form built by magmatic ascensions from the mantle. It comprises the forms of stratovolcanoes, shield volcanoes, domes, or very extensive volcanic complexes or even a simple eruption accident.

W

Würm: This was the last ice age period, corresponding to the fourth European glaciation extending from 70,000 to 12,000 years BP. In the United States it corresponds to the *Wisconsinian* period. The sea level was 460 ft below the current sea level.

X

Xerophyte: This describes vegetation or animals accustomed to the lack of water or prolonged drought.

Bibliography

Agencia Asturiana De Cooperacion, 2009. Ampliación del programa para la regeneración medio ambiental del lago Yojoa. Segunda Etapa, Asturias, España.

Aguilar, T., 1999. Organismos de un arrecife fósil (Oligoceno Superior-Mioceno Inferior) del Caribe de Costa Rica. Rev. Biol. Trop. 47, 453–474.

A.I.D. (USA), 1979. Costa Rica Regional Analysis of Physical Ressources. Washington D.C., U.S.A.

Alfaro, A., 1913. Rocas volcánicas de Costa Rica. Boletín de Fomento, San José, Costa Rica, pp. 123–131.

Alfaro, A., Michaud, G., Biolley, P., 1911. Informe sobre el terremoto de Toro Amarillo, Grecia. Anal. Centro Est. Sismol. de C.R., pp. 35–41. San José, Costa Rica.

Allegre, C.J., Condomines, M., 1976. Fine chronology of volcanic process using 238U-230Th systematics. Earth Planet Sci. Lett. 28, 395–406.

Alvarado, G.E., Carr, M.J., 1993. The Platanar-Aguas Zarcas volcanic centers, Costa Rica: Spatial-temporal association of Quaternary calc-alkaline and alkaline volcanism. Bull. Volcanol 55, 443–453.

Alvarado, G.E., Gans, P.B., 2012. Síntesis geocronológica del magmatismo, metamorfismo y metalogenia de Costa Rica. América Central. Rev. Geol. Amér. Central 46, 7–122. ISSN: 0256–7024.

Alvarado, G.E., Paniagua, S., Tejera, R., 1980a. Mapa del vulcanismo Plio-pleistocénico de Costa Rica. 1:700 000. Instituto Geográfico Nacional Costa Rica.

Alvarado, G.E., Kussmaul, S., Chiesa, S., Gillot, P.-Y., Appel, H., Wörner, G., Rundle, C., 1992. Resumen crono estratigráfico de las rocas ígneas de Costa Rica, basado en dataciones radiométricas. South Am. Earth Sci. J. 6 (3), 151–158.

Alvarado, G.E., Soto, J.G., Taylor, W.D., 1998. Los flujos piroclásticos del volcán Arenal del 5 de mayo de 1998 y sus implicaciones para la amenaza de las obras de infraestructuras cercanas. Bol. OSIVAM 10 (19-20), 1–13.

Alvarado, G.E., Soto, G.J., Schmincke, H.-U., Bolge, L.L., Sumita, M., 2006. The 1968 andesitic lateral blast eruption at Arenal Volcano, Costa Rica. J. Volcanol. Geotherm. Res 157, 9–33.

Alvarado, G.E., Pérez, W., Vogel, T.A., Grüger, H., Patiño, L., 2011. The Cerro Chopo basaltic cone (Costa Rica): An unusual completely reversed graded, pyroclastic cone with abundant low vesiculated cannonball juvenile fragments. J. Volcanol. Geoth. Res. 201 (Issues 1–4), 163–177.

Alvarado, G.E., 1989. Los volcanes de Costa Rica. EUNED, San José, Costa Rica xviii + 175 pp.

Alvarado, G.E., 2000. Los volcanes de Costa Rica. Editorial Universidad Estatal a Distancia, UNED, Costa Rica, p. 269.

Alvarado, G.E., 2009. Geología de la hoja Fortuna, Alajuela, Costa Rica. Rev. Geol. Amér. Central 41, 117–122.

Anderson, D.M., 1987. Mapa Geológico de Honduras: Hoja de Lepaterique. Instituto Geográfico Nacional, Tegucigalpa, Honduras escala 1:50,000.

Arredond, O.S., Soto, J.G., 2006. Edad de las lavas del miembro Los bambinos y sumario científico de la Formación Barva. Revista geológica de América Central 34-35, Costa Rica, pp. 59–71.

Astorga, A., 1987. El Cretácico Superior y el Paleógeno de la vertiente Pacifica de Nicaragua meridional y de Costa Rica septentrional: origen, evolución y dinámica de las cuencas profundas relacionadas al margen convergente de Centroamérica. Universidad de Costa Rica, p. 247. Tésis Lic.

Astorga, A., 1992. Descubrimiento de corteza oceánica mesozoica en el norte de Costa Rica y el sur de Nicaragua. Rev. Geol. Amér. Central 14, 109–112.

Atwater, T., 1970. Implicaciones de las placas tectónicas de la evolución tectónica Cenozoica del oeste de América del Norte. Geol. Soc. Am. Bull. 81, 3513–3556. USA.

Aubrun, Ch.V., 1974. L'Amérique Centrale. Presses Universitaires de France, Paris. Que Sais-je N° 513.

Azambre, B., Tournon, J., 1977. Les intrusions basiques alcalines du Rio Reventazon (Costa Rica). In: Somm, C.R. (Ed.), Soc. Geolo. France, fasc.2, pp. 104–107. Paris, Francia.

Azema, J., Tournon, J., 1980. Un massif ultrabasique charié en marge pacifique de l'Amérique Centrale. Note transmise a la séance du 19.11.1979 a l'Academie des Sciences de Paris. La péninsule de Santa Elena, Costa Rica.

Azema, J., Sornay, J., Tournon, J., 1979. Découverte d'Albien Superieur a ammonites dans le matériel volcano- sédimentaire du Complexe de Nicoya. Somm Soc. GeoL de FR. Tome XXI fasc. 3, Province de Guanacaste, Costa Rica, pp. 129–131.

Azema, J., Glaçon, G., Tournon, J., 1981. Nuevos aportes sobre el Paleoceno con foraminíferos planctónicos del margen Pacífico de Costa Rica. Informe Semestral. Instituto Geográfico Nacional, Costa Rica, pp. 57–69. Jul-Dic. 81.

Azembre, B., Tournon, J., 1977. Les intrusions basiques alcalines du Rio Reventazon (Costa Rica). Som. Soc. Geol. France 2, 104–107.

Barr, K.W., Escalante, G., 1969. Contribución al esclarecimiento de la edad del Complejo de Nicoya, Costa Rica. GeoL ICAITI 2, Costa Rica, pp. 43–47.

Battistini, R., Bergoeing, J.P., 1980. Observations sur le Quaternaire littoral de la côte Caraïbe du Costa Rica. Revista Quaternaria N° XXII, pp. 237–242. Roma, Italia.

Battistini, R., Bergoeing, J.P., 1982a. Volcanisme recent et variations climatiques Quaternaires au Costa Rica. Bull. Assoc. de Géographes Français, N° 485–486, pp. 96–98. Paris, France.

Battistini, R., Bergoeing, J.P., 1982b. Un exemple de côte a structure faillée quadrillée et Néotectonique active: la côte Pacifique du Costa Rica. Bull. Assoc. de Géographes Français, N° 488, pp. 199–205. Paris, France.

Battistini, R., Bergoeing, J.P., 1983a. Características geomorfológicas del litoral comprendido entre bahía Tamarindo y bahía Culebra, Peninsula de Nicoya, Costa Rica. Instituto Panamericano de Geografía e Historia, México D.F., México, pp. 79–90. Revista Geográfica N° 98.

Battistini, R., Bergoeing, J.P., 1983b. Reconnaissance géomorphologique de la façade Pacifique du Costa Rica. Cahiers du CEGET N° 49 3e trimestre 83, pp. 1–73. Bordeaux, France.

Battistini, R., Bergoeing, J.P., 1984. Geomorfología de la costa Caribe de Costa Rica. Instituto panamericano de Geografía e Historia, México D.F. México, pp. 167–188. Revista Geográfica N° 99.

Baxter, S., 1976. Estudio geológico de las Formaciones Mata de Limón y punta Carballo, Costa Rica (Tesis). Escuela Centroamericana de Geología, UCR, Costa Rica, 55 pp.

Beaudet, G., Gabert, P., Bergoeing, J.P., 1982a. Les modelés hérités du Costa Rica. Bull. Assoc. de Géographes Français, N° 488, pp. 183–197. Paris, France.

Beaudet, G., Gabert, P., Bergoeing, J.P., 1982b. La Cordillère de Talamanca et son Piémont (Néotectonique et variations morpho-climatiques dans le Sud-Ouest du Costa Rica. Colloque sur les Piémonts, pp. 121–134. Toulouse 12-15 mai 1982. France.

Beaudez, et al., 1983. In: Introducción a la arqueología de Copán, Tomo I. Proyecto arqueológico Copán, secretaría de estado en el despacho de Cultura y Turismo, Tegucigalpa D.C, pp. 78–92.

Bellon, H., Tournon, J., 1978. Contribution de la géochronométrie K/Ar à l'étude du magmatisme du Costa Rica, Amérique Centrale. Bull. Soc. Geol. France 7 –XX, pp. 955–959. Paris.

Bergoeing, J.P., Brenes, L.G., 1977. Laguna de Hule, una Caldera Volcánica. In: Informe Semestral, julio-diciembre 1977. Instituto Geográfico Nacional, San José, Costa Rica.

Bergoeing, J.P., Brenes, L.G., 1978a. Geomorfología del cantón de Buenos Aires, provincia de Puntarenas, Costa Rica. 1:200 000. Instituto Geográfico Nacional, Costa Rica.

Bergoeing, J.P., Brenes, L.G., 1978b. Mapa Geomorfológico de Costa Rica 1:1 000 000. Instituto Geográfico Nacional, Costa Rica.

Bergoeing, J.P., Brenes, L.G., 2007. Las calderas concéntricas del Platanar. Costa Rica Revista Geográfica IPGH N° 141 Enero-Junio 2007. México.

Bergoeing, J.P., Malavassi, E., 1981a. Mapa geomorfológico del valle Central, Costa Rica. 1:50 000 (9 hojas más texto). Instituto Geográfico Nacional, Costa Rica.

Bergoeing, J.P., Malavassi, E., 1981b. Mapa geomorfológico del valle Central, Costa Rica. 1:100 000 (2 hojas). UCR, Costa Rica.

Bergoeing, J.P., Malavassi, E., 1981c. Síntesis Geológica del valle Central, Costa Rica. 1:100 000 (2 hojas). Instituto Geográfico Nacional, Costa Rica.

Bergoeing, J.P., Malavassi, E., 1981d. Carta Geomorfológica del Valle Central. Escala 1:50.000 (9 hojas más texto) editada en colores por Instituto Geográfico Nacional, Costa Rica.

Bergoeing, J.P., Protti, R., 2006. Geomorfología Paleo-Lacustre del Sur del Lago de Nicaragua. Revista Geográfica, N° 139 IPGH, México DF. México.

Bergoeing, J.P., Protti, M., 2009. Tectónica de placas y sismicidad em Costa Rica. Revista Geográfica del IPGH N°149 julio-Diciembre 2009. México.

Bergoeing, J.P., Brenes, L.G., 2013. El río Tárcoles, desaguadero del Valle Central Occidental de Costa Rica. Revista Geográfica IPGH N° 154, México.

Bergoeing, J.P., Brenes, L.G., 2016. Geomorphology Atlas of Costa Rica, 1:100.000. Full color. Universidad Nacional, Heredia, Costa Rica, 36 pp.

Bergoeing, J.P., Malavassi, E., Protti, R., 1978a. Tres posibles edificios volcánicos del sector Cerros del Aguacate. Informe Semestral Jul-Dic. 78. Instituto Geográfico Nacional, San José, Costa Rica.

Bergoeing, J.P., Mora, S., Jimenez, R., 1978b. Evidencias De vulcanismo Plio-Cuaternario en la Fila Costeña, Térraba, Costa Rica. Informe Semestral Jul-Dic 78. Instituto Geográfico Nacional, San José, Costa Rica.

Bergoeing, J.P., Malavassi, E., Jiménez, R., 1980. Síntesis geológica del Valle Central de Costa Rica (2 mapas). IGN UCR (1:100 000), Costa Rica.

Bergoeing, J.P., Brenes, L.G., Malavassi, E., 1982a. Geomorfología de la hoja Barranca, Costa Rica. Escala 1:50.000. (1 hoja) editada en colores por Instituto Geográfico Nacional, Costa Rica.

Bergoeing, J.P., Brenes, L.G., Malavassi, E., 1982b. Geomorfología del Pacífico Norte de Costa Rica. Escala: 1:100.000. (11 hojas más texto) editada en colores por Instituto Geográfico Nacional, Costa Rica financiado por CONICIT-US-AID.

Bergoeing, J.P., Arce, R., Brenes, L.G., Protti, Q., 2007. La caldera de Barbilla, Costa Rica, investigación preliminar. Revista Geográfica IPGH N° 142, México.

Bergoeing, J.P., Brenes, L.G., Fernandez, M.R., 2010a. Las calderas volcánicas de la Cordillera de Talamanca, Costa Rica. Revista Geográfica, N° 148. IPGH, México DF.México.

Bergoeing, J.P., Brenes, L.G., Fernandez, A.M., Ureña, F.,M., 2010b. Geomorfología de la cordillera Costeña y de los abanicos aluviales en el piedemonte meridional de la Cordillera de Talamanca. Revista Geográfica, N° 148. IPGH, México DF. México.

Bergoeing, J.P., Arce, R., Brenes, L.G., Protti, R., 2010c. Atlas Geomorfológico del Caribe de Costa Rica. Escala 1: 100.000. Editorial SIEDIN Universidad de Costa Rica, San José Costa Rica, 33 pp. Color.

Bergoeing, J.P., Brenes, L.G., Salas, D., 2010d. Atlas Geomorfologico de Costa Rica" Escala 1:350.000. Editorial Instituto Costarricense de Electricidad, ICE, San José, Costa Rica.

Bergoeing, J.P., Brenes, L.G., Salas, D., 2010e. Atlas Geomorfologico de Costa Rica. Escala 1:250.000 Editorial Instituto Costarricense de Electricidad, ICE, San José, Costa Rica.

Bergoeing, J.P., 1977. Modelado glaciar en la Cordillera de Talamanca. Informe Semestral, Jul-Dic 77. Instituto Geográfico Nacional, Costa Rica.

Bergoeing, J.P., 1978a. Geomorfología del sector del Cantón de Talamanca, provincia de Limón, Costa Rica. 1:200 000. Instituto Geográfico Nacional, Costa Rica.

Bergoeing, J.P., 1978b. Geomorfología de Puerto Jiménez, Península de Osa, Costa Rica. In: Informe Semestral enero-junio 1978. Instituto Geográfico Nacional, San José, Costa Rica.

Bergoeing, J.P., 1978c. Geomorfología de los cursos medio e inferior del río Tempisque, provincia de Guanacaste, Costa Rica. 1:200 000. Instituto Geográfico Nacional, Costa Rica.

Bergoeing, J.P., 1979. El volcán Las Nubes. In: Informe Semestral enero-junio 1979. Instituto Geográfico Nacional, San José, Costa Rica.

Bergoeing, J.P., 1981. Fotointyerpretación geomorfológica del sector occidental del Valle Central de Costa Rica. Basada en una imagen del satélite Landsat del 23 de enero de 1971. In: Revista Geográfica IPGH n° 95, Mexico DF, Mexico.

Bergoeing, J.P., 1982a. Rapport sur deux datations radiométriques de quelques échantillons de la Vallée Centrale du Costa Rica. In: Revista Geográfica IPGH n° 95. Mexico DF, Mexico.

Bergoeing, J.P., 1982b. Geomorfología de algunos sectores de Costa Rica basada en la fotointerpretación de imágenes del satélite Landsat en la banda espectral MSS7 (4 cartas color escala 1:500.000e). In: Suplemento Informe Semestral julio-diciembre 1982, vol. 28. Instituto Geográfico Nacional, San José, Costa Rica, pp. 3—15.

Bergoeing, J.P., 1983. Síntesis geomorfológica del sistema fluvial del Tárcoles, Costa Rica. In: Estudios preliminares en géomorfometría y geofactores de base. Cap. IV IGN-CONICIT (1 carta color) San José, Costa Rica, pp. 64—66.

Bergoeing, J.P., 1986. Reconocimiento geomorfológico de la vertiente del Pacífico de Nicaragua. Instituto Panamericano de Geografía e Historia Revista Geográfica 106, México, 30 pp.

Bergoeing, J.P., 1987a. Photo-interprétation géomorphologique du versant Pacifique du Nicaragua, Amérique Centrale. In: Revue Mappe Monde n° 2-1987, pp. 5—8. Montpellier, France.

Bergoeing, J.P., 1987b. Le Costa Rica : contribution à une étude géomorphologique régionale. Thèse d'Etat. Université d'Aix-Marseille-II, France, 437 pp. (Microfilmé par l'Université de Lille).

Bergoeing, J.P., 1987c. L'évolution du Quaternaire au Costa Rica. In: Cahiers Nantais n° 30-31. Hommage au professeur Gras, pp. 167—187. Nantes. France.

Bergoeing, J.P., 1998. Geomorfología de Costa Rica (croquis, estereogramas, cartas, fotos). Instituto Geográfico Nacional de Costa Rica, p. 460.

Bergoeing, J.P., 1999. Geomorfología del Valle de Copán, Honduras (sector comprendido entre Santa Rita y Copán Ruinas). Informe Semestral Instituto Geográfico Nacional de Costa Rica, San José, pp. 47—64, 1999.

Bergoeing, J.P., 2002. Eje terrestre y cambios climáticos Revista Norte Grande N° 29. Universidad Cátolica de Chile, Santiago, Chile, pp. 149—153.

Bergoeing, J.P., 2004. ¿Apocalipsis volcánica? Revista Reflexiones. ISSN: 1021-1209 83 (2), 83—88.

Bergoeing, J.P., 2005. Le cas du Yellowstone une caldeira exceptionnelle. Revista Geografica N° 137, 2005. IPGH, Mexico.

Bergoeing, J.P., 2006. El Cuaternario en Costa Rica, proposición cronológica. Revista Reflexiones. Facultad de Ciencias Sociales Universidad de Costa Rica 85 (2). San José, Costa Rica.

Bergoeing, J.P., 2007. Geomorfologia de Costa Rica. Libreria Francesa, San José, Costa Rica.

Bergoeing, J.P., 2008. Interpretación Geomorfológica del volcán Barú- Panamá. Revista Geográfica IPGH N° 143, México.

Bergoeing, J.P., 2009a. La Trangresión Flandense. Revista Geográfica IPGH N° 144, Mexico.

Bergoeing, J.P., 2009b. Una joven estructura volcánica de Costa Rica, Arenal, volcán turístico volcán letal. Revista Geográfica, N° 145 IPGH, México DF. México.

Bergoeing, J.P., 2009c. Paisajes volcánicos de Costa Rica. Editorial Jadine, San José. Costa Rica.

Bergoeing, J.P., 2011a. Los dos últimos periodos glaciares y la constitución de sackungs en Talamanca, Costa Rica. IPGH, México DF. México. Revista Geográfica, N° 149.

Bergoeing, J.P., 2011b. Riesgo de desaparición de la flecha litoral de Puntarenas, Costa Rica. IPGH, México DF. México. Revista Geográfica, N° 149.

Bergoeing, J.P., 2012a. El antiguo vulcanismo de la Cordillera de Talamanca. Revista Geográfica, N° 151 IPGH, México DF, México.

Bergoeing, J.P., 2012b. La Caldera de colapso de Managua. Revista Geográfica, N° 152, México DF, México.

Bergoeing, J.P., 2013. El río Tárcoles desaguadero del Valle Central Occidental. Costa Rica. Revista Geográfica IPGH N° 154, México DF, México.

Bergoeing, J.P., 2014. Paisajes Fluviales de Costa Rica. EUNED, Costa Rica, bilingüe Inglés–Castellano. Full color, 67 páginas.

Bergoeing, J.P., 2015. Geomorphology of Central América, a Syngenetic Perspective. Elsevier.

Bergoeing, J.P., 2016. Atlas Geomorfológico de Costa Rica, 1: 100.000 Universidad Nacional de Costa Rica, Heredia, Costa Rica, 36 pp.

Berrange, J.P., 1979. Geological map of the Tapanti quadrangle, Costa Rica. In: project sponsored by the Ministry of Overseas Development en collaboration with Dirección de Geología, Minas y Petróleo. Ministerio de Economía Industria y Comercio, San José, Costa Rica.

Berry, E.W., 1939. Contribution to the paleobotany of Middle and South America. In: John Hopkins Univ. Stud. Geol., vol. 13, 168 pp.

Bertrand, M., 1989. Les phénomènes volcaniques et les tremblements de terre de l'Amerique Centrale. Bull. Soc. Geol. France 3 (27), 494–495. Paris.

Bohnberger, O.H., Bengochea, A., Dondoli, C., Marroquin, A., 1966. Report on Active Volcanoes en Central America during 1957–1965. 11a. Reunión geológica de América Central, Guatemala (Bull Erupt. Tokio 9).

Borgia, A., Poore, C., Carr, M.J., Melson, W.G., Alvarado, G.E., 1988. Structural stratigraphic and petrologic aspects of the Arenal-Chato volcanic system, Costa Rica: evolution of a young stratovolcanic complex. Bull. Volcanol. 50, 86–105.

Boudon, G., Rancon, J.Ph., Kieffer, G., Soto, G., Traineau, H., Rossignol, J.C., 1995. Estilo eruptivo del volcán Rincón de La Vieja; Evidencias de los productos de las erupciones de 1966–70 y 1991–92. Roschildia, Costa Rica. http://www.acguanacaste.ac.cr/rothschildia/v2n2/textos/pag10.html.

Boudon, G., Le Friant, A., Komorowski, J.-C., Deplus, C., Semet, M.P., 2003a. Instabilités des volcans de l'arc Antillais: origine et implications sur les risques volcaniques-Rapport Quadriennal 1999–2002. Comité National Français de Géodésie et Géophysique, CNFGG. In: XXIIIéme Assemblée Générale de l'UGGI, Saporro Japon), juillet 2003.

Boudon, G., Semet, M.P., Komorowski, J.-C., Villemant, B., Michel, A., 2003b. Was the last magmatic eruption of la Soufriére, Guadeloupe, in 1440 AD, triggered by partial collapse of the volcano?. In: EGS-AGU-EUG Joint Assembly, Nice, France, 06–11 April 2003, vol. 5. Geophysical Research Abstracts, p. 10398.

Boza, M., 1978. Los Parques Nacionales de Costa Rica. Servicio de Parques Nacionales, San José, Costa Rica.

Brenes-Chaves, G., 1978. Algunas consideraciones sobre posibles problemas biogeográficos en la cuenca del río Sucio (Tesis de Licenciatura en Geografía). UCR, San José, Costa Rica.

Brenes-Monge, M.A., 1968. Contribución a la geología del valle Central Occidental (hoja Turrúcares) (Tesis de Ingeniero Agrónomo). UCR, San lose, Costa Rica.

Brenes-Quesada, L.G., 1976. Análisis geomorfológico de procesos de remoción en masa en parte de la cuenca del río Reventazón, Costa Rica. UCR, San José, Costa Rica.

Brombach, T., Marini, L., Hunziker, J.C., 2000. Geochemistry of the thermal springs and fumaroles of Basse-Terre Island, Guadeloupe. Lesser Antilles. Bull. Volcanol. 61, 477–490.

Brooks, H., 1969. A preliminary report to the Organization of tropical Studies Inc. on lake Izabal. Geol. Hidrol. 19. USA.

Bullard, F., 1956. Volcanic activity in Costa Rica and Nicaragua in 1954. Am. Geophys. Union Trans. 37, 75–82. USA.

Bullard, F., 1957. Active volcanoes of Central America. In: XX Congr. Geol. Intern. Ses. 1, pp. 351–371. USA.

Burkart, B., Deaton, B.C., Dengo, C., Moreno, G., 1987. Tectonic wedges and ofsset Laramide structure along the Polochic fault of Guatemala and Chiapas, Mexico: reaffirmation of large neogene displacement. Tectonics 6 (4), 411–422.

Burkart, B., 1978. Offset across the Polochic fault of Guatemala and Chiapas, Mexico. Geology 6, 328–338.

Butterlin, J., 1956. La constitution géologique et la structure des Antilles CNRS, 453 pp. Paris.

Butterlin, J., 1977. Géologie structurale de la région des Caraibes (Méxique, Amérique Centrale, Antilles et Cordillere Caraîbe). Masson et Cie, 259 pp. Paris.

Calvert, P.P., 1918. Eruptions of the Costa Rican volcano Irazu en 1917–1918. In: Proceedings of the Academy of Natural Sciences of Philadelphia, 73 pp. USA.

Camacho, E., 1992. Volcanes en Panamá. Universidad Tecnológica de Panamá - Cepredenac, Instituto de Geociencias, Panama.

Camacho, E., 1997. Los terremotos en el istmo de Panamá. Laboratorio de Geofísica e Hidrología, Universidad de Panamá, Panamá.

Carballo, M.A., Fischer, 1978. La formación San Miguel. Informe Semestral Ene – Jun 78. Instituto Geográfico Nacional, Costa Rica, pp. 48–144.

Carlut, J., Quidelleur, X., 2000. Absolute paleointensities recorded during the Brunhes chron at La Guadeloupe island. Phys. Erath Planet. Inter. 120, 255–269.

Carlut, J., Quidelleur, X., Courtillot, V., Boudon, G., 2000. Paleomagnetic directions and K/Ar dating of 0 to 1 Ma old lava flows from La Guadeloupe Island (French West Indies): implications for time averaged field models. J. Geophys. Res. 105 (B1), 835–849.

Carr, M.J., 1984. Symetrical and segmented variations of physical and geochimical characteristics of the Central American volcanic front. J. Volcanol. Geoth. Res. 20, 231–252.

Carr, M.J., Saginor, I., Alvarado, G.E., Bolge, L.L., Lindsay, F.N., Milidakis, K., Turrin, B.D., Feigenson, M.D., Swisher III, C.C., 2007. Element fluxes from the volcanic front of Nicaragua and Costa Rica. Geochem. Geophys. Geosyst 8, Q06001. http://dx.doi.org/10.1029/2006GC001396.

Castillo, R., 1978. Geología regional preliminar de una parte del Valle de El General, Costa Rica. In: Bol. Geol, 1. Departamento de Desarrollo Geológico y Recursos Minerales, CODESA, Costa Rica, pp. 289–293.

Castillo, R., Kruschensky, R., 1977. Geologic map and cross secrion of the Abra quadrangle, Costa Rica. In: Misc. Investi Series U.S. Geological Survey Map 1–992, Arlington USA.

Chaves, R., Sáenz, R., 1974a. Mapa geológico de la cordillera de Tilarán 1:200 000. Ministerio de Industria y Comercio, Dirección de Geología, Minas y Petróleo, San José, Costa Rica.

Chaves, R., Sáenz, R., 1974b. Informe técnico y geológico de la cordillera de Tilarán. Ministerio de Econ. Ind. y Com Dirección Geol. Minas y Petrol, San José, Costa Rica.

Chaves, R., 1969. Características físicas, químicas y mineroĺogicas de los materiales eruptados por el volcán Arenal. Informe Semestral Ene- Jun 69. Instituto Geográfico Nacional, Costa Rica.

Chiesa, S., Civelli, G., Gillot, P.-Y., Mora, O., Alvarado, G.E., 1992. Rocas piroclásticas asociadas con la formación de la caldera de Guayabo, cordillera de Guanacaste, Costa Rica. Rev. Geol. Amér. Central 14, 59–75.

Coates, Ag, Obando, 1996. The Geologic evolution of Central American Istmus. In: Jacson, J.B.C., Budd, A.F., Coates, A.G. (Eds.), Evolution an environment in tropical America. Univ. Chicago Press.

Corral, I., Cardellach, E., Gómez-Gras, D., Canals, A., 2009. Contribución al conocimiento de la geología del depósito de Au-Cu de "La Pava" (Península de Azuero, Panamá). Contribution to the knowledge of the geology of the «La Pava» Au-Cu deposit (Azuero Peninsula, Panama) GEOGACETA, 46. Sociedad Geológica de España, Barcelona.

Crossland, C., 1927. The Expedition of the South Pacific of S.S.St. George Marine Ecology and Coral Formations in the Panama Region, the Galapagos and Marquises Islands. Royal Society of Edinburgh.

Cuffey, K., Marshall, S., 2000. Substantial contribution to sea-level rise during the last interglacial from the Greenland ice sheet. Nature 404, 591–594.

De Boer, J.Z., Defant, M.J., Stewart, R.H., Restrepo, J.F., Clark, L.F., Ramirez, A.H., 1988. Quaternary calc-alkaline volcanism in western Panama: regional variation and implication for the plate tectonic framework. J. South Am. Earth Sci. 1, 275–293.

De Boer, J., 1974. Mapa geofísico preliminar de Costa Rica. IGN,-CR, San José, Costa Rica.

De Mets, C., Gordon, R.G., Argus, D.F., Stein, Y.S., 1990. Current plate motions. Geophys. J. Int. 101, 425–478.

Delmelle, P., Stix, J., Baxter, P.J., García-Àlvarez, J., Barquero, J., 2002. Atmospheric dispersion, environmental effects and potential health hazard associated with the low-altitude gas plume of Masaya volcano, Nicaragua. Bull. Volcanol. http://dx.doi.org/10.1007/s00445-002-0221-6.

Delort, R., Walter, F., 2001. Histoire de l'environnement Européen. Préface de Jacques Le Goff. PUF, Paris, 354 p.

Dengo, G., Chaverri, O.H., 1951. Reseña geológica de la región S.O. de la meseta Central de Costa Rica. Revista Universidad de Costa Rica, 5, San José, Costa Rica, pp. 313–326.

Dengo, G., Levy, E., 1970. Mapa Metalogenético de América Central. Public. Geol. ICITII, Guatemala.

Dengo, G., Bohnberger, O.H., Bonis, S., 1970. Tectonics and vulcanism along the Pacific et marginal zone of Central América. Geol. Rdch. Stuttgart. 59, 1215–1232. RFA.

Dengo, G., 1960. Notas sobre la geología de la parte central del litoral Pacífico de Costa Rica. Informe Semestral Jul-Dic 60. Instituto Geográfico Nacional, Costa Rica, pp. 43–58.

Dengo, G., 1961. Notas sobre la geología de la parte central del litoral Pacifico de Costa Rica II. Informe Semestral Jul-Dic 61. Instituto Geográfico Nacional, Costa Rica, pp. 43–63.

Dengo, G., 1962a. Tectonic-igneous sequence in Costa Rica Petrol. Studies Vol. en honor of A Buddington. Geol. Soc. Am. 133–161. USA.

Dengo, G., 1962b. Estudio Geológico de la región de Guanacaste, Costa Rica. Instituto geográfico Nacional, San José, Costa Rica, 112 pp.

Dengo, G., 1962c. Mapa geológico generalizado de la provincia de Guanacaste y zonas adyacentes, preparado con base en el mapa IGN y en los levantamientos geológicos efectuados por la Compañía Petrolera de C.R. por los geólogos Dion; L. Gadner et alt. Instituto Geográfico Nacional, Costa Rica.

Dengo, G., 1967. Geological structure of Central America. In: Studies in Tropical Oceanography Proc. Intern. Conf. on Trop. oc. Univ. of Miami, USA, pp. 56–73.

Dengo, G., 1968. Estructura Geológica, Historia Tectónica y Morfología de América Central. Centro regional de ayuda técnica AID, México.

Dengo, G., 1969. Problems of tectonic relations between Central América and the Caribbean. Trans. Gulf Coast Ass. Geol. Soc. 19, 3121–3320. USA.

Dengo, G., 1973. Estructura Geológica, Historia tectónica y Morfología de América Central. Instituto Centramericano de Investigaciones y tecnología Industrial ICAITI, Costa Rica.

Denyer, P., Kussmaul, S., 2000. Geología de Costa Rica. Editorial Tecnológica de Costa Rica, Cartago, Costa Rica, p. 515.

Denyer, P., Gazel, E., 2009. The Costa Rican Jurassic to Miocene oceanic complex: origin tectonic and relations. J. South Am. Earth Sci 28, 429–442.

Diaz, R.A., 1981. Magnitud de un sismo generado por fallas ubicadas en el valle Central (Tesis de Licenciatura). Facultad de Ingeniería UCR, Costa Rica.

Dondoli, C., Chaves, R., 1968. Mapa adjunto al estudio geológico del valle Central. Dir. Geol. Minas y Petrol, Costa Rica.

Dondoli, C., Torres, M.A., 1954. Estudio geológico y mineralógico pedagógico de la región oriental de la meseta Central. Ministerio de Agricultura e Industrias, 180 pp. Costa Rica.

Dondoli, C., Dengo, G., Malavassi, E., 1968. Mapa Geológico de Costa Rica. 1:700 000 Edic. Prel. Dir. Geol. Minas y Petrol Costa Rica.

Dondoli, C., 1943. La región de El General. Condiciones geológicas y geoagronómicas de la zona. Depto. Nac. de Agric. BoL Tec. 44, 1–16. San Pedro Montes de Oca, Costa Rica.

Dondoli, C., 1950. Liberia y sus alrededores. Suelo Tico, vol. 4, pp. 18–19. San Josê, Costa Rica.

Dondoli, C., 1951. In: Dengo, G., Chaverri, S. (Eds.), Observaciones sobre las andesitas del Virilla encontradas en la perforación del túnel Ing, pp. 324–325. Costa Rica.

Dondoli, C., 1965. Vulcanismo reciente en Costa Rica. Dir. Geol. Minas y Petrol. Minist. Ind., 16 pp., Costa Rica.

Dondoli, C., 1970. Localización de un horizonte laterítico-bauxítico en la zona de Cartago concreciones de gibsita en la laterítica bauxíica de Birrisito de Cartago Dir. Gral. Geol. Minas y Petro. Informes Tec. y Notas Geol., 36 pp. 12 UCR C.R.

Donovan, S.K., Jackson, T.A. (Eds.), 1994. Caribbean Geology: An Introduction. University of the West Indies Publisher's Association, Kingston, Jamaica, pp. 13–39. QE 220 C34.

Echandi, E., 1981. Unidades volcánicas de vertiente norte de la cuenca del río Virilla (Escuela Centroamericana de Geología. Tesis). UCR, Costa Rica, 123 pp.

Elizondo, C., Bergoeing, J.P., 1980. Seasat radar data interpretation for geomorphological research and mapping en Costa Rica. In: (Summary). For the fourtheenth Int. Symp. on Remote Sensing of Environnment, April 23–30, 1980, San José, Costa Rica.

Elizondo, C., 1979. Evolución del uso de sensores remotos en la planificación de los recursos naturales de Costa Rica (Tesis de Licenciatura). UCR, Costa Rica.

Escalante, G., 1965a. Mapa geológico de la parte superior de la cuenca del río Reventazón, Costa Rica. 1:500 000. Instituto geográfico Nacional, Costa Rica.

Escalante, G., 1965b. Geología de la cuenca superior del río Reventazón en Costa Rica. In: Primera Reunión de Geol. De Inst. Gubern de Centro América y Panamá 22–27 feb. 1965. Costa Rica.

Escalante, G., 1968a. Mapa geológico preliminar de la región sureste de Costa Rica. La cordillera de Talamanca y sectores próximos. 1:300 000 inedit.

Espinoza, E., Gutiérrez, C., Cerrato, D., Vázquez-Prada, D., 2008. Cartografía Geológica y Geomorfológica de la Reserva Natural Laguna de Apoyo. Programa Integral por le Ordenamiento Ambiental de Apoyo –AMICTLAN-Geólogos.

Estudio de Caso, 1993. Terremoto del 22 de abril de 1991-Limón. (Pan American Health Organization (PAHO) / Organización Panamericana de la Salud (OPS), Costa Rica.

Fan, G.W., Beck, S.L., Wallace, Y.T.C., 1993. The seismic source parameters of the 1991 Costa Rica aftershock sequence: evidence for a transcurrent plate boundary. J. Geophys. Res. 98, 15,759–15,778.

Felis, T., et al., 2004. Increased seasonality in Middle East temperatures during the last interglacial period. Nature 249, 164.

Fernandez, M., Rodriguez, R., 1952. Estudio geológico del cantón de Atenas (Tesis de Licenciatura). Escuela de Geología UCR, Costa Rica.

Fernandez, M., 2001. Daños, efectos y amenazas de tsunamis en América Central. Centro de investigaciones Geofísicas. Universidad de Costa Rica.

Feuillet, N., Manighetti, I., Tapponnier, P., 2001. Extension active perpendiculaire à la subduction dans l'arc des Petites Antilles (Guadeloupe, Antilles Françaises). C.R. Acad. Sci. Paris, Sciences de la Terre et des Planètes 333, 583–590.

Feuillet, N., Tapponnier, P., Manighetti, I., Villemant, B., King, G.C.P., 2004. Differential uplift and tilt of Pleistocene reef platforms and Quaternary slip rate on the Morne-Piton normal fault (Guadeloupe, French West Indies). J. Geophys. Res. 109, B02404. http://dx.doi.org/10.1029/2003JB002496, 18 pp.

Feuillet, N., 2000. Sismotectonique des Petites Antilles. Liaison entre activité sismique et volcanique. Thése de Doctorat. Université de Paris 7 René Diderot, pp. 1–283.

Field, M.J., December 2001. Sea Levels Are Rising. Pacific Magazine.

Fisher, D.M., Gardner, T.W., Marshall, J.S., Montero, Y.W., 1994. Kinematics associated with late tertiary and quaternary deformation in Central Costa Rica: Western Boundary of the Panamá Microplate. Geology 22, 263–266.

Franco, J.C., 1978. La formación Coris, (Mioceno, valle Central de Costa Rica) (Tesis de Licenciatura. Escuela Centroamericana de Geología). UCR, Costa Rica. http://hal-ens.archives-ouvertes.fr/view_by_stamp.php?&halsid=vgdme5hnn3e97k6e0ken7idj96&label=ens-paris&langue=fr&action_todo=search_advanced&submit=1&search_without_file=yes&f_0=lastname&p_0=is_exactly&f_1=firstname&p_1=is_exactly&l_0=and&halsid=vgdme5hnn3e97k6e0ken7idj96&v_0=franco&v_1=aurore.

Franco, A., 2008. Cinématique Actuelle du Nord de l'Amérique Centrale: Zone de Jonction Triple Amérique du Nord Amérique-Cocos-Caraïbe. Apport des données sismologiques et géodésiques aux modèles régionaux. Université Paris Sud - Paris XI.

Frogley, M.R., et al., 1999. Climate variability in Northwest Greece during the last interglacial. Science 285, 1886–1888.

Gillot, P.Y., Chiesa, S., Alvarado, G.E., 1994. Cronostratigraphy of upper Miocene – quaternary volcanism in northern Costa Rica. Rev. Geol. Amér. Central 17, 45–53.

Girod, M., 1978. Les roches volcaniques, Petrologie et cadre structural. Doin Editeurs, 239 pp. Paris.

Goes, S.D.B., Velasco, A.A., Schwartz, S., Lay, T., 1993. The April 22, 1991, Valle dla Estrella, Costa Rica (Mw=7.7) earthquake and its tectonic implications: a broadband seismic study. J. Geophys. Res. 98, 8127–8142.

Gomez, L.D., Valerio, C.E., 1971. Lista preliminar ilustrada de los moluscos fósiles de la formación río Banano (Mioceno), Limón Costa Rica. Informe Semestral Ene-Jun 71. Instituto Geográfico Nacional, Costa Rica.

Granados, R., 1979. Proyecto geotérmico, investigación geológica en la zona de la Caldera de Guayabo y alrededores. Informe Geol. Inst. Cost. Electricidad, Costa Rica, 45 pp.

Güendel, F., Pacheco, J., 1992. The 1990–1991 seismic sequence across central Costa Rica: evidence for the existence of a micro-plate boundary connecting the Panama deformed belt and the Middle America trench, Eos Trans. Am. Geophys. Union Trans. 73, 399.

Guilcher, A., Berthois, L., Battistini, R., 1962. Formes de corrosion littorale dans les roches volcaniques, particulierement de Madagascar et au Cap vert (Senegal). Cahiers Océanographiques XVIe année 4, Paris.

Guilcher, A., 1966. Les grandes falaises et mégafalaises des côtes sud-ouest de I'Irlande. Ann. Géogr. 407, 26–38. Paris.

Guillermo, E.A., Phillipe, B.G., 2012. Síntesis geocronológica del magmatismo, metamorfismo y metalogenia de Costa Rica. América Central. Rev. Geol. Amér. Central 46, 7–122. ISSN: 0256–7024.

Gutenberg, B., Richter Ch, F., 1942. Seismicity of Central and South América. Geol. Soc. 4, 455. USA.

Gutierrez, 1955. Expedición del doctor Richard Weyl al Macizo del Chirripó, Bosquejo Geológico de la cordillera de Talamanca. Instituto Geográfico Nacional, Costa Rica.

Gutierrez, D., 1962. Apuntes sobre un viaje al volcán Tenorio. In: Informe Semestral Ene-Jun 962. Instituto Geográfico Nacional, Costa Rica, pp. 27–30.

Gutierrez, B.F., 1963a. Actividad volcánica del Irazú. In: Informe Semestral Ene-Jun 63. Instituto Geográfico Nacional, Costa Rica, pp. 33–38.

Gutierrez, B.F., 1963b. Hallazgo de resto de un mamut. In: Informe Semestral Ene-Jun 63. Instituto Geográfico Nacional, Costa Rica, pp. 41–47.

Haas, O., 1942. Miocene molluscs from Costa Rica. USA Paleant. 16, 307–316. USA.

Hall, C., 1984. Costa Rica, una interpretación Geográfica con perspectiva histórica. Editorial Costa Rica, 467 pp. San José, Costa Rica.

Harrison, J.V., 1953. The geology of the Santa Elena península en Costa Rica, Central América. In: Proc. Seventh Scient. Cong, vol. 2, pp. 102–114. USA.

Harrison, J.V., 1956. East and West lines en Costa Rica. In: XX Congreso Geológico Interamericano. Resum, pp. 281–282. México.

Hastenrath, S., 1974. Spuren Pleistozäner vereigsung in den Altos de Cuchumatanes. Guatemala. Eiszeïltalter U. Gegenwert 5, 25–34. R.F.A.

Healy, J., 1969. Notas sobre los volcanes de la sierra Volcánica de Guanacaste, Costa Rica. In: Informe Semestral Ene-Jun 69. Instituto Geográfico Nacional, Costa Rica, pp. 37–47.

Henningsen, D., 1964. Estratigrafía y paleogeografía de los sedimentos del Cretáceo Superior y del Terciario, en el sector Suresre de Costa Rica. Informe Semestral Jul-Dic 64. Instituto Geográfico Nacional, Costa Rica.

Hernitz, S.R., 1981. Landforms under a tropical wet forest cover ont the península Osa, Costa Rica. Zeits fiir Geomorphologie 25. RFA, pp. 259–270.

Hey, R., Johnson, G.L., Lawrie, A., 1977. Recent Plate motions in the Galapagos area. Geol. Soc. Bull. 88, 1385–1403.

Hey, R., 1977. tectonic evolution of the Cocos-Nazca Spreading. Geol. Soc. Am. Bull. 88, 1404–1420.

Hoernle, K., Abt, D., Fischer, K.M., Nichols, H., Hauff, F., Abers, G.A., van den Bogaard, P., Heydolph, K., Alvarado, G., Protti, M., Strauch, W., 2008. Geochemistry of oceanic crust beneath Costa Rica. Supplement to: Hoernle, K et al. (2008): Arc-parallel flow in the mantle wedge beneath Costa Rica and Nicaragua Nature 451 (7182), 1094–1097. http://dx.doi.org/10.1594/PANGAEA.773166.

Hogg, E.H., Bernier, P., 2005. Climate change impacts on drought-prone forests in western Canada. Forest. Chron 81 (5), 675–682. http://dx.doi.org/10.5558/tfc81675-5.

Hoffstetter, R., et al., 1960. Lexique stratigraphique international. In: Amérique Latine: Fasc. 2 Amérique Centrale, vol. 5, 368 pp. 8 cartes, CNRS Paris.

Holdridge, L., 1969. Ecología basada en zonas de vida. Instituto Interamericano de Ciencias agrícolas IICA, Turrialba, Costa Rica.

Horn, S.P., 1990. Timing of deglaciation in the Cordillera de Talamanca. Climate Research, V1, Costa Rica, pp. 81–83. USA.

Horn, S.P., 1993. Postglacial vegetation and fire history in the Chirripó Páramo of Costa Rica, 40. Quaternary Research, USA, pp. 107–116.

Imbrie, J., Palmer Imbrie, K., 1979. Ice Ages, Solving the Mystery. Harvard University Press, Cambridge, Massachusetts and London, England.

Instituto Meteorologico Nacional, 1986. Data climática de Costa Rica 1888–1986. Ministerio de Agricultura y Comercio, San José, Costa Rica.

Instituto Nicaraguense De Estudios Territoriales. Ineter 2002, April 2002. Actualización del Mapa de fallas geológicas de Managua. Informe Técnico, Managua, Nicaragua.

Instituto Panamericano De Geografía E Historia Ipgh, Instituto Panamericano De Geografía E Historia Ipgh. Atlas climatológico e hidrológico del istmo centroamericano Publicación 367 IPGH. Guatemala.

Jacob, K.H., Pacheco, J., 1991. The M-7.4 Costa Rica earthquake of April 22, 1991, tectonic seting, teleseismic data, and consequences for seismic hazard assessment, Earthquake Spectra, 7 B.

Jager, G., 1977. Geología de las mineralizaciones de cromita al este de la península de Santa Elena, provincia de Guanacaste (Tesis de grado). Escuela Centroamericana de Geología UCR Costa Rica, Costa Rica, 136 pp.

James, K.H., 1998. A Simple Synthesis of Caribbean Geology. PDF of Keith James' Caribbean plate model.

Kaspar, F., et al., 2005. A model-data comparison of European temperatures in the Eemian interglacial. Geophys. Res. Lett. 32, L11703.

Komorowski, J.-C., Boudon, G., Semet, M., Coudret, E., Villemant, B., Le Friant, A., 2004. A New Look atthe pyroclastic eruptive history of Soufrière of Guadeloupe (French West Indies) in the past 50,000 years: implications for GIS-based hazard scenario definition. IAVCEI General Assembly 2004, Pucon, Chili, November 14–19 2004, abstract.

Komorowski, J.-C., 2003. Diversité du volcanisme Terrestre : processus et produits. In: De Wever, P. (Ed.), Le volcanisme : cause de mort et source de vie. Editions Vuibert, Paris, pp. 27–112.

Kruschensky, R., Malavassi, E., Castillo, R., 1976. Mapa de reconocimiento geológico y cortes transversales de Costa Rica. Oficina de Defensa Civil y Direc. Geol. minas y PetroL Minist. de Ind. y Com Costa Rica.

Kruschensky, R., 1972. Geology of the Istaru Quadrangle, Costa Rica. Geol. Survey Bull. 1358, 46. USA.

Kukla, G., 2000. The last interglacial. Science 287, 987–988.

Kussmaul, S., Paniagua, S., Gainza, J., 1982. Recopilación, clasificación e interpretación petroquímica de las rocas de Costa Rica. Informe Semestral Jul-Dic 82. Instituto Geográfico Nacional, Costa Rica, pp. 17–79.

Kuyjpers, E.P., 1980. La Geología del Complejo Ofiolítico de Nicoya, Costa Rica. Informe Semestral Ene-Jun 80. Instituto Geográfico Nacional, Costa Rica.

Lachniet, S.M., Roy, J.A., 2011. Quaternary Glaciations-Extent and Chronology a Closer Look. Costa Rica and Guatemala. Developments in Quaternary Science. Elsevier.

Lachniet, M.S., Seltzer, G.O., 2002. Late Quaternary glaciation of Costa Rica. Geol. Soc. Am. Bull. 114 (5), 547–558. USA.

Laserre, G., 1975. América media. México, América Central, Antillas Guayanas. Editorial Ariel, Barcelona.

Lea, D., et al., 2000. Climate impact of late Quaternary Equatorial Pacific sea surface temperature variations. Science 289, 1719–1723.

Lépine, J.-C., Hirn, A., Komorowski, J.-C., 2001. Sismicité de la Soufriére de Guadeloupe, Atelier sur les aléas volcaniques – Les volcans antillais : des processus aux signaux. Programme National des Risques Naturels – PNRN (CNRS) – Institut National des Sciences de l'Univers – INSU, BRGM, CEA, CEMAGREF, CNES, IRD. 18–19 janvier 2001, Paris, abstract volume, p. 24.

Leyrit, H., 2003. Caractéristiques des volcans et provinces volcaniques de France. In: De Wever, P. (Ed.), Le volcanisme: cause de mort et source de vie. Editions Vuibert, Paris, pp. 143–202.

Linkimer, L., Aguilar, T., 2000. Estratigrafía Sedimentaria. In: Denyer, P., Kussmaul, S. (Eds.), Geología de Costa Rica, Cartago, Editorial Tecnológica de Costa Rica, 520 pp.

Linkimer, L., 2003. Neotectónica del extremo oriental del Cinturón Deformado del Centro de Costa Rica (Tesis de Licenciatura). Escuela de Geología. Universidad de Costa Rica.

Lloyd, J., 1963. Historia Tectónica del orogeno Sur Centroamericano. In: Informe Semestral Ene-Jun 63. Instituto Hgeográfico Nacional, Costa Rica, pp. 67–96.

Lockwood, J.P., Benumof, B.T., 2000. Geologic Investigations Field Report, Isla Del Coco, Costa Rica. Prepared for the Government of Costa Rica and the Puffin Investment Company, Ltd, 29 p.

Lorenz, V., 1973. On the formation of maars. Bull. Volcanol. 37, 183–204.

Lovell, W.G., 2005. Conquest and Survival in Colonial Guatemala, third ed. McGill-Queens University Press, Canada.

Macmillan, A.I., Gansa, P.B., Alvarado, I.G., 2004. Middle miocene to present plate tectonic history of the southern Central American Volcanic Arc. Tectonophysics 392, 325–348. Elsevier.

Madrigal, R., Malavassi, E., 1967. Reseña geológica del Area Metropolitana. Inf. Tecn. y Notas Geol. 29, 9 pp. Dir. Gener. de GeoL Minas y PetroL Costa Rica.

Madrigal, R., 1972. Resumen de la estratigrafía de Costa Rica. Escuela Centroamericana de Geología. UCR, Costa Rica.

Madrigal, R., 1976. San Carlos, mapa geomorfológico de Costa Rica: 1:200.000 Color. OFIPLAN, Costa Rica.

Madrigal, R., 1977. Evidencias geomorfológicas de movimientos tectónicos recientes en el valle de El General. Revista Ciencia y Tecnología. UCR, Costa Rica, pp. 97–106.

Madrigal, R., 1978. Terrazas marinas y tectonismo en península de Osa, Costa Rica. Instituto Panamericano de Geografía e Historia, Revista Geográfica 85, México DF México.

Mainieri, P.A., 1976. Proyecto geotérmico de Guanacaste: San José. Informe de probabilidad técnica. Instituto Costarricense de Elecetricidad (ICE), Costa Rica, p. 97.

Malavassi, E., Chaves, R., 1970. Estudio geológico regional de la zona Atlántico norte de Costa Rica. Inf. Tecn. y Notas Geol. Ano 9, p. 35. UCR Costa Rica.

Malavassi, V.E., Suarez, M.A., 1953. Estudio geológico del Macizo de La Carpintera. Universidad de Costa Rica, facultyad de Agronomía, San José, Costa Rica, 57 pp.

Malavassi, E., Madrigal, R., 1967. Mapa geológico del área metropolitana de Costa Rica. Dir. de Geol. Minas y Petrol. Min. de Ind. y com Costa Rica.

Malavassi, E., Madrigal, R., 1970. Reconocimiento geológico de la zona norte de Costa Rica. Dir. de Geol. Minas y Petrol Ano 9, p. 38. Costa Rica.

Malavassi, E., 1960. Algunas localidades de Costa Rica con foraminíferos grandes. Informes. Ciudad Universitaria Anno 1, Costa Rica.

Malavassi, E., 1962. Nota geológica preliminar sobre la región de Tilarán, entre la laguna Cote y Cuevas de Venado. Inf. Depto. de Geol Minas y Petrol, Costa Rica, p. 12.

Malavassi, E., 1965. Reseña Geológica del Valle Central de Costa Rica. Inf. Técn. Y Notas Geol. N° 4, pp. 14–22. UCR. Costa Rica.

Malavassi, E., 1967. Reseña Geológica de la zona de Turrialba. UCR Dirc. Geol. Minas y Petrol. Ano 6, pp. 9–35. Costa Rica.

Malavassi, E., 1975. Explicación de un perfil geológico a través de Costa Rica. Revista Geográfica de América Central 3, Heredia, Costa Rica, pp. 27–31.

Mann, P., 1999. Caribbean sedimentary basins: Classification and tectonic setting from Jurassic to Present. In: Mann, P. (Ed.). In: Hsu, K. (Ed.), Caribbean Basins, Sedimentary Basins of the World Series, vol. 4. Elsevier Science B.V., Amsterdam, The Netherlands, pp. 3–31.

Mapa Geológico De La República De Panamá, 1991. Escala 1: 250.000. Ministerio de Comercio e Industria. Instituto Geográfico Nacional Tommy Guardia, Panamá.

Marko, P.B., Jakson, J.B.C., May 2001. Patterns of morphological diversity among and within arcid bivalve species pairs separated by the isthmus of Panama. J. Paleontol. 590–606. USA.

Marshall, J.S., Fischer, D.M., Gadner, T.W., 1993. Western margin of the Panama microplate, Costa Rica: kinematics of faulting along a diffuse plate boundary. Geol. Soc. Am. Abstracts with programs 25.6: A-284.

Martrat, B., et al., 2004. Abrupt temperature changes in the Western Medierranean over the past 250,000 years. Science 306, 1762–1765.

McCulloch, M., et al., 1999. Coral record of equatorial sea-surface temperatures during the penultimate deglaciation at Huon Peninsula. Science 283, 202.

McNeill, D.F., Coates, A.G., Budd, A.F., Borne, P.F., July 2000. Integrated paleontologic and paleomagnetic stratigraphy of the upper Neogene deposits around Limon, Costa Rica: A coastal emergence record of the Central American Isthmus. Geol. Soc. Am. Bull. 963–981. USA.

Medina, M.C.E., 2009. Modelos numéricos y teledetección en el lago de Izabal, Guatemala (Tesis de doctorado Universidad de Cádiz, España).

Mesa, T., 1979. Consideraciones generales sobre la morfoestructura y el modelado climático de los Cerros de La Carpintera y su relación con el conjunto Irazú, Costa Rica (Tesis de Licenciatura en Geografía). UCR, Costa Rica.

Meza, T.A., 1979. Consideraciones generales sobre la morfoestructura y el modelado climático de los Cerros de La Carpintera y su relación con el conjunto Irazú. Costa Rica. Tésis para optar al grado de Licenciatura en Geografía Física. San pedro Montes de Oca 1979.

Miyamura, S., 1980. Sismicidad de Costa Rica. Universidad de Costa Rica, San José, Costa Rica.

Montero, W., 1975. Estratigrafía del Cenozoico del área de Turrucares, provincia de Alajuela, Costa Rica (Tesis de Bachillerato). Escuela Centroamericana de Geología UCR Costa Rica, p. 40.

Montero, W., 1999. El terremoto del 4 de marzo de 1924 (Ms 7,0): ¿Un gran temblor interplaca relacionado al límite incipiente entre la placa Caribe y la microplaca de Panamá? Rev. Geol. Amér. Central 22, 21–58.

Mora, S., 1977. Estudio geológico del cerro Chopo. Revista Geográfica de América Central 5–6. Universidad Nacional, Heredia Costa Rica.

Mora, S., 1978a. Estudio geológico de los Cerros de Barra Honda y alrededores Tesis de Bachillerato. Escuela Centroamericana de Geología. UCR, Costa Rica, p. 173.

Mora, S., 1978b. Proyecto Talamanca. Informe de reconocimiento geológico de los sitios de presa Bri-Bri y Sheuab. Depto de Geol. ICE, Costa Rica, p. 35.

Mora, S., 1979a. Mapa Geológico de la región Sureste del valle de El General 1:500 000. Escuela Centroamericana de Geología.

Mora, S., 1979b. Estudio geológico de la región sureste del valle de El General (Tesis de Licenciatura, 2 Tomos. Escuela Centroamericana de Geología. UCR, Costa Rica.

Mora, S., 1981. Clasificación Morfotectónica de Costa Rica Informe Semestral Jul-Dic 81. Instituto Geográfico Nacional, Costa Rica, p. 26.

Murata, K.J., Dondoli, 0043, Saenz, R., 1966. The 1963–1965 Eruption of Irazú volcano, Costa Rica. Bull. Volcanol. 29, 765–796. USA.

Newman, A.V., Schwartz, S.Y., Gonzalez, V., De Shon, H.R., Protti, J.M., Dorman, L.M., 2002. Alongstrike variability in the sismogenic zone below Nicoya Peninsula, Costa Rica. Geogr. Res. Lett. 29, n20–77. http://dx.doi.org/10.1029/2002 GL 015409.

Nuhn, H., 1973. Regionalización de Costa Rica para la planificación del desarrollo y la administración OFIPLAN, Costa Rica.

Olsson, A., 1922. The Miocene of Northern Costa Rica. Part 1–2. Bull. Am. Paleont. 9 (39), 179–460. Ithaca USA.

Oviedo-Padrón, E.G., 2005. Análisis geológico-estructural del complejo de maares de Valle de Santiago, Campo Volcánico MichoacánGuanajuato, México: Linares, N.L (Tesis profesional). Universidad Autónoma de Nuevo León, Facultad de Ciencias de la Tierra, 119 p.

Pacheco, J.F., Quintero, R., Vega, F., Segura, J., Jimenez, W., González, V., 2006. The Damas (Mw 6.4) Costa Rica, earthquake of November 20, 2004; aftershocks and slip distribution. Bull. Seim. Soc. Am. 96, 1332–1343.

Paniagua, S., 1984. Contribución al conocimiento de la geología y petrología del vulcanismo Plio- pleistocénico de la cordillera Central de Costa Rica (Tesis de Maestría). Universidad de Chile, Chile.

Paniagua, S., Tejera, R., 1980. Mapa del vulcanismo Plio-Pleistoceno de Costa Rica. Escala 1:700.000. Universidad de Costa Rica, San José, Costa Rica.

Paniagua, S., Soto, G., 1988. Peligros volcánicos en el Valle Central de Costa Rica. Cienc. Tecnol 12, 145–156.

Pelejero, C., Calvo, E., Hoegh-Guildberg, O., 2010. Paleo-perspectives on ocean acidification. Trends Ecol. Evol. 25, 332–344.

Pérez, W., 2000a. Vulcanología y petroquímica del evento ignimbrítico del Pleistoceno medio (0.33 M.a.) del Valle Central de Costa Rica. Tésis de Licenciatura. Universidad de Costa Rica, San José, p. 170.

Perez, W., 2000b. Vulcanología y petroquímica del evento ignimbrítico del Pleistoceno Medio (0.33 M.a.) del Valle Central de Costa Rica. Universidad de Costa Rica, San José xvi + 192 pp.

Pichler, H., Weyl, R., 1975. Magmatism and crustal evolution in Costa Rica. Geol. Rundschau. 64, 457–475. Stuttgart, RFA.

Piedra, J., 1979. Geología del área norte de los Cerros de Escazú, cordillera de Talamanca, Costa Rica. In: Informe Semestral Ene-Jun 79. Instituto Geográfico Nacional, Costa Rica, pp. 99–137.

Pizarro, D., 1993. Los pozos profundos perforados en Costa Rica: aspectos litológicos y bioestratigráficos. Rev. Geol. Amér. Central 15, 81–85.

Ponce, D.A., Case, J.E., 1987. Geophysical interpretation of Costa Rica. In: Mineral resources Assesment of the Republic of Costa Rica, U.S. geological Surv., Misc. Invest. I-1865:8.17.

Protti, M., Schwartz, S., 1994. Mechanics of back ar deformation in Costa Rica. Evidence from an aftershock study of the April 22, 1991, Valle de La Estrella, Costa Rica, earthquake (Mw= 7.7). Tectonics 13-5, 1093–1107.

Protti, M., Güendel, F., McNally, K., 1994. The geometry of the Wadati-Benioff zone under southern Central America and its tectonic significance: results from a high-resolution local seismographic network. Phys. Earth Planet. Inter. 84, 271–287.

Protti, J.M., Schwartz, S.Y., Zandt, G., February 1996. Simultaneous Inversion for Earthquake Location and Velocity Struture Beneath Central Costa Rica. Bull. Seismol. Soc. Am. 86-1A, 19–31.

Protti, M., 1986. Potencial de Instalación de un Sistema en el NW de Costa Rica para la Alerta Temprana al Valle Central al Momento de Ocurrencia de Sismos Fuertes por Debajo de la Península de Nicoya. Observatorio Vulcanológico y Sismológico de Costa Rica, Universidad Nacional. OVSICORI-UNA, Costa Rica.

Protti-Quesada, J.M., 1994. The Most recent Large Earthquakes in Costa Rica (1990Mw 7.0 and 1991 Mw 7.6) and Three-dimensional Crustal and Upper Mantle P-wave Velocity Structure of Central Costa Rica (Ph.D. dissertation). University of California, Santa Cruz, 116 p.

Quiesa, S., Civelli, G., Guillot, Py, Mora, O., Alvarado, G.E., 1992. Rocas piroclásticas asociadas con la formación de la Caldera de Guayabo, cordillera de Guanacaste, Costa Rica. Rev. Geol. Amér. Central 14, 59–75. San José, Costa Rica.

Quintero, R., Segura, J., Jimenez, W., Vega, F., 2004. Evento Principal y réplicas del sismo del 20 de noviembre del 2004 (Mw = 6.4) en Costa Rica Observatorio Vulcanológico y Sismológico de Costa Rica. OVSICORI-UNA, Costa Rica, p. 28. Formato pdf.

Revista Mexicana De Ciencias Geológicas 26 (1), Marzo 2009, 165–176. México.

Rioual, P., et al., 2001. High-resolution record of climate stability in France during the last interglacial period. Nature 413, 293–296.

Ritmann, A., 1963. Les volcans et leur activité. Masson et Cie, Paris, p. 461.

Rivier, F., 1983. Síntesis geológica y mapa geológico del área del bajo Tempisque, Guanacaste, Costa Rica. In: Informe Semestral Ene-Jun 83. Instituto Geográfico Nacional Costa Rica, pp. 30–70.

Rojas, W., Alvarado, E., 1982. Geological and geotectonic framework of Isla del Coco and the marine zone off the central Pacific coast of Costa Rica. Rev. Biol. Trop 60 (Costa Rica).

Rojas, G., 1978. Geomorfologia y uso del suelo. Tesis de grado en Licenciatura en geografía. Universidad de Costa Rica, San José.

Rogers, R.D., 2003. Jurassic-Recent tectonic and stratigraphic history of the Chortis block of Honduras and Nicaragua (Northern Central America) (Ph. D. dissertation). The University of Texas at Austin, 289 p.

Roy, J.A., Lachniet, M.S., July 2010. Late Quaternary glaciation and equilibrium-line altitudes of the Mayan Ice Cap. Guatemala. Central America. Quatern. Res. 74 (1).

Rubio, A., 1949. Notas sobre la Geología de Panamá. Ministerio de Educación, Panamá.

Ruprecht, P., Plank, T., July 31, 2013. Feeding andesitic eruptions with a high-speed connection from the Mantle. Nature.

Rymer, H., Van Wyk De Vries, B., Stix, J., Williams-Jones, G., 1998. Pit crater structure and processes governing persistent activity at Masaya Volcano, Nicaragua. Bull. Volcanol. 59, 345–355.

Sadner, G., 1964. La colonización agrícola de Costa Rica. In: Informe Semestral Ene-Jun 64. Instituto geográfico Nacional Costa Rica, pp. 21–26.

Saenz, R., 1981. Edades radiométricas de algunas rocas de Costa Rica. Depto. de Geología, MIn. E. M. Costa Rica.

Saginor, I., Gazel, E., Carr, M.J., Swisher III, C.C., Turrin, B., 2011. Progress and challenges using 40Ar/39Ar geochronology in Costa Rica and Nicaragua. Rev. Geol. Amér. Central. Costa Rica.

Samper, A., Quidelleur, X., Komorowski, J.-C., Boudon, G., April 25–30, 2004. Timing of effusive volcanism within the whole Basse Terre Island (Guadeloupe, French West Indies) from new K-Ar Cassignol-Gillot Ages, 6. European Geosciences Union, 1st General Assembly, Nice, France. Geophysical Research Abstracts.

Sanchez, M., 1979. Consideraciones generales sobre la biogeografía y su aplicación a la cuenca inferior y media del río Mosca, Costa Rica (Tesis de Licenciatura en Geografía). UCR, Costa Rica.

Sandoval, F., 1971. Geología de la región Noreste del valle Central (Hoja Grecia). Inf. Tecn. y Notas Geol. (Reimp. 9). Dir. Geol. Minas y petrol, Costa Rica, p. 44.

Sapper, K., 1925. Los volcanes de la América Central. Max Niemeyer Halle, Deutschland, p. 144.

Sapper, K., 1943. Viajes a varias partes de la República de Costa Rica 1899–1924. In: Trejos Quirós, J.F. (Ed.), Imprenta Universal, p. 140. Costa Rica.

Savage, J.M., 1966. The origin and history of the Central American herpetofauna. Copeia 719–766. USA.

Sayles, R.W., 1931. Bermuda during the Ice Age. Am. Acad. Arts Sci. 66, 381–468. USA.

Schaufelberger, P., 1935. Un estudio geológico de la meseta Central Occidental. Rev. Inst. de Defensa del Café 1 (2), 148–160. Costa Rica.

Schimels, B., 1901. Recent decline en the level of lake Nicaragua. Am. Geol. 28, 396–398. USA.

Schmidt-Effing, R., 1975. El primer hallazgo de amonitas en América Central Meridional y notas sobre las facies Cretácicas en dicha región. Informe Semestral Ene-Jun 75. Instituto Geográfico Nacional, Costa Rica, pp. 53–61.

Schmidt-Effing, R., 1980a. Geodynamic history of oceanic crust en Southern Central América. Berliner Gesuroesteiss. ABH (A) 19, 201–202. Berlin.

Schmidt-Effing, R., 1980b. Rasgos fundamentales de la historia del Complejo de Nicoya (América Central meridional). Brenesia 18, 231–252. San José, Costa Rica.

Schmincke, H., Kutterolf, S., Perez, W., Rausch, J., Freundt, A., Strauch, W., 2008. Walking through volcanic mud: the 2,100 year-old Acahualinca footprints (Nicaragua). I stratigraphy, lithology, volcanology and age of the Acahualinca section. Bull. Volcanol. 51 (5), 479–493.

Schubert, C., 1984. Investigaciones sobre el Cuaternario de la República Dominicana. Instituto Panamericano de Geografía e Historia, revista Geográfica n° 99, México. DF. México, pp. 69–92.

Shakun, J.D., Clark, P.U., Shaun, A.M., Feng, H.E., Mix, A.C., Marcott, A.C., Liu, Z., Otto-Bliesner, B., Schmittner, A., Bard, E., 2012. Global warming preceded by increasing carbon dioxide concentrations during the last deglaciation. Nature 484, 49–54. USA.

Siebert, L., Alvarado, G.E., Vallance, J.W., Van Wyk De Vries, B., 2006. Large-volume volcanic edifice failures in Central America and associated hazards. In: Rose, W.I., Bluth, G.J.S., Carr, M.J., Ewert, J.W., Patino, L.C., Vallance, J.W. (Eds.), Volcanic Hazards in Central America, Geol Soc Soc Amer Spec Pap, vol. 412, pp. 1–26.

Silva Romo, G., Mendoza Rosales, C., 2009. Evaluación geológica de los modelos para El truncameinto cenozóico Del sur de México: Erosión por subducción y detachment del bloque Chortis.

Silver, E.A., Reed, D.L., Tagudin, J.E., Heil, D.J., 1990. Implications of the north and south Panama thrust belt for the origin of the Panama orocline. Tectonics 9, 261–281.

Simkin, T., Siebert, L., 1994. Volcanoes of the World, second ed. Geoscience Press for the Smithsonian Institution, Tucson, ISBN 0-945005-12-1 (hard covers), xi + 349 pp.

Sirocko, F., et al., 2005. A late Eemian aridity pulse in central Europe during the last glacial inception. Nature 436, 833–836.

Soto, G.J., Alvarado, G.E., 2006. Eruptive history of Arenal Volcano, Costa Rica, 7 ka to present. J. Volcanol. Geotherm. Res 157, 254–269.

Sprechmann, P., et al., 1979. Estratigrafía de Costa Rica (informe preliminar). Escuela Centroamericana de GeoL UCR Costa Rica.

Sprechmann, P., 1982. Estratigrafía de Costa Rica I – Unidades Estratigráficas sedimentarias V. congreso Lat. de Geol. Actas I-55–71 Argentina.

Tarling, M.P., Et, D.H., 1973. La derive des continents. Doin Editeurs Paris.

Taylor, G.D., 1975. The geology of the Limón Area of Costa Rica. Louisiana State Univ. Dept. of Geol., USA.

Thompson, W.G., Goldstein, S.L., April 15, 2005. Open-System Coral Ages Reveal Persistent Suborbital Sea-Level Cycles. Science 308 (5720), 401–404.

Thompson, L.G., 1980. Glaciological investigations of the tropical Quelccaya ice cap, Peru. J. Glaciol. 25 (91), 69–84.

Tournon, J., Alvarado, G., 1995. Mapa Geológico de Costa Rica. Escala1: 500.000. más texto. Ministerio de Relaciones Exteriores de Francia. Delegación regional de la cooperación Científica y Tecnica – ICE, Francia. Impreso por La Vigie, Dieppe.

Tournon, J., 1971. Apuntes sobre la geología de la península de Santa Elena. Manuscrito publicado po la Universidad de Costa Rica, p. 4.

Tournon, J., 1972. Présence de basaltes alcalins récents au Costa Rica. (Amérique Centrale). Bull. Volcan. XXXVI-1, 140–147. Paris. France.

Tournon, J., 1980. Contribución a la morfología de la parte occidental de la cordillera Central. In: Informe Semestral Ene-Jun 80. Instituto Geográfico Nacional, Costa Rica.

Tournon, J., 1983. La cadena volcánica cuaternaria de Costa Rica, Composiciones químicas de las lavas, presencia de dos series. Informe Semestral Ene-Jun 83. Instituto geográfico Nacional, Costa Rica.

Tournon, J., 1984a. Magmatismes du Mesoïque à l'Actuel en Arnérique Centrale L'exemple du Costa Rica, des ophiolites aux andesites. Thèse doctorat d'Etat. Université de Paris VI, 318 pp.

Tournon, J., 1984b. Magmatismes du Mesozoique a l' Actuel en Amerique Centrale: L'example de Costa Rica. Des Ophiolites aux Andesites (Tesis doctoral). Univ. Pierre et Marie Curie, París, 335 pp.

Trewartha, G.T., 1970. The Earth's Problem Climates. The Univ. of Wisc. Press, Madison, Milwaukee and London, p. 334.

Tricart, J., 1975. Types de lits fluviaux en Amazonie Bresilienne. Ann. Geogr. 473, 1–54. Paris.

Tricart, J., 1981. Apercu sur le Quaternaire au Salvador. Bull. Soc. Geol. de France. Tome III 81, pp. 59–68. Paris.

Tuzo, Wilson J., 1959. Geophysics and continental growth. Am. Sci. 47, 1–24.

Tuzo, Wilson J., 1960. Some consequences of expansion of the earth. Nature 185, 880–882.

Tuzo, Wilson J., 1963. A possible origin of the Hawaiian Islands. Can. J. Phys. 41, 863–870.

UNEP-OAS, 1997. Cooperation between United Nations Environmental Program and the Organization of American States for Costa Rica.

Vargas, G., 1978. Diagnóstico y recomendaciones para el manejo y ordenamiento de los recursos naturales en la cuenca del río San Lorenzo, Alajuela, Costa Rica (Tesis de Licenciatura en Geografía). UCR, Costa Rica.

Vargas, G., 1981. La chaine volcanique de Tilarán et le bassin inférieur du fleuve Bebedero: conditions écologiques, végetation et mise en valeur, Costa Rica (Thèse 3ème Cycle en Géographie). Université de Bordeaux-III, Bordeaux.

Vargas-Ramirez, J.E., 1978. Geología de una parte de la hoja Naranjo. Dir. Geol. Minas y Petrol. Minist. Econ. Ind. y Com Costa Rica.

Venzke, E., Sennert, S.K., Wunderman, R., April 2008. Smithsonian Institution's Global Volcanism Network. Bull. Volcanol 70 (6), 771–772.

Villemant, B., Hammouya, G., Michel, A., Semet, M.P., Komorowski, J.-C., Boudon, G., Cheminee, J.-L., 2005. The memory of volcanic waters: shallow magma degassing revealed by halogen monitoring in thermal springs of La Soufrière volcano (Guadeloupe, Lesser Antilles). Earth Planet. Sci. Lett. http://dx.doi.org/10.1016/j.epsl.2005.05.013.

Viramonte, J.G., Incer-Barquero, J., October 1, 2008. Masaya, the "Mouth of Hell", Nicaragua: volcanological interpretation of the myths, legends and anecdotes. J. Volcanol. Geoth. Res. ISSN: 0377-0273 176 (3), 419–426. http://dx.doi.org/10.1016/j.jvolgeores.2008.01.038.Elsevier Science.

Weyl, R., 1955a. Vestigios de una glaciación del Pleistoceno en la cordillera de Talamanca, Costa Rica. Informe Trimestral Jul-Sept 55. Instituto Geográfico Nacional, Costa Rica, pp. 9–32.

Weyl, R., 1955b. Viaje al cerro Chirripó en la cordillera de Talamanca. Informe Trimestral ene-marzo 55. Instituto Geográfico Nacional, Costa Rica, pp. 7–13.

Weyl, R., 1956. Vulcanismo y plutonismo en el sur de Centro América. Informe Trimestral jul-Sep 56. Instituto Geográfico Nacional, Costa Rica, pp. 9–17.

Weyl, R., 1957. Contribución a la Geología de la cordillera de Talamanca de Costa Rica, Centro América. Instituto Geográfico Nacional, Costa Rica.

Weyl, R., 1960. Las ignimbritas* centroamericanas. Informe Semestral Ene-Jun 60. Instituto Geográfico Nacional, Costa Rica, pp. 39–59.

Weyl, R., 1971. La clasificación morfotectónica de Costa Rica Informe Semestral Jul-Dic 71. Instituto Geográfico Nacional, Costa Rica, pp. 107–125.

Weyl, R., 1980. Geology of Central América. Gebruder Borntraeger, Berlin- Stuttgart. RFA.

Williams, H., 1952. Volcanic history of the meseta Central Occidental, Costa Rica. Univ. of California Pub. Geol. Soc. 29 (4), 145–180. Berkeley, Los Angeles, USA.

Williams-Jones, G., Rymer, H., Rothery, D.A., 2003. Gravity changes and passive degassing at the Masaya caldera complex, Nicaragua. J. Volcanol. Geoth. Res. 123 (1-2), 137–160.

Winter, A., et al., 2003. Orbital control of low-latitude seasonality during the Eemian. Geophys. Res. Lett. 30 (4), 12.

Woodring, W.P., Malavassi, E., 1961. Miocene foraminifera mollusks and barnacles form the valle Central, Costa Rica. J. Paleont. 35 (3), 489. Tulsa USA.

Wug, L.A., De La Cruz, S., 2008. Historia de la Actividad Petrolera en Guatemala. Sociedad Geológica de Guatemala 1. La Geología en Guatemala.

Consulted Maps

Alvarado, G.E., Paniagua, S., Tejera, R., 1980b. Mapa del vulcanismo Plio-pleistocénico de Costa Rica. 1:700 000. Instituto Geográfico Nacional Costa Rica.

Bergoeing, J.P., Brenes, L.G., 1978c. Mapa Geomorfológico de Costa Rica 1:1.000.000. Instituto Geográfico Nacional, Costa Rica.

Bergoeing, J.P., Brenes, L.G., 1978d. Geomorfología del cantón de Buenos Aires, provincia de Puntarenas, Costa Rica. 1:200 000. Instituto Geográfico Nacional, Costa Rica.

Bergoeing, J.P., Brenes, L.G., 2010a. Atlas Geomorfológico del Caribe de Costa Rica 1:1 000 000. Editorial Universidad de Costa Rica.

Bergoeing, J.P., Brenes, L.G., 2010b. Atlas Geomorfológico de Costa Rica 1:350.000. Instituto Costarricense de Electricidad, ICE, Costa Rica.

Bergoeing, J.P., Brenes, L.G., 2016. Atlas Geomorfológico de Costa Rica: 1:100.000 Edit. UNA, Universidad Nacional, Heredia, Costa Rica.

Bergoeing, J.P., Brenes, L.G., 2016. Geomorphology Atlas of Costa Rica. 1:100.000. Full color. Universidad Nacional, Heredia, Costa Rica, 36 pp.

Bergoeing, J.P., Malavassi, E., 1981e. Mapa geomorfológico del valle Central, Costa Rica. 1:50.000 (9 hojas más texto). Instituto Geográfico Nacional, Costa Rica.

Bergoeing, J.P., Malavassi, E., 1981f. Mapa geomorfológico del valle Central, Costa Rica. 1:100 000 (2 hojas). UCR, Costa Rica.

Bergoeing, J.P., Malavassi, E., 1981g. Síntesis Geológica del valle Central, Costa Rica. 1:100 000 (2 hojas). Instituto Geográfico Nacional, Costa Rica.

Bergoeing, J.P., Brenes, L.G., Malavassi, E., 1982c. Geomorfología del Pacífico norte de Costa Rica. 1:100 000 (11 hojas más texto). Instituto Geográfico Nacional, Costa Rica.

Bergoeing, J.P., Brenes, L.G., Malavassi, E., 1982d. Geomorfología de la hoja Barranca, Costa y. 1:500 000. Instituto Geográfico Nacional, Costa Rica.

Bergoeing, J.P., 1978d. Geomorfología de los cursos medio e inferior del río Tempisque, provincia de Guanacaste, Costa Rica. 1:200 000. Instituto Geográfico Nacional, Costa Rica.

Bergoeing, J.P., 1978e. Geomorfología del sector del Cantón de Talamanca, provincia de Limón, Costa Rica. 1:200 000. Instituto Geográfico Nacional, Costa Rica.

Bergoeing, J.P., 1982c. Geomorfología de algunos sectores de Costa Rica basada en imágenes del satélite Landsat MSS7 1:500 000 (4 mapas más texto). Instituto Geográfico Nacional, Costa Rica.

Chaves, R., Saenz, R., 1974c. Mapa geológico de la cordillera de Tilarán 1:200 000 Ministerio de Industria y Comercio. Dirección de Geología, Minas y Petróleo, San José, Costa Rica.

Escalante, G., 1965c. Mapa geológico de la parte superior de la cuenca del río Reventazón, Costa Rica. 1:500 000. Instituto geográfico Nacional, Costa Rica.

Escalante, G., 1968b. Mapa geológico preliminar de la región sureste de Costa Rica. La cordillera de Talamanca y sectores próximos. 1:300 000 inedit.

Ice, 1976. Mapa geológico regional (Guanacaste) 1:400 000. Departamento de Geología, ICE, Costa Rica.

Instituto Costarricense De Electricidad (Ice), 1976a. Boletín Hidrológico 10, 349. Sna José, Costa Rica.

Instituto Costarricense De Electricidad Ice, 1976b. Mapa Geológico regional (proyecto geotérmico) 1: 400 000. Depto. Geol. ICE, San Josê, Costa Rica.

Instituto Geográfico Nacional Mapas de base, topográficos a escalas: 1/50 000; 1/200 000; 1/500 000 y 1/1 000 000 así como numerosas fotografías aéreas verticales y oblicuas entre las cuales algunas se reproducen en el texto.

Instituto Panamericano De Geografía E Historia Ipgh. Atlas climatológico e hidrológico del istmo centroamericano Publicación 367 IPGH. Guatemala.

Mora, S., 1979c. Mapa Geológico de la región Sureste del valle de El General 1:500 000 Escuela Centroamericana de Geología.

Index

'*Note*: Page numbers followed by "f" indicate figures, "t" indicate tables.'

Kolibri 360 GmbH
Ferdinand-Jühlke-Straße 7
99095 Erfurt, Deutschland
produktsicherheit@kolibri360.de